INDUSTRIAL and Manufacturing WELLNESS

The Complete Guide to
Successful Enterprise
Asset Management

Mike Sondalini

INDUSTRIAL PRESS, INC.

Industrial Press, Inc.
32 Haviland Street, Suite 3
South Norwalk, Connecticut 06854
Tel: 203-956-5593, Toll-Free: 888-528-7852
E-mail: info@industrialpress.com

Library of Congress Cataloging-in-Publication Data

Name: Sondalini, Mike, author.
Title: Industrial and manufacturing wellness: the complete guide to successful enterprise asset management / Mike Sondalini.
Description: 2nd edition. South Norwalk, Connecticut: Industrial Press, Inc., [2016]. Includes bibliographical references and index.
Identifiers: LCCN 2016008105; ISBN 9780831135904 (hardcover: alk. paper); ISBN 9780831194031 (ePUB); ISBN 9780831194048 (eMOBI); ISBN 9780831194024 (ePDF)
Subjects: LCSH: Plant maintenance—Management. Industrial equipment—Maintenance and repair. Production management. Production planning.
Classification: LCC TS192 .S675 2016; DDC 658.2/02—dc23
LC record available at http://lccn.loc.gov/2016008105

ISBN print: 978-0-8311-3590-4
ISBN ePUB: 978-0-8311-9403-1
ISBN eMOBI: 978-0-8311-9404-8
ISBN ePDF: 978-0-8311-9402-4

Sponsoring Editors: Taisuke Soda and Judy Bass
Managing Editor: Laura Brengelman
Developmental Editor: Deborah A. Ring
Interior Text: Janet Romano-Murray
Cover Design: Ryan Maxwell and Janet Romano-Murray

industrialpress.com
ebooks.industrialpress.com

10 9 8 7 6 5 4 3 2 1

DEDICATION

In memory of quality management guru W. Edwards Deming.
Your vision of what is possible continues to guide and inspire.

TABLE OF CONTENTS

For additional online materials referred to throughout the text, visit Industrial Press's website at http://industrialpress.com and go to the page for *Industrial and Manufacturing Wellness*, where a link to these dowloadable files is provided.

PREFACE

This book contains the complete Industrial and Manufacturing Wellness (IMW) asset management solution for creating and sustaining outstanding production plant and equipment reliability. It evolved from the Plant and Equipment Wellness concepts for perpetual machinery health. The methodology is also known by the equally meaningful and shorter name Plant Wellness Way. The full explanation and examples make this book a user manual. The book is accompanied by spreadsheet templates and support documents that are available online. **You can access these resources at the publisher's website, http://industrialpress.com, then go to the page for *Industrial and Manufacturing Wellness*, where a link is provided to the online materials.**

The Plant Wellness Way methodology is based on sure materials science, reliability engineering facts, good financial management sense, and risk management best practices. Its purpose is to achieve world-class physical asset performance so that all industrial and manufacturing operations can reach Operational Excellence. Because IMW is based entirely on the laws of reliability engineering, risk elimination, and microstructure physics, you can get Operational Excellence success without using the methods of Lean or Six Sigma.

In January 2014, the ISO 55000 physical asset management series was released. It is gratifying to see the faultless match between the methods and outcomes of the Plant Wellness Way and the requirements of the ISO 55001 asset management standard. The Plant Wellness Way methodology lets you design and build a highly successful enterprise asset management system that will also fully satisfy ISO 55001 certification. It does so using low-cost, comprehensive methods that will ensure you make the best choices for your operation.

Heartfelt thanks go to Taisuke Soda, former editorial director at Industrial Press, Inc., for his decision to publish this book, and to his successor, Judy Bass. Thanks go to Laura Brengelman, managing editor at Industrial Press, for her professional and practical publishing advice. It's been a pleasure to work with project editor Deborah A. Ring. She turned a rough draft manuscript into a highly readable book and, in that process, made me a better writer. Thank you as well to Janet Romano-Murray for creating a user-friendly book layout. An author's family is always impacted by the writing of a book, and deep gratitude goes to my wife Sue for her patience and support over the months of composing and then editing the manuscript.

It is unlikely that all that is written in this book is perfectly right. Most of what we think is true today will be shown to be not quite so in the future—there are always better, more complete, and more effective answers to be found. That's the way it has been since the dawn of civilization. Humanity advances gradually by doing, failing, and learning. The path to Industrial and Manufacturing Wellness will continually improve and become simpler and more effective as the asset management and reliability body of knowledge grows ever truer and our methods and techniques get ever more successful. Right now, you have in your hands the best possible method for getting outstanding physical asset reliability success in industrial and manufacturing operations: the Plant Wellness Way.

Mike Sondalini
www.lifetime-reliability.com
April 2016

INTRODUCTION

There is no shortage of asset management strategies and methods in use, yet incredibly few companies have great asset management performance. Something is wrong with the way most organizations go about managing their operating assets. What this book does that no other has done before is to focus on building a complete business system to get world-class industrial asset performance.

Industrial and Manufacturing Wellness (IMW) lets you design and build a holistic life-cycle system for enterprise asset management that delivers world-class plant and equipment reliability and locks it into place forever. Using this methodology, you optimize production plant performance through the correct selection and application of financial, project design, operations, maintenance, risk management, and reliability engineering. You use effective methods, techniques, and tools that get outstanding equipment reliability and maximum life-cycle operating profits by creating risk-free operation. You pick the best strategies, tactics, and practices to give your plant and equipment failure-free life. The Plant Wellness Way's methods and techniques use science, accounting, and math so that its answers deliver sound, highly profitable, and defendable business choices. Yet the methodology is straightforward to follow so that every operation can use it.

Building a System of Reliability

Outstanding physical asset reliability is a formulaic result that you can deliver in any operation. We know what to do to make any industrial asset outstandingly reliable—don't let its parts fail. Getting highly reliable equipment is a certainty because reliability is simply the result of meeting achievable engineering outcomes. When you use the methods that keep machinery parts' microstructures healthy, you'll get world-class plant and equipment reliability. It's impossible to do that unless you build a holistic, integrated, reliability creation system in your business and across the asset's life cycle—this understanding is what has been missing in the way organizations have been going about physical asset management. You'll get outstanding reliability when your business uses a business-wide, life-cycle-long system for creating reliability. World-class plant reliability cannot be achieved any other way. To do so, you'll need a methodology that is sure to give you a lifetime of equipment health and wellness.

An operation that is built to be a system for creating reliability is structured so that it optimally achieves that purpose. When your company is a "system of reliability," you engineer and use your business processes to get utmost operating success. A Plant Wellness Way (PWW) system of reliability learns how to perform better. It actively generates new and better solutions and ensures that they become the new company standards that your people use. All successful methods and solutions are taken into your business so that you accelerate ahead of your competition with powerful innovations, new knowledge, and the excellent skills of a masterly workforce. With IMW, there is no limitation on the achievement of world-class maintenance, reliability, and operational performance. Your day-to-day struggles disappear as your processes create the world-class reliability that you need to achieve Operational Excellence.

The Plant Wellness Way is used to design and build a business-wide, life-cycle system for world-class physical asset reliability. It puts a systematic solution to create exceptional operating asset performance into companies. An IMW operation makes world-class reliability a part of its business DNA. Its operating assets are highly reliable because all of its people understand, do, improve, and sustain a system of reliability that can only produce outstanding production from outstandingly reliable assets.

Plant Wellness Way Overview

A great enterprise asset management methodology will rapidly deliver sure physical asset management success. This book introduces and explains how to create world-class reliability by designing and then building the necessary processes. You'll be able to build a combined business and life-cycle system to keep your plant and equipment highly reliable and generate optimal operating profit.

The simple PWW techniques use regular professional office software and computers. You need to know only the basics of engineering, risk management, and financial management to use the Plant Wellness Way. You do not require proprietary RAMS (reliability, availability, maintainability, safety) modeling software that only mathematicians, degreed engineers, and specialists can understand. When operations do have access to reliability analysis and maintenance and operations optimization tools and software, they can be used to better understand which options will be most effective. But you do not need such programs and software to benefit from using the Plant Wellness Way methodology. With this book and its supplementary materials (available online), you'll always develop sound and powerful operations, maintenance, and reliability improvement strategies and make great choices for outstanding equipment performance.

Three premises guide us to Industrial and Manufacturing Wellness. The first is that equipment only stops after a critical part fails. The health of every critical part determines your equipment performance. Keep those components strong and healthy, and your equipment will be exceptionally reliable. This premise is the cornerstone of all enterprise asset management success. Its achievement is what the Plant Wellness Way delivers, and it is fundamental to creating Operational Excellence and generating profits.

The second premise is that humans make errors. We destroy equipment reliability by our misunderstandings and mistakes. If people are necessary for your business processes, then their degree of competence, interest in doing better, level of curiosity, level of dedication, desire to learn more, and many other normal human traits cannot be left to chance. As you eliminate and control more human errors, your plant and equipment will run more successfully and failure-free. Whenever possible, use technology to replace humans. When people are necessary, use technology to help them achieve excellence and assist them in doing great work. When people do tasks where technological assistance is not available, teach them to be experts in their duties and organize them into a collaborative team of experts so they use right knowledge and make fewer mistakes.

The third premise is that your organization wants to create a world-class company, built of highly reliable processes producing desired results that stakeholders and customers are delighted to have. Poor plant and equipment reliability is a business process failure that prevents business success. As plant and equipment are built, used, and maintained more precisely, the quality will be better, the risk of failure will be lower, production costs will fall, and delivery time will be shorter. Customers like that and will buy your product, thereby making the business a success.

Parts, people, and processes—these are what make your products and services. Each one is important to business performance and must be nurtured to achieve excellence if a company is to be a leader in its field. Figure I.1 situates PWW among the methodologies that are currently available for asset reliability growth and shows the direction in which it will take a business.

The sole purpose of the Plant Wellness Way is to deliver outstanding physical asset reliability and operational performance. Zero failures in your processes and work is the aim. You do not seek

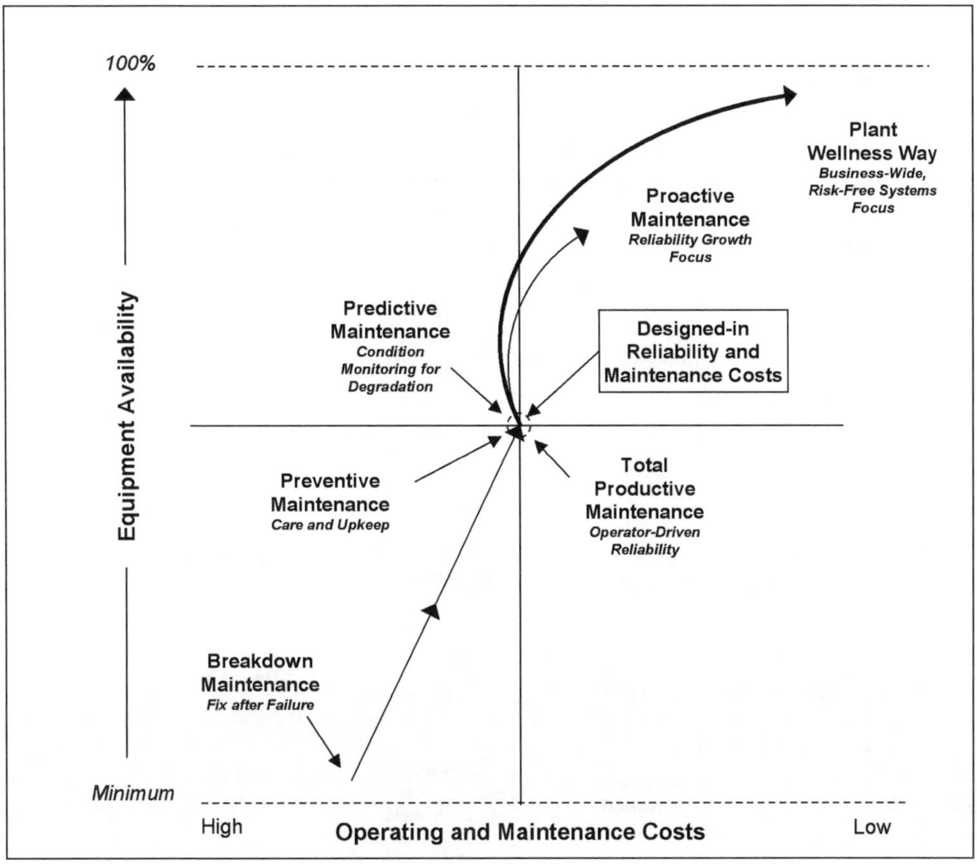

Figure I.1—Climb to the Heights of Enterprise Asset Management Success

acceptable failure rates for plant and equipment—the objective is to have no failures whatsoever during the service life of the asset. You identify and implement effective asset management strategies that optimize operating profits using the fewest resources for the least cost. A business that needs its plant to operate trouble-free for five years will make different choices than a business that wants a 30-year trouble-free service life. The methodology caters to all scenarios while helping each operation tailor the best solutions to their specific situation.

The Six IONICS Processes

The asset management strategies and solutions that you apply are the outcomes of the six IONICS processes[1] of the Plant Wellness Way methodology:

- Identify business and operational risks
- Order risks by importance
- Numerate risk elimination options

- Introduce risk control solutions
- Control operational processes
- Synthesize new ideas to continually improve

Figure I.2 is a simple flowchart of the IONICS processes used to create a lifetime of highly reliable plant and equipment. Work through them one by one. None is difficult.

By adopting the Plant Wellness Way as your company standard, you bring an enterprise asset management philosophy and methodology into your organization that helps you do the following:

- Fix the problem, not minimize its symptoms
- Optimize the operational value-added streams by eliminating plant and equipment risk
- Remove all physical asset failures with best failure prevention practices

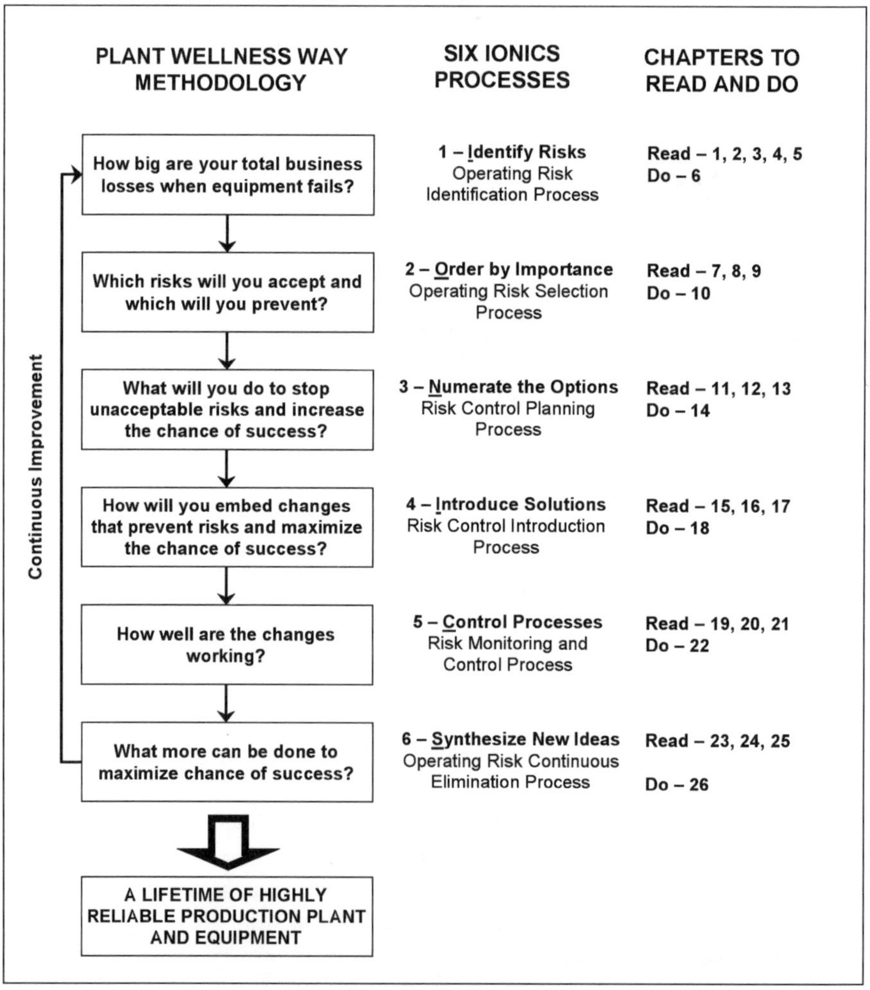

Figure I.2—The Plant Wellness Way Asset Management Methodology

- Target world-class quality results with accuracy-controlled tasks
- Use precision maintenance and precision operation on all physical assets
- Normalize exactly how to achieve outstanding reliability and production success
- Establish a culture of reliability in your company

These seven points form the acronym FORTUNE because that is what you get when you create asset reliability that delivers least operating costs and outstanding operational performance. You improve your asset maintenance and plant operations processes and practices so they always deliver world-class production asset performance. When a company adopts IMW, it cannot stay the same—it becomes a world-class operation because it follows world-class practices.

Plant Wellness Way Vision

- The Plant Wellness Way system of reliability methodology…
- Maximizes life-cycle profit by…
- Creating a holistic, life-cycle-long and business-wide system to…
- Proactively eliminate equipment failures through…
- Reducing process variation to only high-quality results that…
- Deliver failure-free parts and components by…
- Creating and sustaining ideal parts' microstructure health conditions and…
- Markedly reducing human error, thereby...
- Ensuring high precision quality results that…
- Produce world-class plant and machinery reliability for…
- The least engineering, operations, and maintenance commitment to…
- Give the lowest plant and equipment risk profile and life-cycle cost, so that…
- You get world-class operating performance and profits forever.

All of the Industrial and Manufacturing Wellness concepts and techniques needed to build a complete system for successful enterprise asset management are covered in this book. It gives you the know-how to get world-class asset management and reliability results. Each IONICS process is introduced and explained in the six sections of the book. They each cover relevant key business, engineering, and reliability concepts and principles of the Plant Wellness Way methodology. At the end of each section, the learning is put to use to help you climb a sure and safe path to the pinnacle of outstanding plant and equipment performance. The methodology is built on solid science, physics, and math, so your decisions are grounded in facts and reality.

Figure I.3 is shows how using the Plant Wellness Way[2] puts people, processes, capital and culture together in a never-ending cycle of innovation and learning that lifts your organization to the heights of Operational Excellence by making excellence the normal way of doing business. Use this book on your climb to world-class performance and learn how to create, design, and build a holistic, integrated, lifetime system for world-class reliability the Plant Wellness Way—good fortune awaits you!

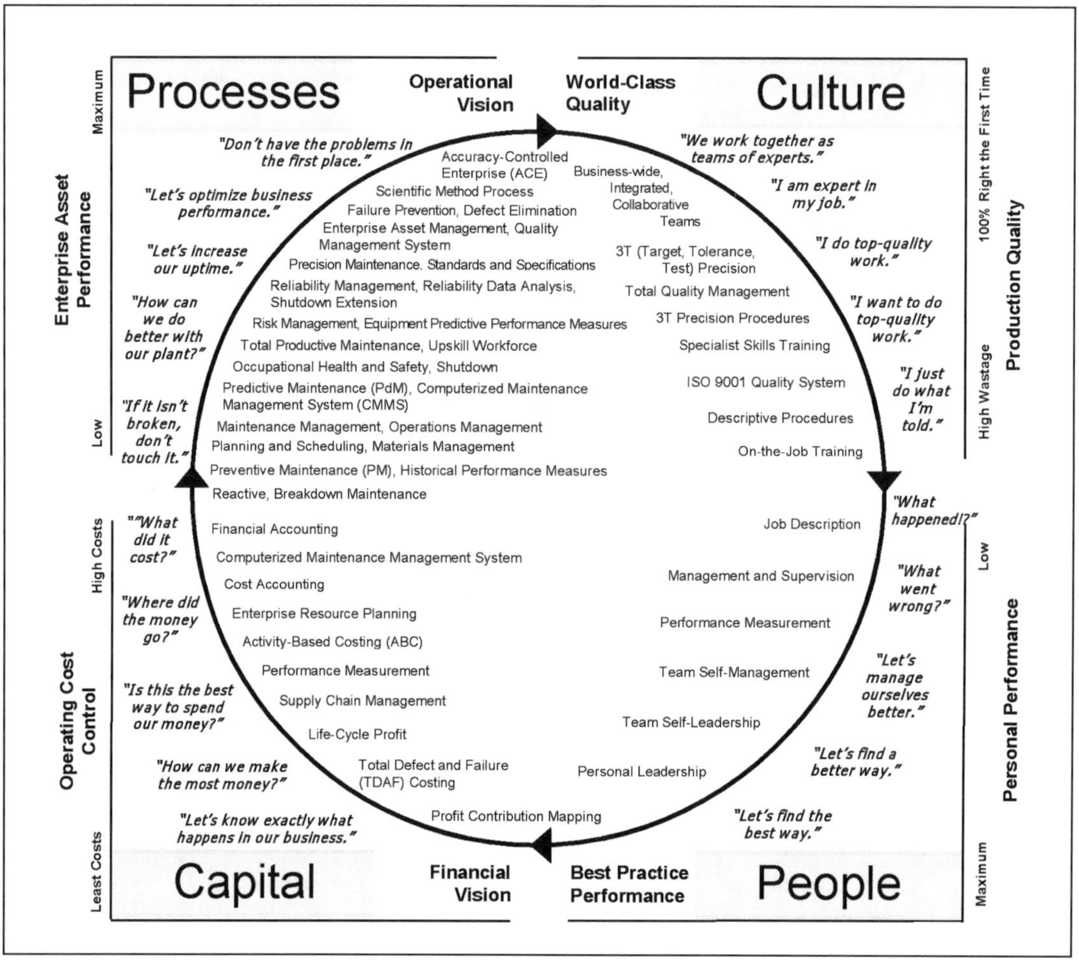

Figure I.3—The Plant Wellness Way Cycle

References

1. Thank you to Bruce McLaughlin for his contribution to the development of the IONICS and FORTUNE acronyms.

2. Thank you to Peter Brown, formerly of Industrial Training Associates in Perth, Western Australia, for suggesting the "Plant Wellness" name.

INDUSTRIAL and Manufacturing WELLNESS

The Complete Guide to
Successful Enterprise
Asset Management

1

Reliability of Work, Processes, and Machines

A business, a job, or a machine must work right by design before it can work right in reality. A business produces products and services from a collection of interacting processes. Build a business of world-class processes, and you'll get a world-class business. Follow a well-built, exact work procedure with properly organized and planned tasks and activities, and the job will get done right. Do work without using a designed procedure to control and coordinate the job, and you won't know what you are going to get. Inside a machine, parts work in a prescribed arrangement to carry their loads, stresses, and strains. When the design is poorly engineered or poorly built, then poor performance is what you will get from the machine. If the design is robustly engineered and well built, you will get a reliable machine that returns the investment.

If you want a company in which great results are natural and excellence abounds, you need to ensure that your processes, jobs, and machines are built to deliver excellence. Every step in every process, every task in every job, and every part in every machine needs to work right all the time. That can only happen in the real world when your processes, work, and equipment are designed to deliver the right outcomes every time they are used.

Creating a more successful business means designing, then building, more successful processes. A successful process comprises correct inputs, effective tasks, knowledgeable people, and reliable machines working in concert. With the activities, equipment, and processes in your company performing at world-class quality, world-class business results become natural.[1] Measuring the chance of business process or work success requires statistics and probability math. Such math can be difficult, but you need only simple multiplication to see what chance you have of getting work and process success in your organization.

Job and Work Process Reliability

Every job is a link in a work process chain. The results of the process depend on how well each job and its activities are done. An activity done wrong introduces errors and defects that jeopardize job and process success—each process failure damages company performance. Figure 1.1 is a process map depicting a five-task job. From such a flowchart, you can gauge how successful the work, the job, and the process will be.[2]

To determine work task success rates, you collect data on work task failures. This lets you determine the likelihood of doing each task right, after which you can calculate the chance of doing the whole job right. If Task 1 has a 100% chance of perfect work, its probability of success is 1. If it is done right 50% of the time, it has a 0.5 probability of success. Formula 1.1 is used to calculate job reliability, or the chance of doing the whole job successfully. The underscore distinguishes work task reliability ($\underline{R}$) from system reliability (R), which does not use the underscore.

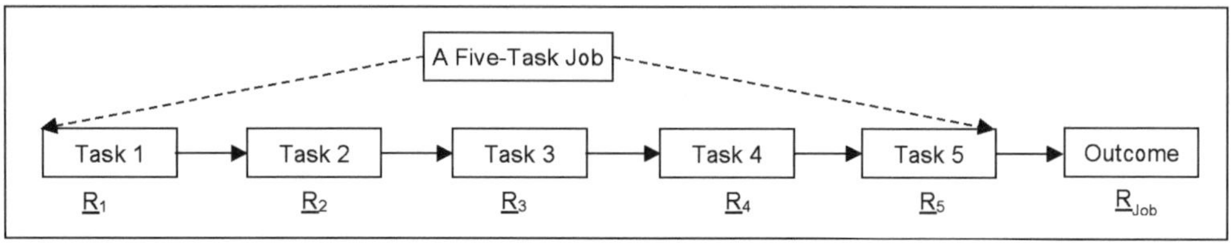

Figure 1.1—A Series of Tasks in a Work Process

Formula 1.1

$$\underline{R}_{Job} = \underline{R}_1 \times \underline{R}_2 \times \underline{R}_3 \times \ldots \underline{R}_n$$

We can use this formula to see the effect of mistakes on the chance of success in our five-task job. A short list of human error rates applicable to industrial plant operating and maintenance functions is given in Table 1.1.[3] Routine simple inspection and observation tasks incur 100 times fewer errors than complicated work done nonroutinely. Equipment and machinery repair tasks belong to the "complicated, nonroutine" category. Usually repairs are done irregularly on complex machinery, and human error rates during maintenance of 1 in 10, or more, are common (which means that 9 out of 10 times, a task will be done right—a 0.9 chance of success).

Table 1.1—Selected Human Error Rates

Situation and Task	Error Rate (per task)	No-Error Rate (success rate)
Routine simple tasks		
Read checklist or digital display incorrectly	0.001	0.999
Check for wrong indicator in an array	0.003	0.997
Fail to correctly replace printed circuit board	0.004	0.996
Incorrectly carry out visual inspection for a defined criterion (e.g., a leak)	0.003	0.997
Select wrong switch among similar	0.005	0.995
Read 10-digit number incorrectly	0.006	0.994
Routine tasks with care needed		
Incorrectly replace a detailed part	0.02	0.98
Put 10 digits into a calculator incorrectly	0.05	0.95
Do simple arithmetic incorrectly	0.01–0.03	0.99–0.97
Read five-letter word with poor resolution incorrectly	0.03	0.97
Dial 10 digits incorrectly	0.06	0.94
Punch or type character incorrectly	0.01	0.99
Complicated, nonroutine tasks		
Fail to notice incorrect status in roving inspection	0.1	0.9
New work shift—fail to check hardware, unless specified	0.1	0.9
High-stress, nonroutine work	0.25	0.75
Fail to notice wrong position of valves	0.5	0.5
Fail to act correctly after one minute in emergency situation	0.9	0.1

If every task in Figure 1.1 has a 0.9 chance of success, the whole job reliability is calculated as follows:

$$\underline{R}_{Job} = 0.9 \times 0.9 \times 0.9 \times 0.9 \times 0.9 = 0.59\ (59\%)$$

Even at 90% certainty for each of the five tasks, the chance that the whole job will be done without error is a poor 59%. In other words, the job will be wrong 41 times for every 100 times it is done. To get a 90% success rate for the whole job, the calculation below warns us that each task will need a 98% chance of success—only 2 errors in every 100 times it is done.

$$\underline{R}_{Job} = 0.98 \times 0.98 \times 0.98 \times 0.98 \times 0.98 = 0.9\ (90\%)$$

As a job gets longer, each activity in it is another opportunity for mistakes. The more activities a job comprises, the greater the number of opportunities to make errors and leave defects, and the fewer times the job will be done right. For a job that is 12 tasks in length, with each task having a 90% chance of success, reliability is calculated below as 0.28—the job will contain defects and errors 72 times out of every 100 times it is done. To get the job success rate up to 90 out of 100, every task will need to be 99% perfect—no more than 1 error in every 100 times it is executed.

$$\underline{R}_{Job} = 0.9 \times 0.9 \times 0.9 \times 0.9 \times 0.9 \times 0.9 \times 0.9 \times 0.9 \times 0.9 \times 0.9 \times 0.9 \times 0.9 = 0.28\ (28\%)$$

If every task in our five-task job is done right except Task 3, which is done correctly 60% of the time, the reliability of the job is as follows:

$$\underline{R}_{Job} = 1 \times 1 \times 0.6 \times 1 \times 1 = 0.6\ (60\%)$$

The chance of the whole job being done right is just 60%. All operating and maintenance work consists of tasks done in series, all of which have far more than the five steps in our simple example. Maintenance jobs of 40 to 50 tasks long, and often longer, are common. Unless every task is done right, the job will leave behind defects and mistakes. The high human error rates for repair work make breakdown maintenance and overhaul repairs very risky if you want maximum equipment reliability and utmost production uptime. Depending on the industry, early-life failure of plant and machinery represents 50% to 70% of all equipment failures. Failure early in equipment life is most often caused by bad work quality control.[4] Is it any wonder that many companies suffer from poor-performing operations when their managers, engineers, maintenance crews, and operators use failure-prone series processes?

To do a job perfectly, every task must be 100% right. In a series process, such as doing a repair job, operating a production line, using a supply chain, or running a business, when there is a mistake in one step, a defect is made or a problem is created, and the final outcome will also be wrong. This makes for a simple work reliability rule: the chance of job success is never greater than the chance of success for the worst-performed task. It's the same with every series arrangement: "One poor, all poor; one bad, all bad" is a reliability mantra to remember. It explains why you can have constant production quality problems—make one error anywhere in a series work process, and the finished item will be defective.

Today's aircraft industry has been outstandingly successful at controlling the outcomes of maintenance processes. It has developed highly reliable work techniques to maintain aircraft in extremely safe flying conditions. It is instructive and insightful to know what these companies do.

When you buy an airplane from a manufacturer, you also get a large set of maintenance manuals explaining in great detail exactly how to maintain the aircraft. The manuals are written by the designers. Every aircraft part is specified by a set of engineering parameters, right down to the formulation of its materials of construction. The designers define and explain the details of the ideal way to install and care for each component in the aircraft. Every maintenance activity is prescribed, including the drawings to use, the job procedures to follow, the techniques to apply, any special tools required, the parts to be replaced, and all work record forms. When independent double checks are needed, the procedure specifies where and how the checks are to be done. The industry is highly regulated worldwide, and it is a universal requirement when doing any aircraft maintenance to precisely follow the manufacturer's manuals.

The first question that aircraft mechanics ask before starting a job is, "Where is the manufacturer's maintenance procedure?" They know they can only do their work right if they follow the aircraft's designers approved manuals. Aircraft maintenance technicians are trained, tested, and certified competent on a model of plane before they can get their license to work. They can only work on the specific aircraft models they are licensed for and no others. Throughout their careers, aircraft technicians' work is regularly monitored for consistency of quality and accuracy. When new and improved methods are introduced by the aircraft maker, the technicians must be retrained and recertified. No matter where an airplane is maintained in the world, everyone working on it must be licensed for the currently approved maintenance procedures. If they are not up to the standard, they must stop working on aircraft until their competency is restored.

These are some of the processes the global airline industry uses to maintain planes and make air travel as safe as it is today. The industry has found, from many decades of experience and continuous improvement, that faultless aircraft maintenance requires processes to ensure that every job and all tasks are exactly specified and perfectly achieved every time they are done.

Transferred Defect Inheritance and Quality Inheritance

Every defect in a process step has the potential to impact numerous future steps. A defect in an item or work done in a prior step that causes trouble in a later step is termed an "inherited" defect. It is an error or fault that travels along with the item or job and becomes a future problem in the process or another process. One defect may only become a minor irritation, while another could turn into a severe business-destroying disaster. Transferred defect inheritance is involved in many business and operational problems and industrial equipment failures.

A common example of defect inheritance found in machinery is the adverse impact on parts from bad machining practices during manufacture.[5] Three groups of alloy coated steel parts were machined with differing surface roughness, Group 1 was coarsely rough machined, having a surface roughness of 80 microns (μm) between topographic peaks, another group was rough machined with 20 μm roughness, and the final group was given 0.32 μm roughness by grinding. All groups were heat treated to harden the surface coating and ground to a finish surface roughness of 0.16 μm, then put into wear trials to find their resistance to abrasion. The coating of the Group 1 specimens wore out the quickest and suffered the greatest number of surface cracks. Group 2 specimens had less wear and fewer cracks than Group 1, and Group 3 had little wear with no cracks at all. Under the microscope a difference in the coating microstructure was observed. The Group 1 rough machining had generated greater heat and produced high internal stresses that had caused many crevices, defects, and microcracks in the coating, but these were not present in the Group 3 specimens. A quality characteristic of a prior process step had changed the behavior of a subsequent process step. Surface hardness is important for machine parts that wear during service. If a machine had Group 1 rough machined parts installed, its maintenance costs and production downtime would be far more than if Group 3 parts had been fitted. The quality characteristics of a manufacturer's machining process have dire consequences for the businesses using their machines.

Another example of defect inheritance is a shaft journal machined out-of-round in a rough turning step that is later turned or ground to the finished size in a fine machining step will have retained its initial oval trait. The ovality is inherited for the life of the journal. If the oval journal is within the design tolerance for its size and shape, it will pass dimensional inspection and be used in service, but the ovality produces higher localized stress in the rolling bearing mounted on the journal. During operation, the higher local stress combines with other stresses to increase the probability of early bearing failure. To prevent the fine-turning step from making oval shapes in journals, it is necessary to go back to the prior manufacturing process steps to find the faults that caused the oval shape. The problems uncovered in the previous manufacturing steps would have come from earlier failures in the process. Those early failures would have still earlier defects. You would find that there are ever repeating steps of transferred defects followed by the troubles they cause.

Defect inheritance occurs in all processes. Any time an error, a misjudgment, a bad decision, a fault, a deficiency, or any other possible adverse outcome that can occur in a process step happens, it will create the opportunity for problem after problem to arise later. The problems cannot be stopped when they arise—they can only be fixed, replaced, or lived with. Problems stop when there is no defect present in the first place to cause the problem. The same data and examples of defect inheritance apply equally to the exact opposite—quality inheritance. Top-quality results achieved earlier in a process also transfer to future process steps. Doing fine-quality work brings its own satisfaction and success, but also it brings more success later in the process because quality items perform far better than poor-quality items when used in service. High quality results always contribute to the production of good results later, but poor-quality work will only harm future success. The better the quality you produce in each process step and job task, the higher the chance of success in all the subsequent steps of the future processes that use that quality characteristic.

Business Process Reliability

Figure 1.2 shows a simple production process used to make a product.

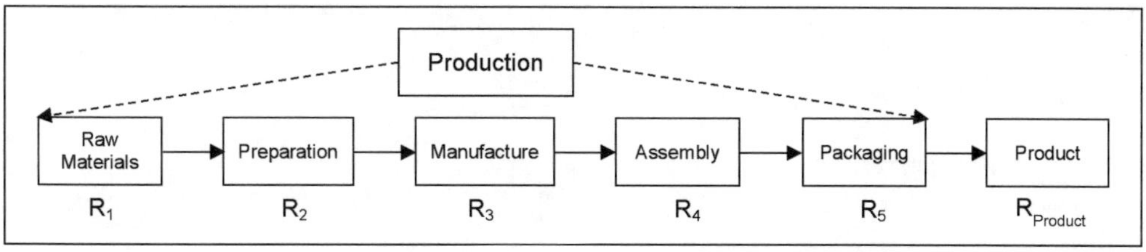

Figure 1.2—A Series of Steps in a Production Process

Within each process step, there are many subprocesses. The Raw Material step will have numerous processes within it and impacting it, the Preparation step will have its own processes, as will the Manufacture step, and so on for all of them. Figure 1.3 shows some of the processes in the Manufacture step for making a mechanical machine part. When tallied together, there are hundreds of activities in dozens of processes impacting an industrial operation.

Production plants experience many processes in their lifetimes.[6] The design, manufacture, supply chain, warehousing, installation, operation, and maintenance processes comprise numerous tasks that must be done right. From time to time, mistakes and poor choices are made in all of them. Those defects eventually lead to equipment or production failures. To understand how business and work processes impact equipment performance, you need to see the interconnectivity of all processes used across the life cycle to engineer, buy, make, and run the equipment.

Figure 1.4 is a representation of the many supply chain and operational processes involved in making a product. Process after process connects with others in a tangled web of interaction across

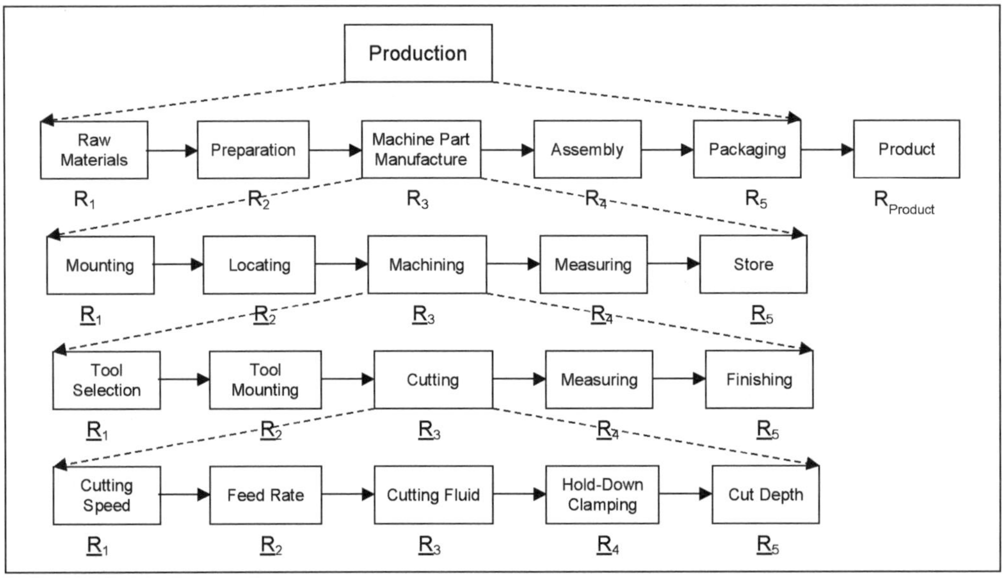

Figure 1.3—There Are Many Work Subprocesses in Every Production Process

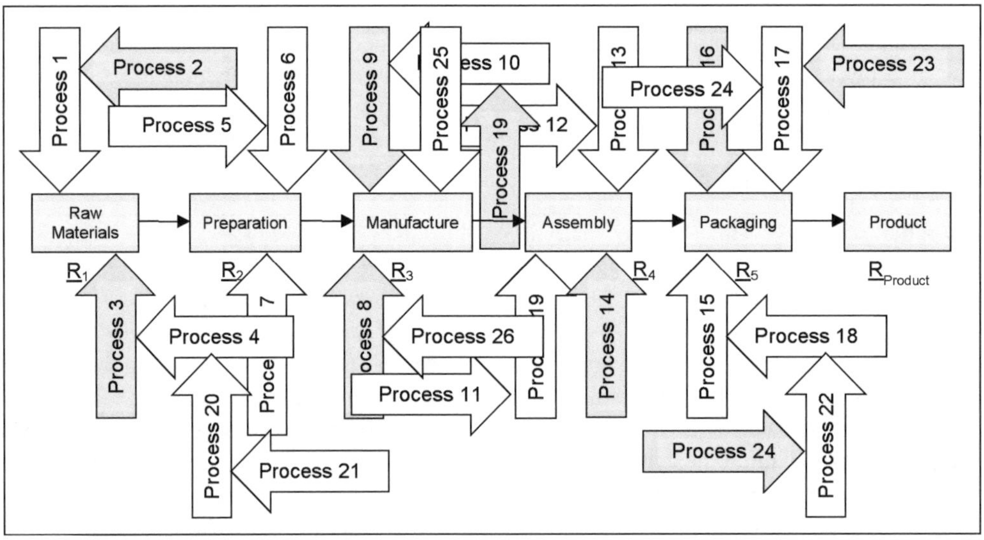

Figure 1.4—Numerous Processes Interact across Every Process Chain

time and space. There are dozens and dozens of processes containing task upon task. There are hundreds of tasks in most businesses; many companies have thousands of them. Companies with highly complex operations, such as building spaceships or airplanes, have tens of thousands of activities to control. Each one presents an opportunity for things to go wrong.

Because each process feeds many other processes, any error in one has a ripple effect that harms those downstream. A process that goes wrong that is not corrected can impact numerous others in the future. For example, a poor maintenance repair will cause a future production failure; an operator error that overloads a machine will lead to a future breakdown; the wrong choice of materials of construction by the designer of a gas-processing plant will contribute to a future explosion and possibly the death of people. That is why it is important for every step in a series process to go right every time—the future consequences are unforeseeable and can be devastating.

Getting the individual tasks in every process 100% right the first time is a seemingly impossible challenge in running a business. Guaranteeing that every activity is done correctly cannot be left to chance. Doing dozens of processes and thousands of activities perfectly requires a standardized system of excellence. Without ensuring excellence in every process step, you cannot get excellent products or services. World-class operations recognize the interconnectedness and holistic nature of their business and work hard to ensure that everything is right at every stage in every process across the entire business life cycle.

Industrial Equipment Reliability

Figure 1.5 on the following page shows how series processes are used in operating plants. It highlights that series processes abound throughout the lifetime of every piece of equipment. During design, manufacture, assembly, operation, and maintenance, multitudes of risks exist that can adversely impact equipment and business performance.

A machine is a series of parts configured to move and act in an organized sequence. One part functions on another, which then causes the next part to act, and so on. The parts that suffer operating stresses during use are known as the *critical working parts.* If a critical part in a machine fails, the machine stops. That is why production plants and industrial operations can have many breakdowns—it only takes one failure in one part of one machine to stop the whole plant. In plants with thousands of equipment items, there are millions of opportunities for plant and equipment failures.

The segmented centrifugal pump-set assembly shown in Figure 1.6 on the following page, is used as an example to help explain and understand equipment reliability. The electric motor turns a rotor that is connected by a drive coupling to the pump shaft, on which is mounted an impeller. In order for the pump impeller to spin and pump liquid, the pump shaft must rotate, as must the coupling, as must the motor rotor, as must the magnetic field in the motor. All of these requirements for the impeller to turn form a series arrangement. If the diagram displayed every piece of equipment needed to make liquid flow from the impeller, the whole process would start at the power provider's generator and show dozens of process steps. If any process step in the chain fails, the impeller will not turn, and no liquid will flow.

The reliability of a series configuration is calculated by multiplying the reliability of each item in the arrangement, using the following formula:

Formula 1.2

$$R_{Series} = R_1 \text{ x } R_2 \text{ x } R_3 \text{ x } ... R_n$$

As soon as the reliability of any item in the series drops to zero, the whole series goes to zero, and the entire system stops working. If the shaft coupling of the pump-set fails, its reliability becomes zero. The impeller mounted on the pump shaft cannot turn, and the pump-set is failed. If the electric motor cannot rotate, the pump-set is again failed. An Internet search by the author for causes of centrifugal pump-set failures found 228 ways for the wet end components to fail, 189 ways for a me-

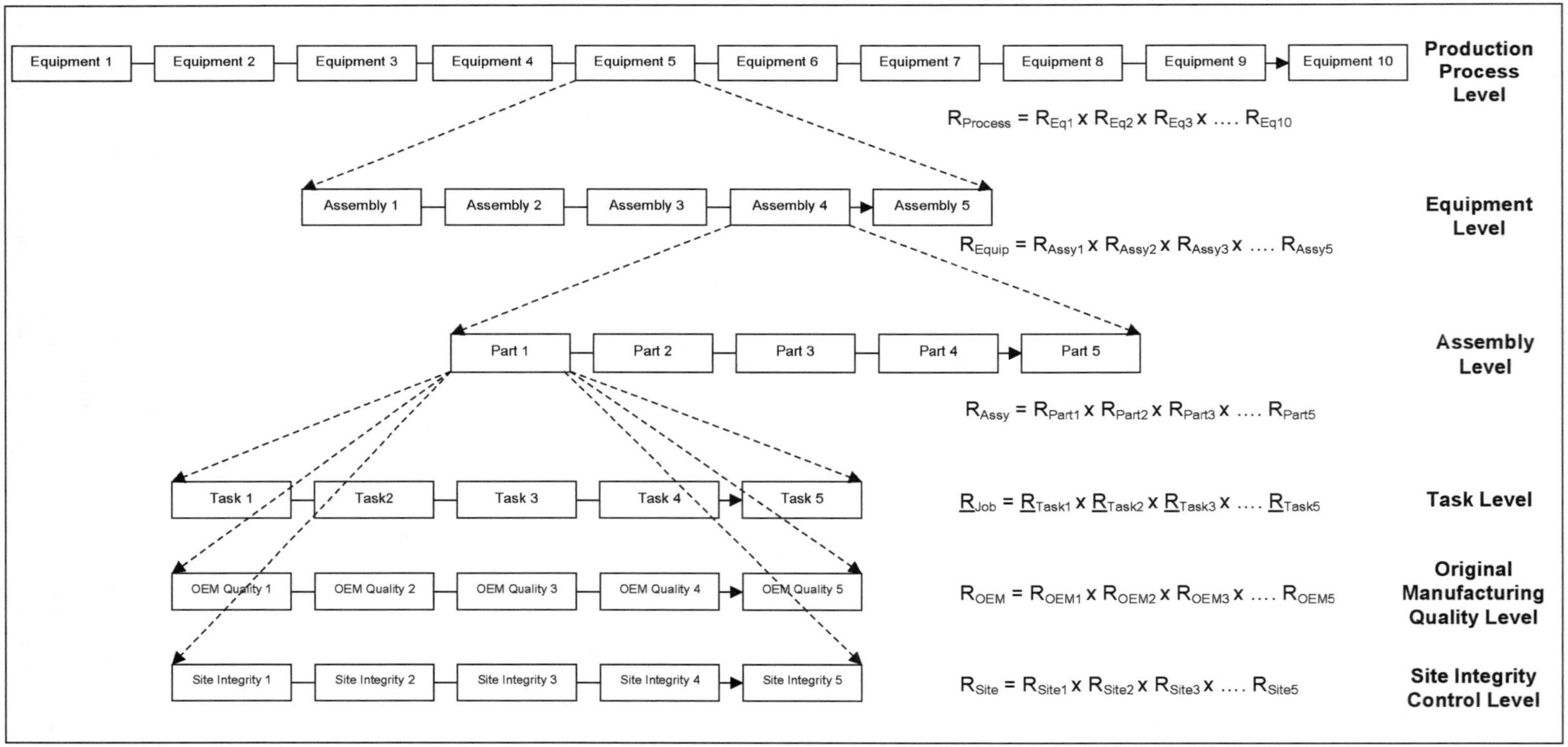

Figure 1.5—Impacts on Reliability during an Operating Equipment's Lifetime

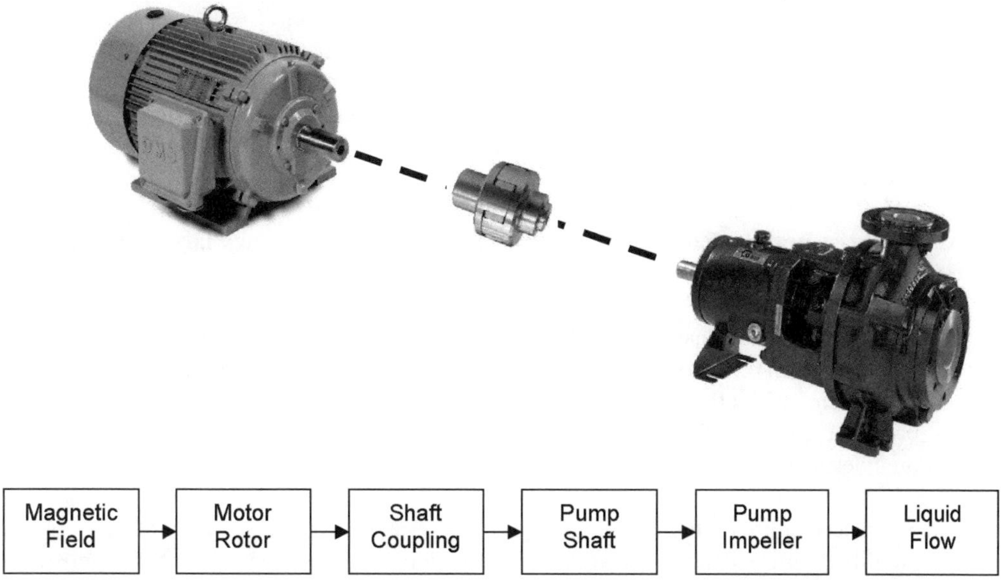

Figure 1.6—Series Arrangement of Assemblies in a Centrifugal Pump-Set

chanical seal to fail, 33 ways for the shaft drive coupling to fail, and 103 ways for the electric motor to fail. This totals 553 ways for one common item in a plant to stop functioning. In operations with many equipment items, there is a constant struggle against mountainous odds to keep them working. Improving the reliability of your series-constructed equipment is critically important for reducing operating plant failures.

A series arrangement has three Series Reliability Properties.

1. **The reliability of a series system is no more reliable than its least reliable component.**

   The reliability of a series of parts (a machine is a series of parts working together) cannot be higher than the reliability of its least reliable part. If the reliability of each part in a two-component system is 0.9 and 0.8, the series reliability is 0.9 x 0.8 = 0.72, which is less than the reliability of the least reliable item. Even if work is done to lift the 0.8 reliability to 0.9, the best the system reliability can be is 0.81.

   Series Reliability Property 1 means that anyone who wants high reliability from a series process must ensure that every step in the series is even more highly reliable.

2. **Add "k" items into a series system of items, and the probability of failure of all items in the series must fall by an equal proportion to maintain the original system reliability.**

   Say one item is added to a system of two. Each part has 0.9 reliability. The reliability with two components is originally 0.9 x 0.9 = 0.81, and with three it is 0.9 x 0.9 x 0.9 = 0.729.

To return the new series to 0.81 reliability, all three items must have a higher reliability, for example, 0.932 x 0.932 x 0.932 = 0.81. In this case, each item's reliability must rise 3.6% in order for the system to be as reliable as it was with only two components.

Series Reliability Property 2 means that if you want highly reliable series processes, you must remove as many steps from the process as possible so your opportunities for failure decrease—simplify, simplify, simplify!

3. **An equal rise in the reliability of all items in a series causes a much larger proportionate rise in system reliability.**

   Say a system-wide change is made to a three-item system, such that the reliability of each item rises from 0.932 to 0.95. This is a 1.9% individual improvement. The system reliability goes from 0.932 x 0.932 x 0.932 = 0.81 to 0.95 x 0.95 x 0.95 = 0.86, which is a 5.8% improvement. For a 1.9% effort, there is a gain of 5.8% from the system. This is a 300% return on investment.

   Series Reliability Property 3 seems to give big system reliability growth for free. Series Reliability Property 3 means that system-wide reliability improvements deliver far more payoff than making individual step improvements. It is the principle that delivers the most operating profit most quickly.

These three reliability properties are key to great enterprise asset management and Operational Excellence.

The Control of Series Process Reliability

Reliability engineering principles also give us the answer to series process problems—the parallel arrangement. Figure 1.7 shows a parallel layout. The second and higher-numbered items form a redundant configuration with the first item. Should the first item fail, the second item continues in operation, and the outcome from the system is maintained.

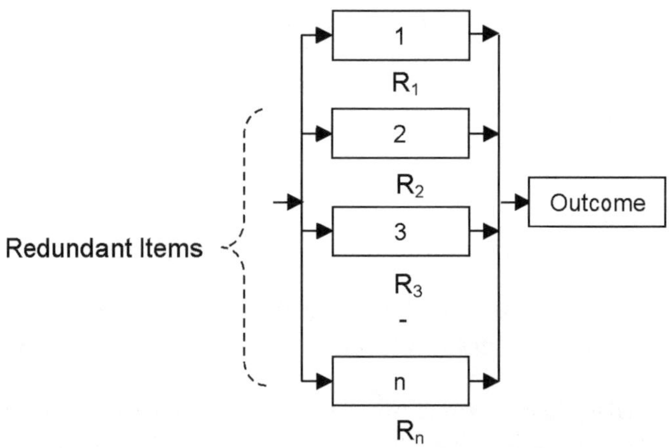

Figure 1.7—A Parallel Process

Reliability behavior in parallel arrangements is very different from that in series arrangements. Formula 1.3 is used to calculate the reliability of a parallel arrangement in which each element is in use and any one of them can do the full duty (known as fully active redundancy).

Formula 1.3

$$R_{Parallel} = 1 - [(1 - R_1) \times (1 - R_2) \times \ldots.(1 - R_n)]$$

Other system configurations of redundancy are common, such as a unit on duty and one on standby, two out of three, or three out of four, such that one unit is a standby for the other concurrently operating units in the system. Each type of parallel configuration has its own reliability formula that applies to the specific arrangement.

In a fully active parallel arrangement of four items, each with a terrible 0.6 reliability (a 40% chance of failure), the whole system reliability is represented as follows:

$$\begin{aligned} R &= 1 - [(1 - 0.6) \times (1 - 0.6) \times (1 - 0.6) \times (1 - 0.6)] \\ &= 1 - [(0.4) \times (0.4) \times (0.4) \times (0.4)] = 1 - [0.0256] = 0.9744 \end{aligned}$$

This arrangement gives a 97% chance of system success even though each item has a 40% chance of failure. We can use this fact to redesign series processes to get high reliability from them. Putting things in parallel gives you a way to lift production uptime. It is also a powerful strategy used to get greater job reliability and to build robust, antifragile business processes.

There is a natural economic limit to how many redundant items you can justify in a parallel arrangement. Each extra item requires money to acquire, install, and support. Each item needs regular maintenance and incurs ongoing operating expenditure by its presence. You want as few redundancies as possible in a process, but you can justify a redundancy when the risk of not having it is too high to accept.

Risk is the deciding factor when choosing plant, equipment, or work process redundancy. When the consequence of failure for an item in a series arrangement is excessive, it becomes practical to install parallel redundancy whenever the savings resulting from the redundant item more than pay for its cost, future upkeep, and eventual disposal. Adding a redundancy does not mean you can dismiss the risk. Providing a standby unit does not give you the right to allow anything to go wrong with the working equipment because you have covered its failure with the backup item. Once the duty equipment fails and you start the standby, you lose the benefit of redundancy. Without the standby item, the operating risk instantly jumps to total production loss. When a duty unit in a redundant arrangement stops and the standby is used, it is important to get the failed item fixed in an organized and timely manner—but do it immediately.

Parallel Tasks and the Carpenter's Creed

An example of high-reliability work is the Carpenter's Creed: "measure twice, cut once." Carpenters have known for millennia that a double check will save problems and trouble later. We can turn the adage into the parallel task shown in the reliability block diagram of Figure 1.8, in which a second measurement is done to confirm the first. By using a proof test activity to verify that the original task has been done right, we create a highly reliable "task system." Although the measurements are sequential, the logical purpose of the proof test measurement is to check the first one. This forms the parallel task arrangement shown in the block diagram.

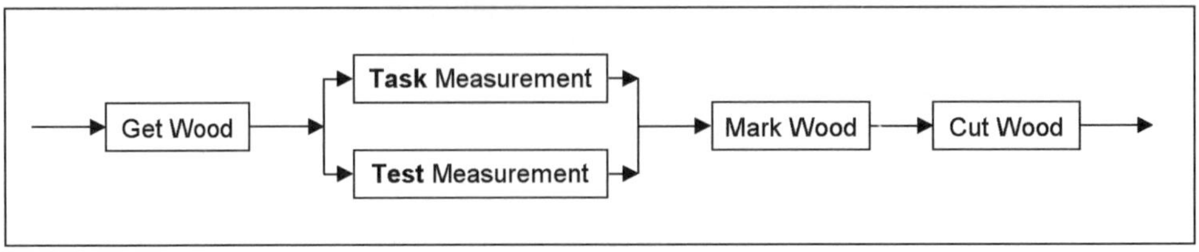

Figure 1.8—Carpenter's Creed: "Measure Twice, Cut Once" Is a Parallel Redundant Activity

The effectiveness of the Carpenter's Creed can be shown mathematically. A typical error rate in reading a tape measure is 0.005—that is, 5 times in every 1,000 it will be misread, or 995 times out of 1,000 it will be read correctly (a task reliability of 0.995). This means the average carpenter will mark the wood in the wrong spot about 1 time in every 200 measurements. It is not hard to imagine a carpenter averaging 40 to 50 cuts a day. About once each working week, the carpenter will mark and cut the wood in the wrong place and have to throw the job away. When he adds the proof test required by the Carpenter's Creed, he creates a parallel arrangement in which both tasks must fail before the system of two measurements together is failed. He would have to measure incorrectly twice in a row. With the chance of making one measurement wrong being 0.005, the reliability of the two measurements combined into a "measuring system" is found using Formula 1.3.

$$R = 1 - [(1 - 0.995) \times (1 - 0.995)] = 1 - 0.000025 = 0.99998$$

With the proof test added, the chance of getting the cut position right rises to 0.99998, which is an error rate of 2 in every 100,000 times. At 50 cuts a day, a measurement error is made once every 200 working days, or about every 40 working weeks. Doing a check test means 40 times fewer scrapped jobs. That is the advantage of adding parallel proof test activities to work tasks: to ensure that each activity is done right before the next step is started. Note that it is the proof test alone that protects against error. It is only by doing the check test that human error is prevented and high task reliability is achieved. Without the test, you have no error prevention.

There is one vital requirement for any proof test to reduce the chance of a common cause error. Common cause error is a shared error in which the same mistake is done in both the original and the test tasks. For the proof test, you must use a different measuring device than was used to make the original measurement. It is unlikely to have two measuring devices out of calibration at the same time unless there are systematic calibration problems within the organization. You should also have a different person do the proof test. The person and the measuring equipment form a system. Changing only the measuring device for the proof test and not the person doing the test leaves your business exposed to common cause problems from shared misunderstandings and wrong beliefs existing among your people. Having two totally independent measuring devices greatly reduces the chance of a common cause error. Similarly, by using two competent people to perform independent proof tests, you protect against common misunderstandings, incorrect information, and wrong training. It is unlikely for two knowledgeable, competent people to share the same mistaken education and bad work practices unless they were both wrongly educated and trained.

Figure 1.9 shows the five-task job depicted in Figure 1.1, with each task having a parallel inspect-and-measure proof test to confirm that it is correct. By adding test activities to all tasks in the five-step maintenance job, you create a high-reliability work process.

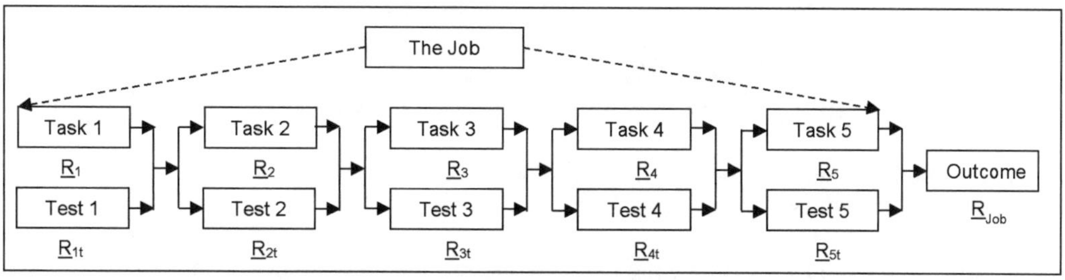

Figure 1.9—A Job with Parallel Test Tasks

If the test has 0.99 reliability—testing is carefully performed using high-quality tools and procedures—then each parallel-tested step reliability is as follows:

$$R_{Task} = 1 - [(1 - R_1) \text{ x } (1 - R_{1t})]$$
$$= 1 - [(1 - 0.9) \text{ x } (1 - 0.99)] = 1 - [(0.1) \text{ x } (0.01)] = 1 - 0.001 = 0.999 \text{ (99.9\%)}$$

The reliability of the whole job is represented by the following equation:

$$R_{Job} = 0.999 \text{ x } 0.999 \text{ x } 0.999 \text{ x } 0.999 \text{ x } 0.999 = 0.995 \text{ (99.5\%)}$$

A job that began at 0.59 reliability without any proof tests rises to 0.995 probability of success with proof-tested tasks. But even 0.995 reliability means that 5 times out of every 1,000 opportunities, the job will be wrong. In a large, busy operation with many people, 1,000 opportunities for error accrue rapidly. Similarly, when numerous processes are used to make a product, there are hundreds, even thousands, of opportunities a day for error to happen along the process chains. We need job and process reliabilities of great certainty if we want excellence in our businesses. You can achieve this by adding another parallel activity to each "task system." Figure 1.10 is an example. The test, which involves careful inspection and/or measurement, has a reliability of 0.99, while 0.9 reliability is used for each of the other parallel activities because "human factors" causing human errors are present when they are performed.

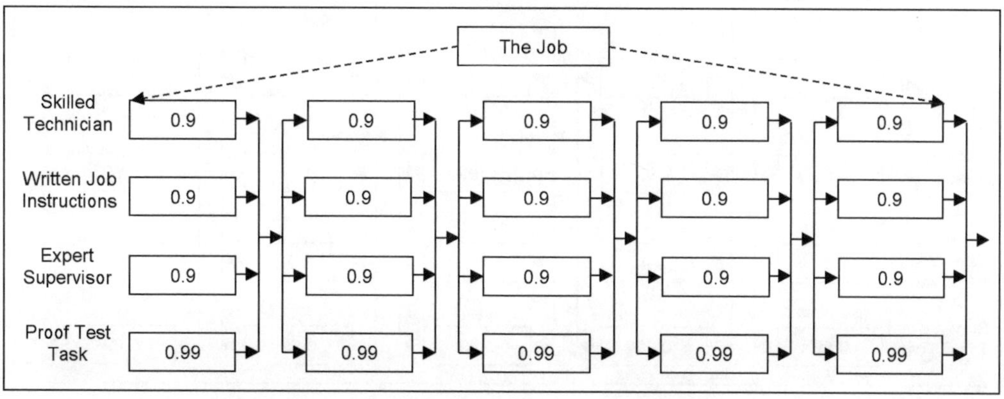

Figure 1.10—A Multiparalleled Task Work Process

The reliability equation for each of the multiparalleled work tasks is as follows:

$$R_{Task} = 1 - [(1 - 0.9) \times (1 - 0.9) \times (1 - 0.9) \times (1 - 0.99)]$$
$$= 1 - [(0.1) \times (0.1) \times (0.1) \times (0.01)] = 0.99999 \text{ (1 error per 100,000 opportunities)}$$

The reliability of the entire job of five tasks with each task paralleled in this error-preventing configuration is as follows:

$$R_{Job} = 0.99999 \times 0.99999 \times 0.99999 \times 0.99999 \times 0.99999 = 0.99995 \text{ (99.995\%)}$$

The error rate for the whole job is very low: 5 errors per 100,000 opportunities. This is the way to drastically reduce human error and get highly reliable work. To have high-reliability work processes, build parallel inspection activities into the job tasks.

My brother-in-law, who used to work for Japan Airlines (JAL), tells a story of watching Japanese aircraft maintenance technicians overhaul a JAL airplane jet engine. He tells this story because it is so unusual. During his visit to the maintenance hangar, he was enthralled by the extraordinary maintenance procedure that the JAL technicians followed. He watched as a man on a podium in front of a jet engine being worked on read from a manual. Once he'd finished speaking, the technicians at the engine began working on the equipment. The man on the podium went and looked carefully at the work being done. When the technicians finished, they stepped away from their work, and the man, who seemed to be the supervisor, tested and checked their workmanship. As he went through the double-checking process, he would, from time to time, note comments on a form that he carried. Once his inspection was completed and the technicians had also signed off on their work, he returned to the podium and read the next instruction from the manual. The whole process was repeated while my brother-in-law watched in astonishment.

What he saw was JAL's stringent policy of rebuilding its jet engines by following standard operating procedures paralleled to verbal instruction and supervisory monitoring. The expert supervisor read each task step, explained it, and then monitored the fully qualified and experienced aircraft tech-

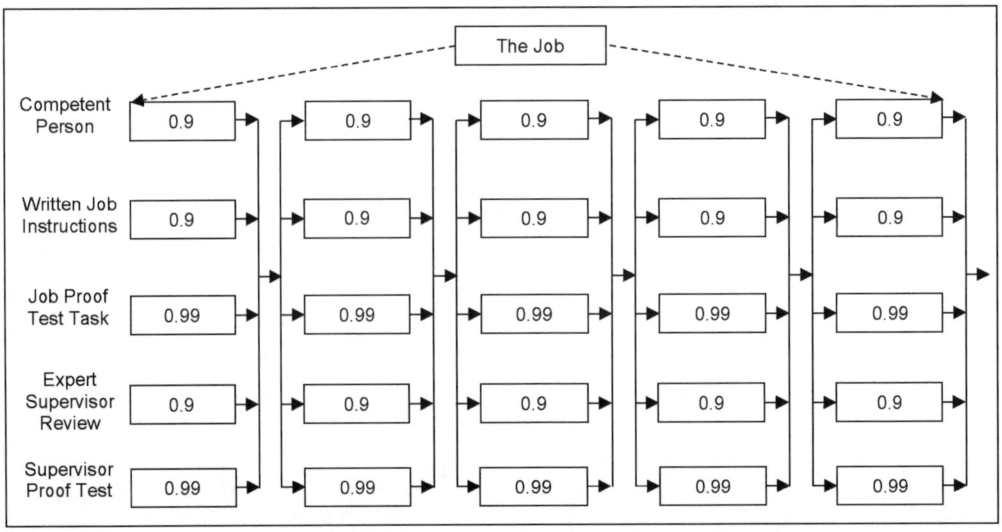

Figure 1.11—A Super-Sure Error-Prevention Work Process

nicians as they did the task. As the technicians performed the work, the supervisor watched and checked their workmanship. The task was completed only when the technicians and the supervisor confirmed that the work met the required standard and a record of proof was made of its successful completion. Then the next task step of the job was performed in the same way. By this method, JAL absolutely ensured that its jet engines were correctly rebuilt and fully meet specification. If you fly Japan Airlines, it is reassuring to know the rigors their aircraft mechanics go through to ensure that their jet engines and planes are in top order. Figure 1.11 shows how adding proof tests as done by JAL to create five-level parallel tasks gets amazing reliability.

The reliability of each five-level paralleled error-preventing step is as follows:

$$R_{Task} = 1 - [(1 - 0.9) \times (1 - 0.9) \times (1 - 0.99) \times (1 - 0.9) \times (1 - 0.99)]$$
$$= 0.999999 \text{ (99.9999\%, or 1 error per 1,000,000 opportunities)}$$

The reliability of the entire job of five super-sure tasks is as follows:

$$R_{Job} = 0.999999 \times 0.999999 \times 0.999999 \times 0.999999 \times 0.999999 = 0.999995$$
$$\text{(99.9995\%, or 5 errors per 1,000,000 opportunities)}$$

Performing each task independently of the other parallel tasks is a vital condition to meet in order to get these levels of work reliability. In the Supervisor Proof Test, the supervisor must use different test equipment from that used by the technician in the Job Proof Test.

Getting the maximum reliability from processes should drive all production management thinking and business risk decision making. The design of the work ought to ensure that high reliability is the natural outcome. You want your people to know for themselves when a thing is "done right." With the use of parallel-tested tasks, human error is detectable and controllable to any level of risk by setting commensurate task quality standards to achieve and by independently double-checking that they are met. Make proof-testing a standard practice in your system of work—make double checks and proof tests "the way we do things around here." When the reliability is insufficient for a situation or the risk is too high, add a parallel test activity to guarantee a higher chance of success. At least parallel critical tasks with very specific and certain error-preventing tests so you can be very sure that a work process is able to deliver the needed results and quality.

There is a sure way to fail the rigorous security of multiple-level parallel test activities—don't do the proof tests! An example comes from the 1960s, when an international airline adopted five-level parallel inspections as part of doing critical maintenance work. The joystick and flaps on aircraft wings of the era were connected by wires and pulleys within the wing. The arrangement let the pilots control the position of the wing flaps. After a maintenance task involving the wires, the job record form was returned to the office with five signatures on it, signifying proper work completion and full five-level inspection. The aircraft was needed immediately, and it was pulled out of the hangar and handed over to the pilot ready to fly. It is a necessity that aircraft pilots conduct their own independent tests to confirm that an airplane is in a safe state to use. When the pilot worked the joystick, the flaps jammed and would not move. The urgently needed plane was pulled back into the hangar, and the wires were reinspected. A wire was found to be off its pulley, yet the completion form advised five times that the work had been done properly.

The technician who had done the work had signed the maintenance record form, as had four other people. The five signatures on the maintenance record indicated that each one had personally seen the job and agreed it was right. This situation is not uncommon—people not checking a thing for themselves. They are mistakenly confident in the capability of the person before them doing their work well, and, seeing that it has already been approved and passed, they don't do their own checks.

They misunderstand their role in the work process, which is to be a proof test to protect against human error. Thank goodness the pilot was not one of the five. The aircraft pilot's inspection and test of the maintenance department's work is an error-proof activity that is intentionally designed into the handover process because the risk from aircraft maintenance failure is too great to accept.

When a company culture values expediency over accuracy, or the organization's management practices bullying to rush work to completion, people will take shortcuts and tell lies. This is how you can fail the best security and safety plans even with four extra parallel tests stipulated. Such ethics live in a company for decades, regularly causing unintended, systematic failures throughout an organization's processes.

The Best Answer Is to Mistake-Proof by Design

Human error cannot be prevented. It is human nature to make mistakes. We will always make errors because our brains and bodies have physical limits.[7] But that does not mean a mistake must lead to a failure. There is a better way to control failure than paralleling test activities. It is to mistake-proof an equipment's design to ensure that human error cannot cause failure. "Mistake-proof" means changing the design of a thing so the design itself ensures that mistakes have no effect on the outcome. A simple example is replacing stairs with long, inclined ramps. A fall on a ramp is unlikely to happen because there is no place to catch one's toes and trip. Even if you fell, the injury would be only a bruise, whereas if you tumbled down a set of stairs, you could break a bone.

Figure 1.12 shows our five-task job designed so that each task is mistake-proofed. The job is always completed with perfect reliability.

You get 100% reliability in a mistake-proof process. In such situations and circumstances, no human error leads to failure. Mistake-proofing does not mean errors are not made—they are inevitable. Rather, when mistakes happen, they do not fail the job or the machine. Examples of mistake-proofing equipment include changing designs of parts so they assemble in only one way and providing parts with a telltale indication of correct positioning. We cannot stop mistakes, but we can stop human error from causing failure. Use mistake-proof designs, and the right outcomes will result the first time and every time. When you can design the effects of human error away with mistake-proofing, there is no better way to guarantee utmost reliability.

The Asset Life Cycle

Figure 1.13 shows the typical life cycle of physical assets in a facility. It, too, is a series—concept, feasibility, detailed design, procurement, installation, commissioning, operation, and decommissioning. There are multitudes of interconnected work processes in every phase, providing innumerable opportunities for error. By now you should not be surprised to learn that a great number of them become latent problems that play out over time to cause future equipment failures. People can make mistakes and errors anywhere, at any time. Investigations into safety incidents confirm that the root

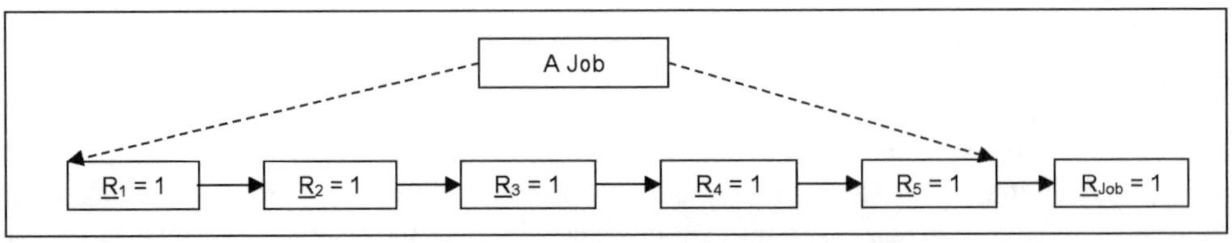

Figure 1.12—A Series Work Process with Each Task Mistake-Proofed

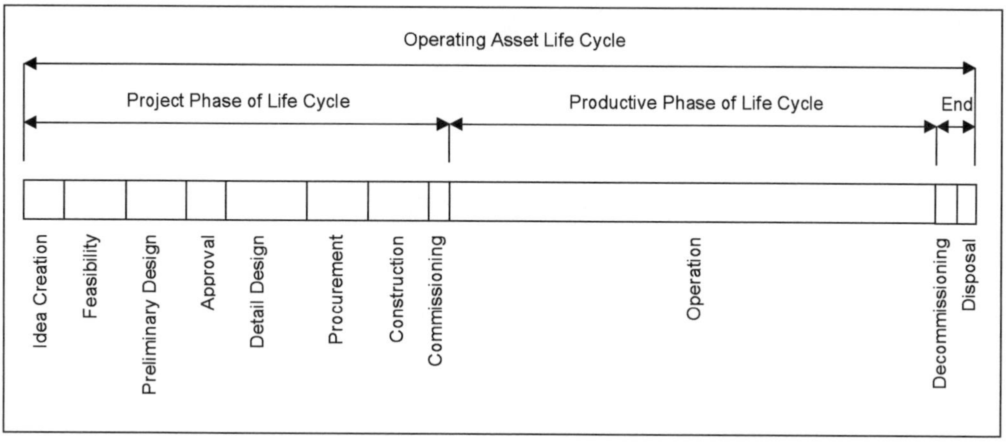

Figure 1.13—The Life Cycle of an Industrial Facility Involves Multitudes of Series Process

causes of failure occur at all stages of a facility's life cycle.[8] This is why you will regularly hear plant and equipment maintainers cursing equipment and plant designers for their hidden design "traps." The reliability of the operating phase is totally dependent on the reliability of all of the numerous human-dependent activities performed in the prior phases.

Getting high reliability from any series process, whether it is a business life cycle, a job, or a machine, is a decision you make, and then you put into place the necessary quality practices, error-prevention methods, and mistake-proofing techniques to deliver it with certainty.

References

1. A. V. Feigenbaum, *Total Quality Control,* 3rd ed. (New York: McGraw-Hill, 1993).
2. Mike Sondalini, "Total Control over Human Error" (paper presented at the ICOMS Asset Management Conference, Fremantle, Australia, May 26–30, 2008).
3. David J. Smith, *Reliability, Maintainability, and Risk: Practical Methods for Engineers,* 7th ed. (Boston: Elsevier, 2005), appendix 6.
4. Stanley F. Nowlan and Howard F. Heap, *Reliability-Centred Maintenance* (San Francisco: Dolby Access Press, 1978).
5. John Osarenren, *Integrated Reliability Condition Monitoring and Maintenance of Equipment* (Boca Raton, FL: CRC Press, 2015), chap. 4, section 4.1.
6. Benjamin S. Blanchard, Design and Management to Life Cycle Cost (Forest Grove, OR: M/A Press, 1978).
7. Ron Moore, RM Group, Inc., "Correlation of Injuries with PM & PdM Maintenance Work Orders and with Corrective & Reactive Work Orders" (presentation, Knoxville, TN, 2007).
8. Malcolm Gladwell, *Blink: The Power of Thinking without Thinking* (New York: Little, Brown, 2005).
9. A. G. Foord and G. Gulland, "Can Technology Eliminate Human Error?" Trans IChemE, Part B, *Process Safety and Environmental Protection,* vol. 84, no. B3 (2006): 171–173.

2

The Physics of Failure

Figure 2.1 represents a distorted conveyor pulley in overload condition. If this happens to a piece of equipment, the parts fail fast. The equipment will run when you press the start button, but not for long. There is little forgiveness when machines are pushed or distorted beyond their design capability. If you want reliable plant and equipment, parts must stay well within their stress limits. Once the available microstructure is overstressed, there is sure to be a failure sooner or later. Plant and equipment are only reliable if their parts' materials of construction are fit and healthy. When loads push a part beyond its capability, it fails, and the machine that it is a part of breaks down.

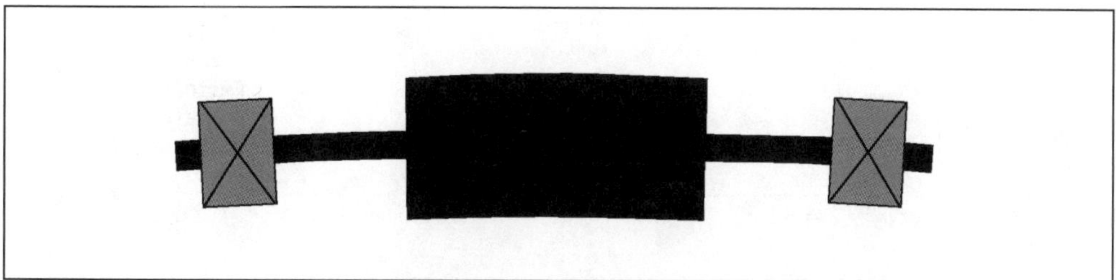

Figure 2.1—Machine Distortion Overloads Parts

There is a retired professor of maintenance and reliability who tells a story in his reliability engineering seminars about the financial outcomes for two organizations with different strategic views on equipment reliability. Some years ago, a maritime operation bought three diesel engines for a new ship. At about the same time, in another part of the world, a railway bought three of the same model diesel engines for a new haulage locomotive. The engines went into service on the ship and the locomotive, and no more was thought about either selection.

Several years later, the opportunity arose to compare the costs of using the engines. The ship owners had three times less maintenance cost than the railway. The size of the discrepancy raised interest. An investigation was conducted to find out why there was such a large maintenance cost difference for identical engines in comparable duty. The engines in both services ran for long periods under steady load, with occasional periods of heavier load when the ship ran faster "under-steam" or the locomotive went up inclines.

In the end, the difference came down to one factor: the shipping operation had made a strategic decision to de-rate all engines by 10% of nameplate capacity and never run them above 90% of design rating. The railway ran its engines at 100% duty because they had been designed for that duty and so, it was thought, they should be worked at that duty. That single decision reduced the shipping company's maintenance costs by 200%. Such is the size of the financial impact of a seemingly minor difference in the load carried by equipment parts.

The Cause of Machinery and Equipment Parts Failure

A force put on a machine part stresses the part's physical structure. The loads come from the use and operation of plant and equipment under service conditions. Machines break down because their parts' atomic structures can no longer take the imposed forces. Microstructures fail for two reasons: as a result of deformation, when stress or fatigue causes the atomic bonds to separate, or degradation, when the atomic bonds are attacked and removed.

When stress is applied, the atomic matrix reacts by deforming to absorb the load. If the stress is too great, the bonds across the load-carrying section separate and the material breaks—this is *overload.* If a load is excessive but rapidly removed, only a few bonds separate. The unbroken bonds that remain carry the load but with less structure available—this is *fatigue.* The equipment designer's role is to select material for a part that has adequate strength for the expected stresses.

Figure 2.2 shows why parts fail from stress and fatigue. It indicates how operating stresses can overload a part's microstructure or weaken it so that it can no longer take the load. The material-

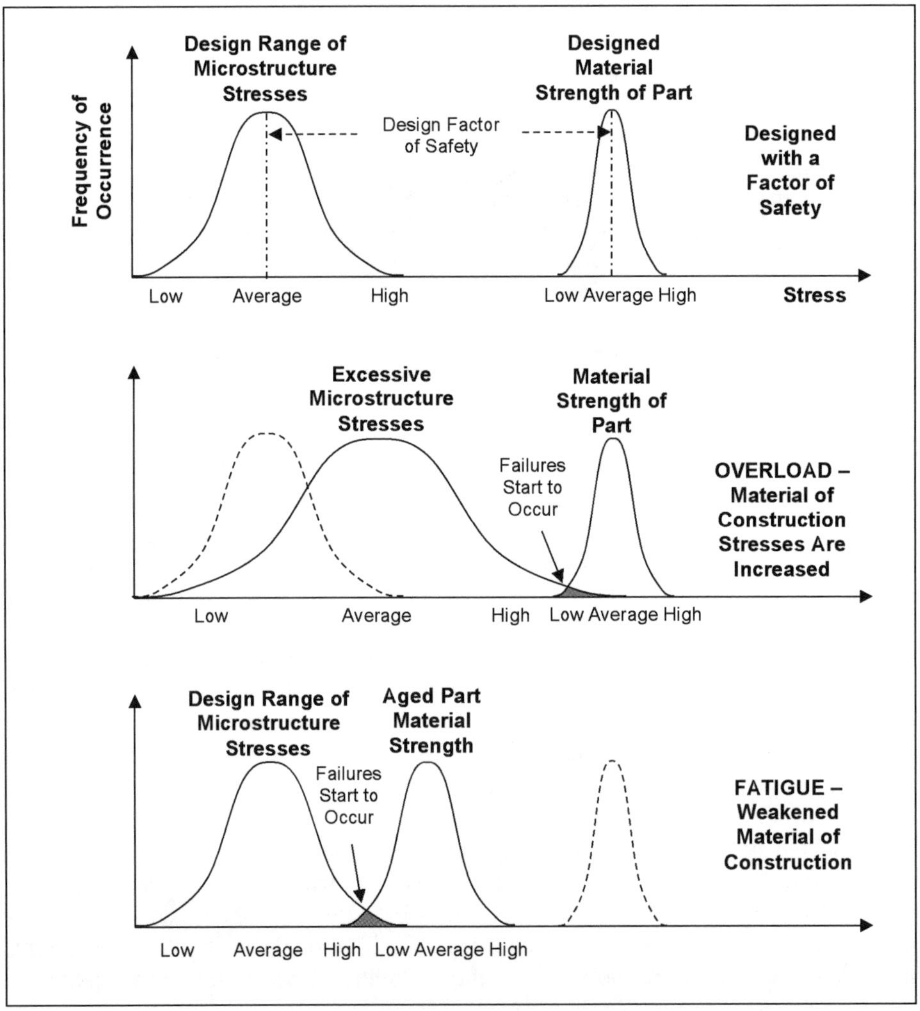

Figure 2.2—Parts Fail When the Stress on Parts Is Greater than the Strength of Parts

strength distribution curves on the right side of the figure represent the stress levels at which materials of construction fail. The curves are known as density functions of probability versus stress/strength. They show the natural spread of variation in the stress-carrying capacity of identically specified material bought from different suppliers. The load-bearing capacity of a material of construction is dependent on the manufacturer's formulation and how well its chemical and microstructure properties are controlled during formation. This variation is probabilistic; when you buy material from different suppliers to make parts, there will be a range of stress-carrying capabilities for seemingly identical material. The curves on the left side of Figure 2.2 are probability distributions showing the extent of operating stresses that a part is expected to suffer. They vary from negligible when equipment is at rest to maximum under occasional extreme loading. The stress values used in designing machinery parts are those expected when running at the specified operating conditions defined in the service duty scope for the machine.

The top set of curves show the equipment operated and maintained as the designer intends. The strength of material used in a part and the range of expected operational stresses are wide apart. There is no chance that the part will fail, and it can expect a long working life because the highest operating stress is well below the least-strength part's capacity to handle the stress. The gap between the distributions is a factor of safety that the designer gives us to accommodate the unknown and unknowable.

The middle distribution curves represent a situation in which a part's microstructure stresses rise beyond the factor of safety allowance. Some stresses in the part grow so large that they exceed the remaining material strength, and the item overloads and fails. If the designer chooses material with a low-stress capability for a part when the chance of having overloads is great, then at a future time, a stress will arise that exceeds the capacity of the part. The weakest parts will fail early; the strongest will take more stress before they, too, fail.

Equipment failure can be attributable to aging of parts when accumulated stress or chemical attack over time weakens the materials of construction. This is shown in the bottom curves, where the part's material properties are degraded until the part is too weak to carry high loads, and it fails.

Figure 2.3 displays how suffering an excessive stress destroys the material of construction. A portion of the material strength is lost with each high-stress incident as atomic bonds break. Figure 2.3 also highlights the failure prediction dilemma—the timing and severity of overload incidents is

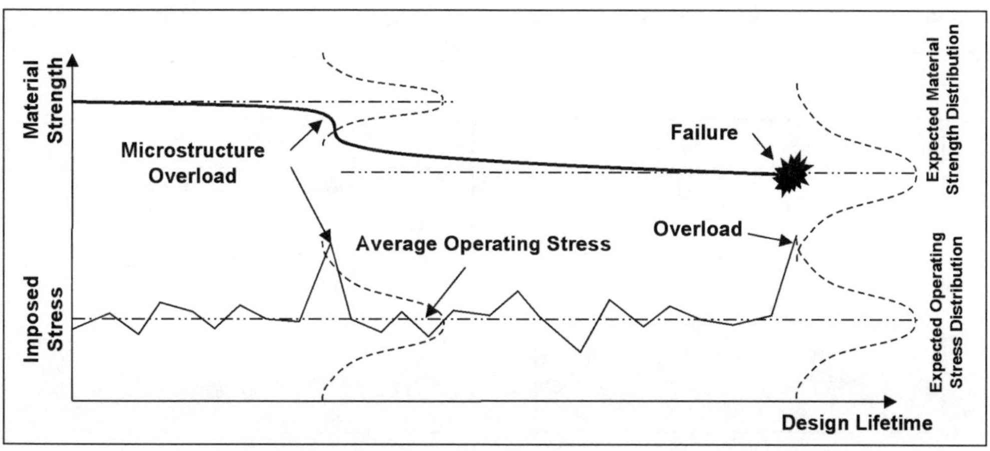

Figure 2.3—Effects of Overload Stresses on the Failure of Parts

random. Every component failure has its causes, but when a part is exposed to high-risk situations that could cause it to fail, failure becomes a matter of chance. Expecting to get a full working life from parts is impossible if they suffer out-of-design conditions.

Figure 2.4 shows what happens to fatigued parts over the long term. The microstructure gradually weakens, either from the accumulated damage of occasional overload conditions or from the gradual aging by regular fluctuating stresses. Fatigued parts eventually fail because a fateful load occurs one day that destroys the remaining microstructure. These excessive stresses are not necessarily the fault of bad operating practices. They are often attributable to wrong engineering choices or the selection of poor maintenance quality standards that cause stressful situations and increase the probability of failure. If you can prevent distortion of parts so that stress levels on their materials of construction are kept far below the values that break their microstructures, then the parts will not fail, and your machines will be highly reliable—and remain so.

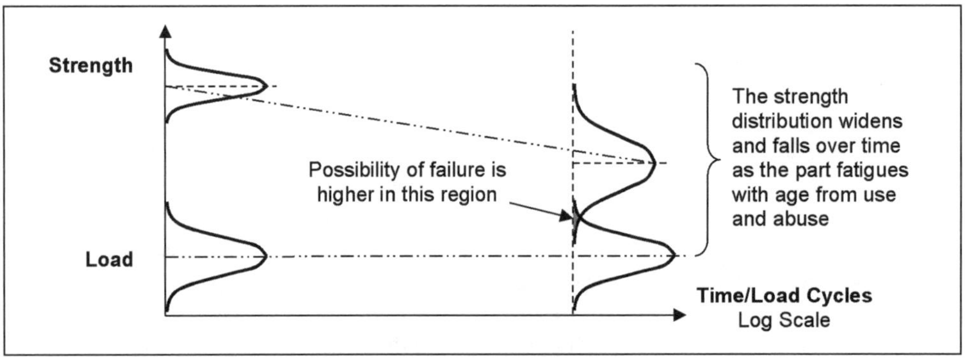

Figure 2.4—Time-Dependent Load and Strength Variation as Stress Damage Accumulates

Parts' microstructures will also fail if chemicals contacting their surface attack the atomic matrix and destroy the material of construction. Examples include oxygen in the air degrading rubber; hydrogen ions in water causing steel to corrode through mechanisms such as pitting and crevice corrosion, shown in Figure 2.5; and aggressive chemicals attacking intergranular phases of the microstructure in alloy metals. In these situations, atomic structures fail by degradation. If you prevent degrading environments from touching your equipment parts, then their atomic structures cannot be attacked and failed.

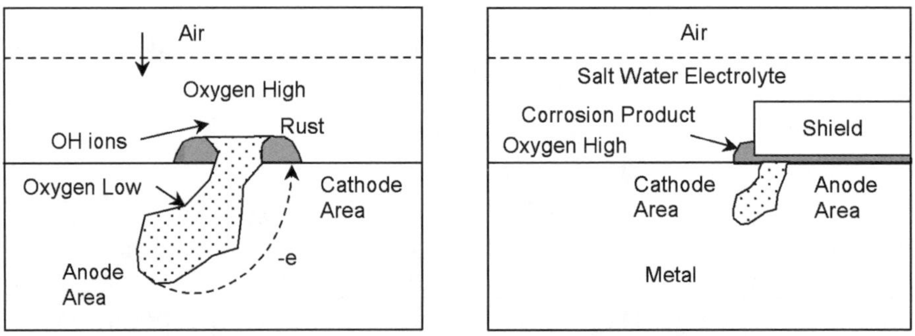

Figure 2.5—Pitting Corrosion and Crevice Corrosion in Seawater

The Physics of Failure

The study of the mechanisms and processes of failure in parts and machines is known as the Physics of Failure. It is the foundation of today's best-practice equipment design methodology used to engineer and build reliable machines. Figure 2.6 shows the Physics of Failure approach for designing machinery and equipment.[1] It recognizes the influences and effects of the Physics of Failure mechanisms on parts. Equipment components are computer-modeled (or prototypes are laboratory tested) to simulate the performance of the materials of construction in a range of operating situations. The computer-generated model is analyzed for weaknesses. The part is put through various operating conditions, such as overloads, temperature effects, geometry changes, and distortion. The modeling identifies the part's likely behavior in the simulated situations and indicates what loads it can take before failing. The results warn of the design limit and operating envelope for the material of construction. During operation, we must ensure that parts are never loaded and stressed to those levels or allowed to degrade to the point that they cannot take their service loads. It is the role of project engineering, maintenance management, and reliability engineering to ensure that parts do not fail and that machines do not stop during operation.

We know what causes parts to fail and equipment to break down—sudden excessive stress, accumulated stress, or removal of structural material. During the design of plant and equipment, we apply our knowledge of the Physics of Failure to select the right materials and designs to deliver af-

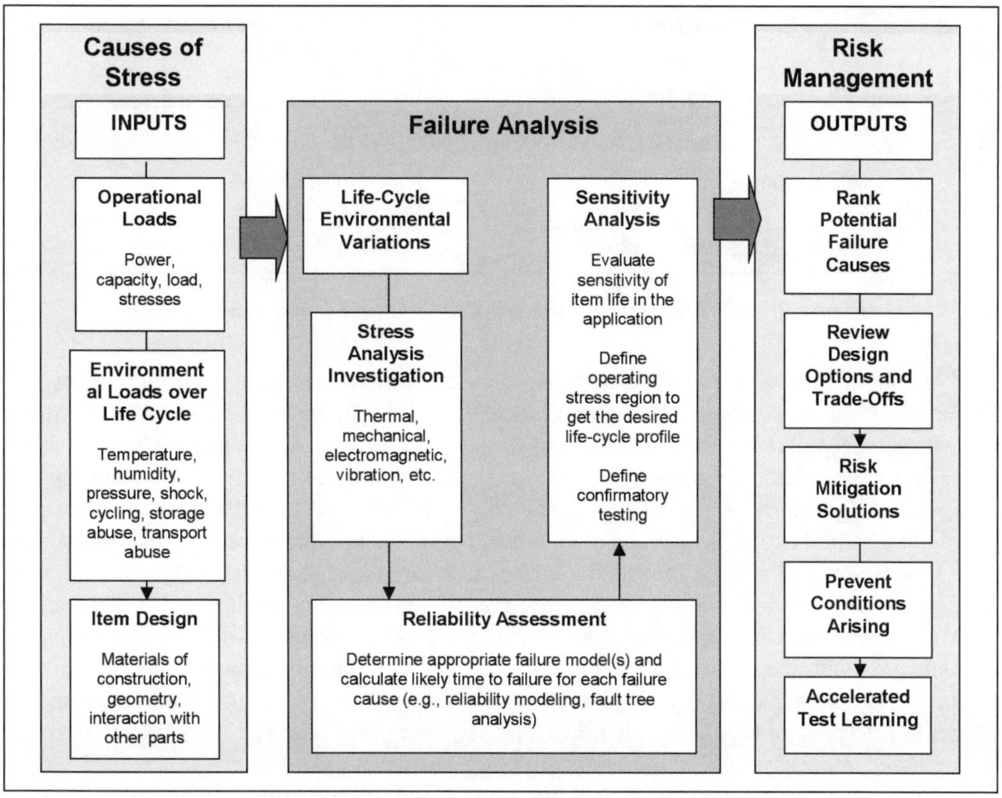

Figure 2.6—Physics of Failure Approach to Reliability Improvement

fordable reliability during operating life. The design envelope sets the service limits of a part's working life. To maximize reliability, we first must have parts that can comfortably take extreme service loads. Second, we must ensure that the microstructure stresses that occur during operation are kept well within the design envelope. Third, the part's physical structure cannot be attacked by damaging chemicals or elements from the contacting environment. If you want reliable equipment, don't strain parts' microstructures to the point of breaking or let their materials of construction be harmed.

Limits of Material Strength

Apply force to an object, and it deforms. Its atomic structure is strained. The greater the force that is applied, the more the deformation. In metals, this relationship is known as Hooke's law. Figure 2.7 shows the stress/strain relationship for some common types of engineering metals under constant tensile load. Metals have an elastic region where load and strain are proportional (the straight lines on the graphs). In this region, the metal acts like a spring. Remove the load and the deformation (strain) reduces, and the metal returns to its original shape. If the load increases, the strain rises to a point that the microstructure can no longer sustain the load, and the metal yields like plasticine. The yielding can be gradual, as in the left-hand plot, or it can be sudden, as in the right-hand plot.

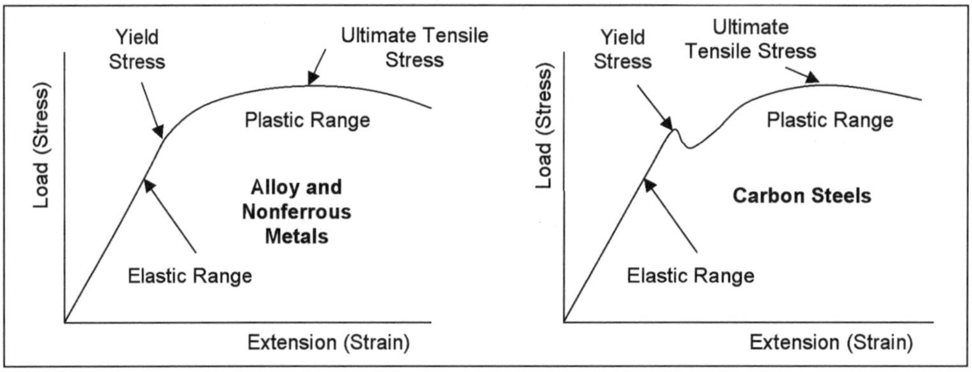

Figure 2.7—When Metal Materials Reach Load Limits, They Deform

Material overstress happens to electronic, electrical, and mechanical parts in equipment that is put under excessive operational loads or suffers environmentally induced stress or forces, such as from vibration, temperature fluctuation, or physical distortion. Parts do not know what causes them stress. They know not the reasons they are overloaded or what caused them to be degraded. Opinions and limited budgets carry no weight with them—they only suffer the facts. Parts react to the stresses and conditions they experience. When the contact region is large, the forces are distributed over a broad surface area and stress is low. As the surface area is reduced, the stresses at the contact points rise. The greatest stress occurs when loads are concentrated on a very small area of microstructure.

After a failure starts in a part, it progresses and grows larger whenever sufficient stress is present. If the stress exceeds the material capability, the part will deform until the microstructures collapse.[2] If a constant stress on a metal part stays well within the straight elastic region of the curve, the item will not fail. To prevent tensile failure in machine parts, we design for maximum operating stresses that are in the lower portion of the stress/strain line by using factors of safety on yields of 3, 4, and even more for high-risk equipment such as crane hooks and lifting chains. High-tensile bolts are an exception, as they are designed to work in the upper half of the stress/strain line.

Have you ever bent a metal wire or paper clip back and forth until it broke? If you have, you have performed a fatigue stress test. A wire that is bent 90 degrees one way and then 90 degrees the other way does not last long. Each bend produces an overstress, causing damage to accumulate in the microstructure until eventually the wire fatigues and fails.

Fatigue failure is caused by cycling loads. In cyclic stress situations, the microstructure behaves differently than it does under constant tensile or compressive load. When cyclic loading is added into a part's working situation, it markedly reduces the part's life expectancy. This is a confusing phenomenon, as the cycling does not require the load to go from positive to negative—for example, bolt fatigue occurs when the bolt is cycled under positive tensile load. Nor does a cycling load need to exceed factor of safety limits for a part to fail. There is also a relationship between a metal's ultimate tensile strength (UTS) and hardness and its ability to handle fatigue loads. The higher the tensile strength and hardness, the more likely it is that the metal will fatigue if it is subject to high fluctuating loads.

There has been a great deal of fatigue load testing done with many materials. Fatigue curves are created in a metallurgical testing laboratory by putting a standard sample bar under cycling tensile load. Eventually the tension combined with cycling stretches the bar to yield by plastic deformation and, ultimately, failure. These tests produce graphs of tensile strength versus number of cycles to failure. They help us understand how much fluctuating load a material can take and still survive. Figure 2.8 is an example of a fatigue curve for wrought (worked) steel, which is used in many industries. Under fluctuating loads just above 90% UTS, the sample lasts 2,000 cycles. Under loads of around 60% UTS, the sample lasts 200,000 cycles before failure—100 times longer. But when loads are below half of the UTS, the sample has an indefinite life.

Note that not all metals have a defined fatigue limit as steel does. Nonferrous metals, such as aluminum and brass, have also been fatigue-tested, and their curves are readily available. The fatigue curves for those metals eventually go to failure. Machine parts made of aluminum or brass metals fatigue with use at even low stress levels, and they will need to be replaced well before the part approaches fatigue failure. Fatigue life limits are the reason commercial aircraft frames made of alu-

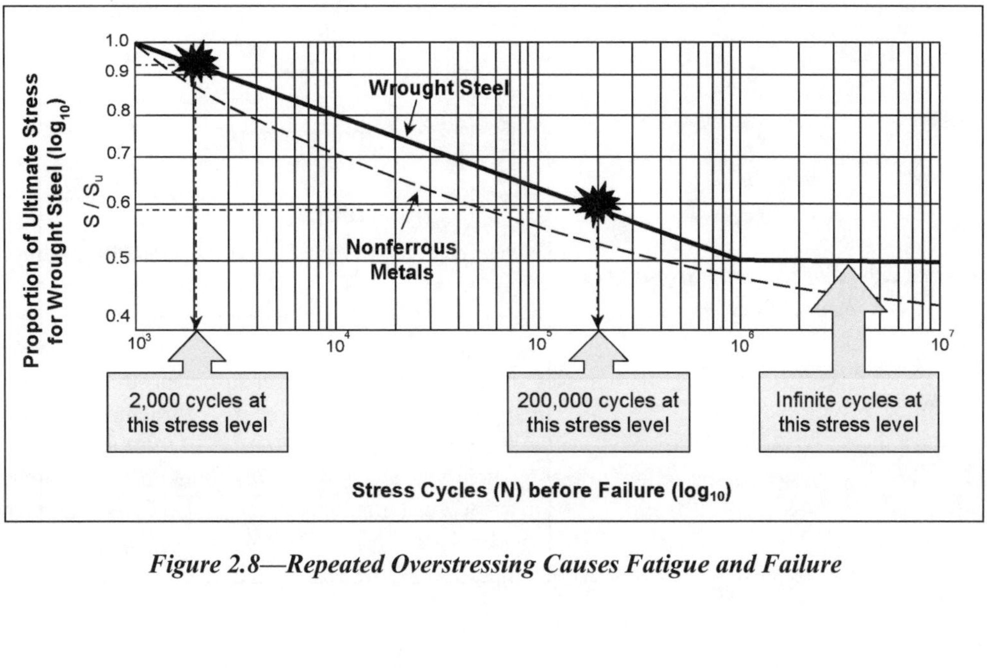

Figure 2.8—Repeated Overstressing Causes Fatigue and Failure

minum are retired from service. The replacement of parts before failure from operational age and use is known as *preventive maintenance.*

In every case, whether as a result of extreme burden or cyclic fatigue, excess stress eventually leads to material microstructure failure. If that happens to your parts, the equipment will break down.

Engineering Limitations Are a Part of the Design

The loads under which a machine works produce the forces that cause microstructure stresses, which, in turn, cause a part to fail. Change the working loads, and the forces change. Change the forces, and the stresses change. We know the engineering formulas of these force/stress relationships. The amount of stress at a particular point of a part's microstructure depends on its shape. To calculate whether there is too much stress on a part, you must know its material properties, its geometry, and the distance between that point and the point where the force is applied. The equations and calculations can get complicated for intricate shapes, but for a round bar, the formulas for the relationships between applied force and the three resultant stresses—axial, torsional, bending—are well documented.

Axial Stress = $F/\pi d^2$
Torsional Stress = $16\ [F \times l/\pi d^3]$
Bending Stress = $32\ [F \times l/\pi d^3]$,

where F is applied force, d is the bar's diameter, and l is distance to the force from the point being analyzed.

Figure 2.9 shows the change in microstructure stress for a change in applied force on a round bar. All three stresses add together at every point on the bar. With an increase in force, the axial stress rises by an equal proportion. For a torsional load, the microstructure stress rises by 16 times. For a bending load, the stress rises by 32 times. Compounding these stresses is the influence of shape changes in the microstructure, such as the keyways and the shaft fillet, shown in Figure 2.10. Fatigue

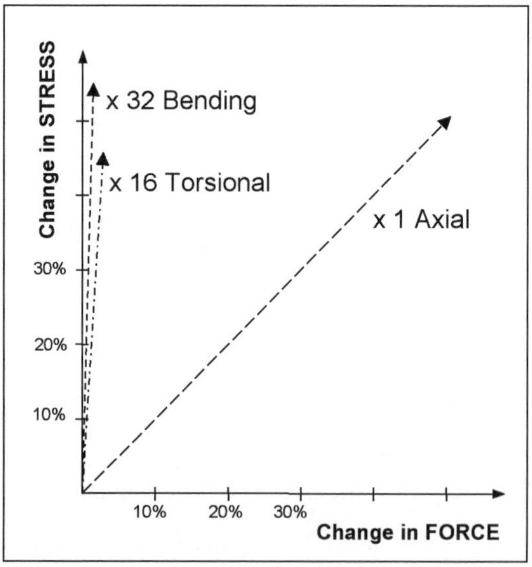

Figure 2.9—Force versus Strain in Round Bars

stresses also increase when parts have sharp contours and when stress raisers, such as notches, holes, and keyways, are present in the part. The contour multiplies the effect of the stress. The stress concentration factor K represents the proportionate rise in the microstructure stress caused by the type of shape change.

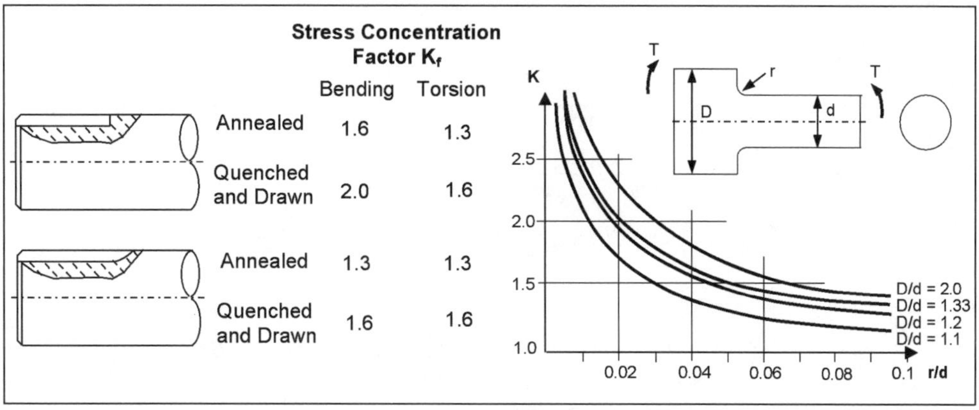

Figure 2.10—Stress Concentration from Keyways and from a Fillet

Because all stresses act concurrently throughout a part, the combination of stresses during high loads may surpass the part's material yield stress at the weakest point and plastically deform the microstructure. If the stress continues past the UTS, a crack will form in the microstructure and dislocations will be created within the material of construction as it separates. Once that happens, things get much worse very fast. The stress needed to propagate an existing failure in the microstructure is significantly less than the stress needed to start the failure. The atomic crack front of a dislocation is razor sharp (imagine you're on the cutting edge of a sharp knife slicing through cheese). Any load that is applied at a crack stress concentration point is multiplied by orders of magnitude.[3] When you change the load conditions on parts by a few percentage points, you can alter the stress levels by thousands of percent. Once there are cracks in the microstructure, the resulting failure can be instant. Parts with stress concentration fractures within the material of construction can break from fatigue even under normal operating loads.

Production Limitations Are a Part of the Design

The metallurgical and engineering limits inherent in a machine always dominate production requirements. Machines create fatigue situations for their own parts. As a machine operates, its components are put under tensile and cyclic load, and stress is created in the parts' microstructure. Figure 2.11 is the previous "stress versus cycles to fatigue" failure plot for wrought steel (Figure 2.8). The dotted lines from the y-axis to the S–N Curve are spaced 10% apart. Where they cross the fatigue curve, the dotted lines drop to the x-axis. At 80% UTS, a sample piece is expected to last about 10,000 cycles before failing. At 70% UTS, it should last about 40,000 cycles. At 60% UTS, its life is about 200,000 cycles before failure. Each 10% drop in stress increases operating fatigue life by four or five times. In fatigue situations, every change in stress greatly impacts the service lifetime of equipment and the resulting maintenance costs.

For plant and equipment, it is the size of the stresses within parts that dictates whether they retain their integrity. Under load, a part can only survive up to its stress limits; passing that limit leads

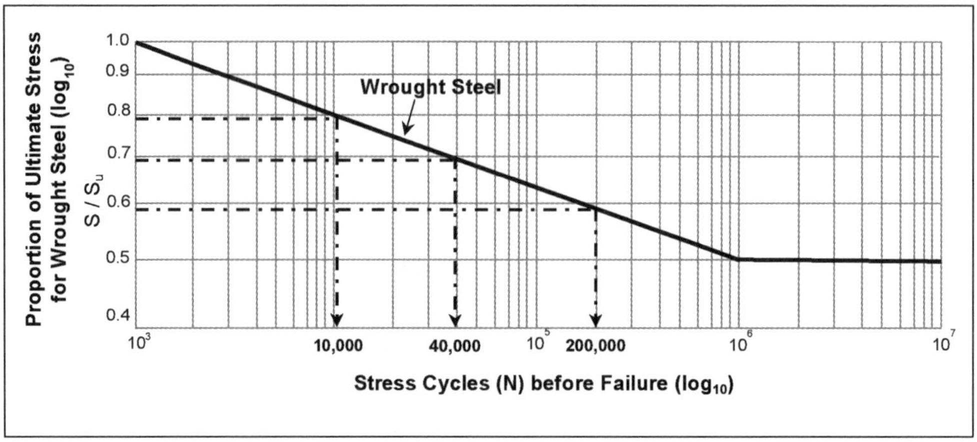

Figure 2.11—Ferrous Metal Fatigue Stress versus Life

to its destruction. Somewhere in between disaster and dawdle, you must decide the stress levels at which to run your machines. That decision has surprisingly huge business cost implications because it determines how many times your equipment will break down during its life.

Although machine designers limit the effects of stress by using factors of safety in choosing the materials of construction, they also presume that the machine is precisely assembled, that it is installed without extra preimposed stresses, and that it is run within its design envelope for its whole life. Situations such as shaft misalignment, soft foot, unbalance, incorrect fits and tolerance, wrong lubricant, insufficient lubricant, water in lubricant, excessive temperatures, and hundreds of other causes of failure are not expected in normal operation. Designers do not engineer equipment to run with parts already heavily preloaded high up the stress/strain line. When machine parts are badly deformed during installation, the microstructure can be so severely stressed that the parts already carry a high percentage of their UTS before the machine even starts operation. Prestressing machine parts as a result of bad installation will drastically shorten their life—10% additional stress can cost you 10 breakdowns. When machine installers and maintainers put parts and machines into place deformed, they guarantee the early death of the equipment.

During World War II, British airmen referred to ongoing trouble with aircraft, in spite of aircraft mechanics' best efforts, as "gremlins at work." A gremlin is an imaginary creature that lives inside machines and equipment, breaking parts and causing trouble in systems and devices. We are the gremlins. Our machines and equipment are failed by us in our ignorance. Failure is not an unlucky accident—we murder our machines.[4] They die by being killed in the ways listed below.

- We twist them
- We squeeze them
- We buckle them
- We hit them
- We poison them
- We burn them
- We shake them
- We snap them
- We choke them
- We boil them
- We crush them

Parts do not care why the stress they suffer was imposed on them—they do not know that a delivery was late, or that they were overloaded to make up time, or that the repair technician did not have the tools or the skills to do the job right. Parts simply fail when the stress on them becomes excessive, regardless of the cause. Machinery "murderers" are found among equipment manufacturers, among constructors and installers, among equipment users, among maintenance crews, and among business decision makers. People do not intentionally cause failures, but the parts and machines still end up broken.

All machines have production limitations. Run them with parts distorted by overloads or deformed out of shape during installation, and they will soon fail. Overloading a plant to try to make up for lost production time or to fulfill late orders may seem like a heroic management choice, but really it is a destructive asset management and business risk decision. Managers who approve overload levels of operation cause the early death of their machinery. To maximize plant uptime and throughput, run equipment at production rates at which the working parts' microstructure is always below design stresses. Install your machines so they are deformation-free. Make sure the surfaces of parts under load have no stress-raising damage. You will be a wise asset manager if you never run your plant harder than 90% of its rated load.

Equipment Reliability Cliffs

Equipment and machinery reliability grows as machines are made more accurately, installed precisely, and components kept in better condition and health. Once you reach superb operating and maintenance precision with accurate work quality control, your reliability growth will bound forward by years at a time. Conversely, when assembly accuracy is poor or parts' environments are degraded, you will lose reliability at a rate of years at a time. In fact, the rate at which you lose reliability when component health and strength are lost is so steep that it is as if you'd flung your machines over a cliff. Many industrial companies blindly commit industrial suicide daily by leaping off "reliability cliffs." Reliability cliffs show up in graphs and plots of equipment service life as rapid declines into reliability disaster.

Oil Particulate Contamination Reliability Cliff

The table on the left side of Figure 2.12 located on the following page, is famous in the world of equipment maintenance and reliability.[5] It shows the results found by the British Hydromechanics Research Association in the 1990s from a three-year controlled field study of 117 hydraulic machines (injection molding, machine tools, material handling, mobile equipment [earth moving, etc.], marine hydraulics, and test stands) conducted with the purpose of correlating hydraulic fluid cleanliness with machine breakdown frequency. Using a life factor equal to 1 achieved with the ISO 4406 18/15 level of particle contamination often recommended by equipment manufacturers, you can see how quickly operating life falls once hydraulic oil contamination rises above that value. It is clear that oil that is dirty with solid particulates rapidly fails hydraulic systems. When you put the information in the table onto the graph on the right side, you see a reliability cliff.

Metal Fatigue Reliability Cliff

The graph on the left side of Figure 2.13 is a laboratory fatigue curve for a wrought steel. The cycles to failure, N, are charted against applied stress, S, producing an S–N fatigue plot. When the data are graphed on the linear scales of the right-hand chart, another reliability cliff appears.

Ave, ISO Code	Ave, Hrs between Breakdowns	Relative Life Factor
24/21	200	0.19
23/20	250	0.24
22/19	325	0.31
21/18	430	0.41
20/17	600	0.57
19/16	800	0.76
18/15	1,050	1
17/14	1,400	1.33
16/13	1,900	1.81
15/12	2,600	2.48
14/11	3,800	3.62
13/10	5,000	4.76
12/9	6,500	6.19
11/8	9,000	8.57
10/7	20,000	19.05

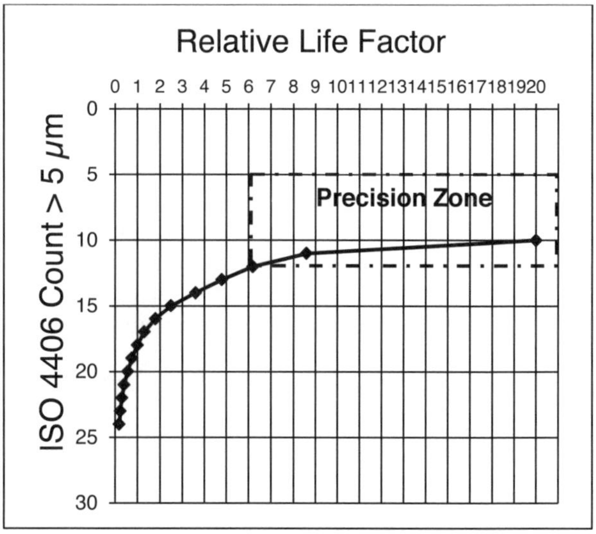

Figure 2.12—Hydraulic Oil Contamination Reliability Cliff

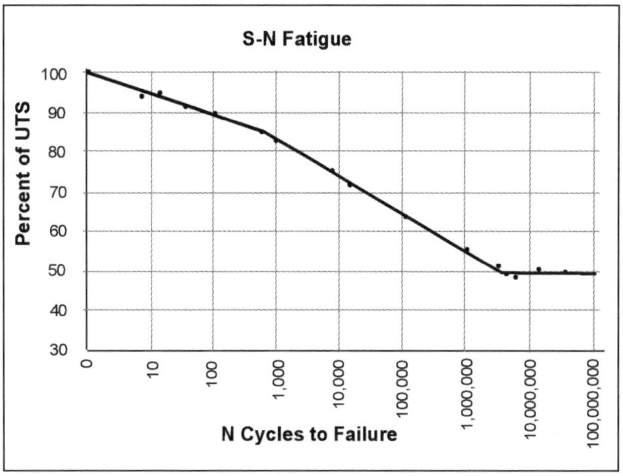

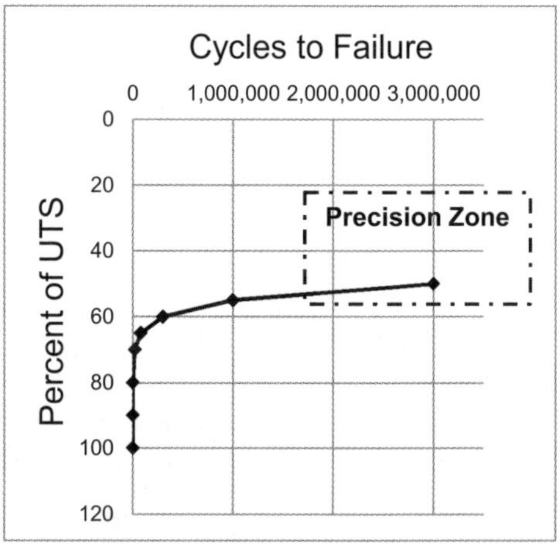

Figure 2.13—
Ferrous Metal Fatigue Reliability Cliff

Roller Bearing Internal Clearance Reliability Cliff

The curve depicted in Figure 2.14 is from laboratory research. It shows the ratio of 6310 ball bearing service life compared with design life versus bearing internal clearance (25.4 μm = 0.001 inch). It presents the effect of altering bearing internal clearance on bearing operating life span. You can easily see a reliability cliff to the left and the right. For a 6310 ball bearing, a small preload actually improves service lifetime above designed life, but once that point is passed, or if it is not achieved, service lifetime falls off a reliability cliff into rapid bearing failure.

Figure 2.15 derives from the 6310 ball bearing life curve and shows how much change will destroy each 10% of bearing life. Starting at the full life preload of –20 μm, the change in clearance for each 10% loss of life is gauged and graphed. Another reliability cliff—an abyss, actually—appears. Every 2 μm preload error away from the ideal preload will destroy 10% of the bearing's service

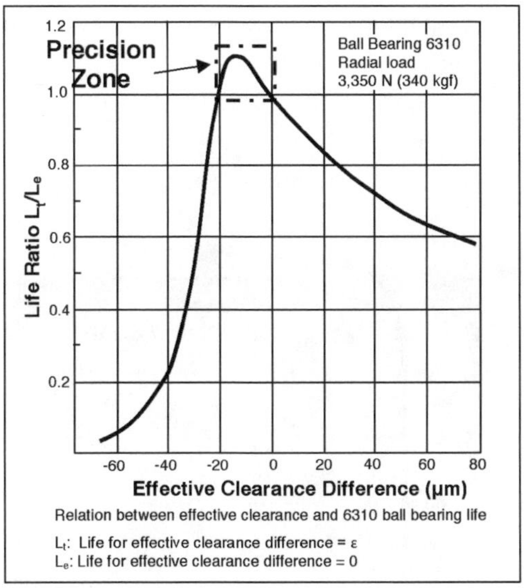

Figure 2.14—Bearing Life versus Bearing Clearance Reliability Cliff

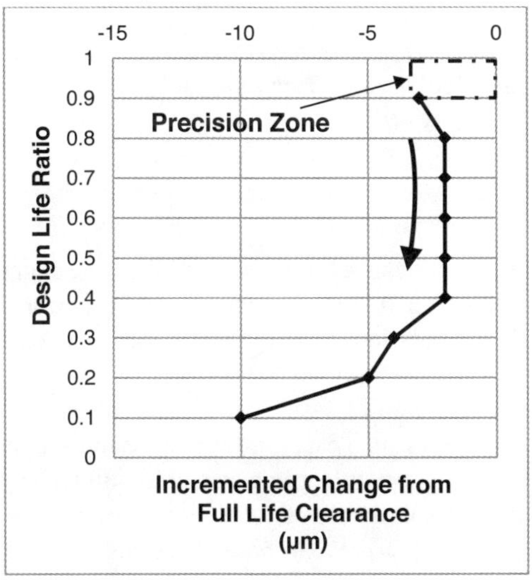

Figure 2.15—Bearing Life versus Change of Bearing Clearance Reliability Cliff

life. Once the bearing is already horribly preloaded, it takes another 5 μm preload to lose 10% more service life. Two microns is about the size of a germ.

Getting roller bearing clearance right is a vital part of machinery assembly. The machine designer specifies suitable manufacturing sizes and tolerances for optimum new assembly fits. Keeping internal clearances right—that is, within 10μm of perfect—during operation and rebuild is enormously hard. You must prevent component distortion, prevent temperature rise (and fall), and minimize fatigue stress cycling: that is, you must control all operating conditions masterfully. It is hard to do, but it is vital for high roller bearing reliability if you want zero-breakdown machinery.

Shaft Misalignment Reliability Cliff

The left-side graph in Figure 2.16 is adapted from the *Shaft Alignment Handbook* by John Piotrowski.[6] It is a plot of the typical operating life span of rotating machinery subjected to various amounts of shaft misalignment. The data are based on a large number of case studies in which shaft misalignment was found to be the root cause of rotating equipment breakdown. Graph the bottom curve of the failure zone in the chart, and the reliability cliff on the right reveals itself.

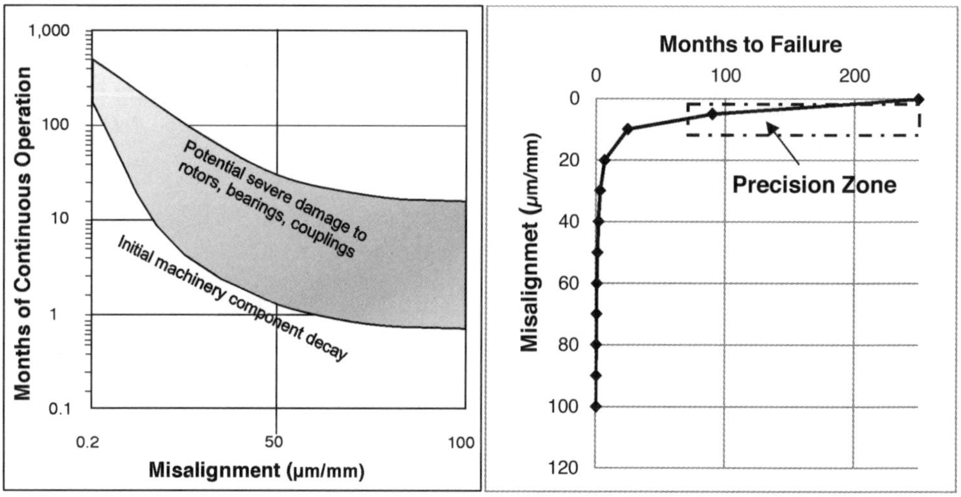

Figure 2.16—Shaft Misalignment versus Machine Life Reliability Cliff

Equipment and Parts Degradation Cycle

Overstressed parts get damaged and then fail. Every overload steals a little of a part's life. The degradation cycle depicted in Figure 2.17 shows the influence of a part's failure sequence on its performance. All parts go through the recognizable stages of degradation. First the component works as expected, but then abnormal situations arise and it gradually starts to fail.

Some parts fail without exhibiting warning signs of a coming disaster. They show no evidence of degradation—there is just a sudden, catastrophic failure. In such cases, all we see is the sudden death of the part. This commonly happens to electronic parts. It is worth noting that almost all failures, even to electrical and electronic parts, are ultimately mechanical (involving relative movement), contaminant, or over-temperature related.[7] For the most part, we can prevent those situations.

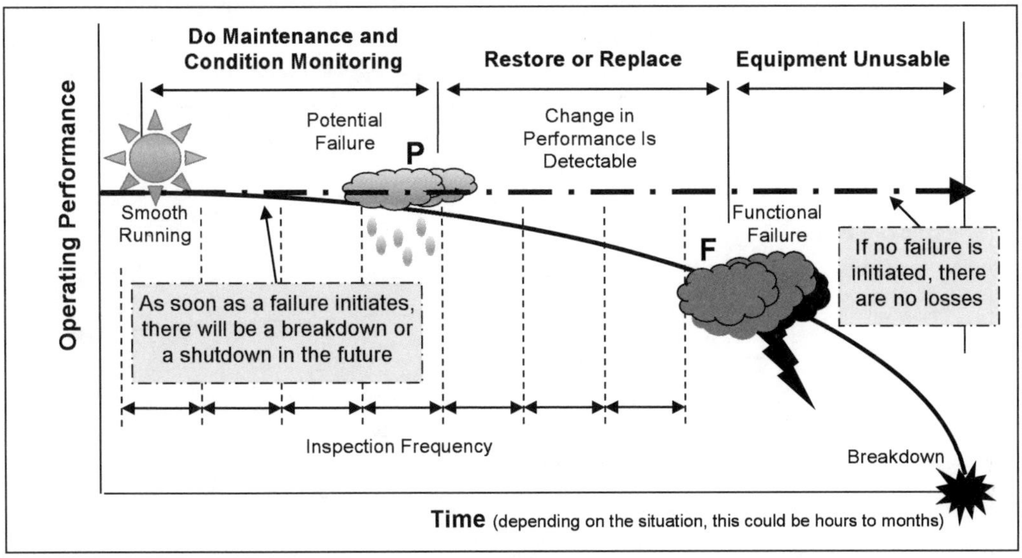

Figure 2.17—Component Failure Degradation Sequence

The point at which degradation is first possible to detect is the potential failure point,[8] P. After this point, failure will eventually happen; how soon breakdown occurs depends on the size of the stresses imposed on the failing part. The point at which degradation has progressed beyond salvage and the equipment performance is critically affected is the functional failure point, F.

The degradation cycle is the basis of predictive maintenance strategy and condition monitoring. The degradation curve explains why and when to use condition monitoring on plant and equipment. Knowing that many equipment parts show evidence of developing failure, it is sensible to inspect them at regular intervals for signs of approaching failure. Once you select an appropriate technology to detect and measure degradation, a part's condition can be trended and the impending failure monitored until it is time to make a repair.

You set the inspection frequency to detect the onset of a problem so there is time to address the failure before it happens. Condition monitoring can be as simple as regular "feel and listen" observations of parts and equipment performance by the operator, or as complicated as using continuous on-board monitoring with sensors and instrumentation feeding computer-controlled diagnostic and prognostic software to judge the likely remaining life. When repair work takes a long time to plan, organize, and schedule, that justifies using more complex technology and giving yourself a long P–F interval to prepare for the restoration or rectification. If you choose to use low-tech condition monitoring solutions with short P–F windows, there will be little time to the breakdown after a failure is detected. Such a predictive maintenance strategy necessitates having quick access to spare parts or having redundant equipment to take over when the duty machine or plant fails.

During operation, every company using physical assets faces the dilemma of probabilistic parts failure, shown in Figure 2.18. The figure presents a series of degradation curves for selected working parts in a centrifugal pump-set. Each part has varying P and F points. The longevity to the P point and the drop rate of the P–F curve for each part depends on the stresses on the materials of construction and the physical degradations suffered by the part. When you use plant and equipment, you live in a situation in which any part can fail at any time if its operating stresses exceed the microstructure's load-carrying capacity. Our machines and equipment typically contain dozens, and often hundreds, of parts under stress, and each of them presents an opportunity to have a breakdown. Every physical asset lives in a probabilistic uncertainty as a result of multiple opportunities for failure caused by all of its degrading and deformed parts. The best protection against failure is to proactively keep every critical part in your plant and equipment fit, healthy, and well for its lifetime.

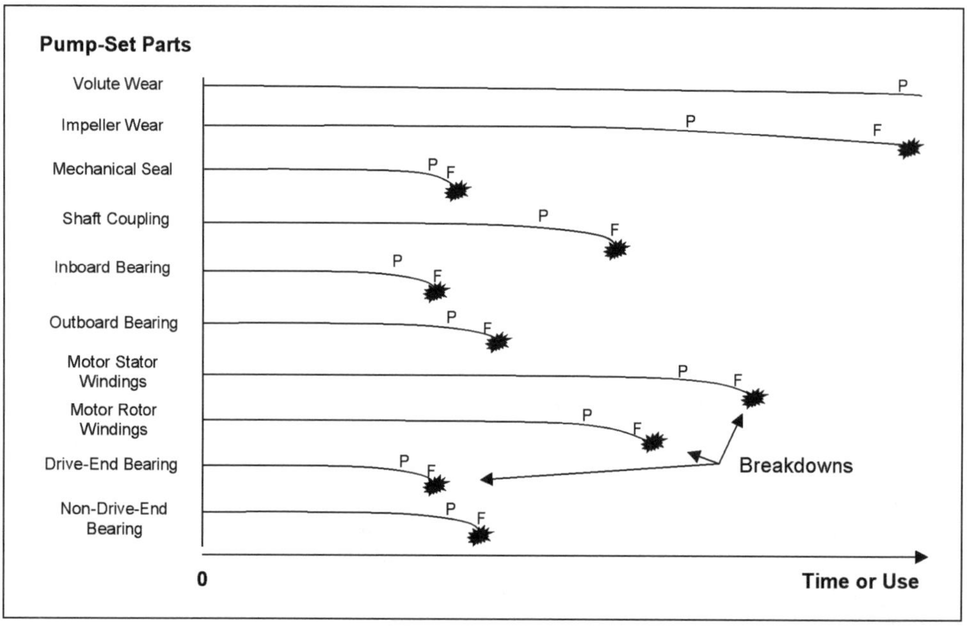

Figure 2.18—Each Part within Equipment Degrades on Its Own Curve

Reach the Standards That Cause Reliability

Reliability success is often said to be a journey, but that is the wrong metaphor—reliability success is a climb! World-class machinery reliability is only found at the peaks of machinery precision zones. It cannot survive anywhere else: long-lived, trouble-free reliability is a mountainous climb to the place where lasting precision quality lives.

Reliability is a step change. Either you already have reliability, or you are doing one of the following: (1) constantly correcting problems or (2) fixing broken things. There is only one way to make machines reliable: their parts must be even more reliable. The health of your parts produces the reliability of your machines. High machine reliability requires asset life-cycle processes and supply chains that deliver dependable precision quality to working parts. The path away from reliability cliffs starts by setting health quality targets for parts that, when achieved, will result in reliability.

Plant and equipment failure factors such as vibration, fits and tolerances, deformation, unbalance, misalignment, lubricant condition, and fastener tension need to be quality controlled to the values that produce highly reliable machines. Let one go out of control, and world-class reliability is unachievable. Step out of the precision zone with any factor, and you fall over its "reliability cliff" and your machines die. Reliability requires developing design, engineering, manufacture, warehousing, installation, operating, and maintenance procedures with quality standards to prevent deformation and degradation and then training managers, engineers, operators, and maintainers to follow them with great certainty.

Life is wonderful at the top of the reliability cliffs. The view is magnificent. The air is clear and fresh. You get the sunshine from dawn to dusk. You have time and money to enjoy yourself and pursue happiness. But take one step too far in the wrong direction, and success ends at the edge of the cliff.

References

1. Michael Pecht, "Why the Traditional Reliability Prediction Models Do Not Work—Is There an Alternative?" (Center for Advanced Life Cycle Engineering, Electronic Product and Systems Center, University of Maryland).

2. J. E. Gordon, *The New Science of Strong Materials, or, Why You Don't Fall through the Floor,* 2nd ed. (New York: Penguin, 1976).

3. Robert C. Juvinall, *Engineering Considerations of Stress, Strain, and Strength* (New York: McGraw-Hill, 1967).

4. Rod Bennett, "Machines Don't Die, They're Murdered" (National Condition Monitoring Forum Australia, August 2006).

5. Accessed at http://www.oiltransfer.com/files/british.pdf, June 26, 2015.

6. John Piotrowski, *Shaft Alignment Handbook,* 3rd ed. (Boca Raton, FL: CRC Press, 2007).

7. First heard from retired professor David Sherwin in his three-day "Introduction to Reliability Engineering" course, Perth, Western Australia, June 2007.

8. John Moubray, *Reliability-Centred Maintenance* (Oxford, UK: Butterworth-Heinemann, 1991).

3

Variability in Outcomes

Probability, likelihood, chance—the more we learn about them, the more we realize how much they impact our lives, our businesses, and our machines.[1] All around us, things happen. People make choices and act. We only see the effects of those choices in the future. Often we can't differentiate one effect from another because past choices interact to cause mysterious and unknowable events. Operators, maintainers, manufacturers, engineers, managers, purchasing officers, suppliers, and many others make choices all the time that impact the lives and reliability of our plant and equipment.

All unknowns and vagaries introduce variability—the cause of most of our operating and business problems. Variability is "the range of possible outcomes." A business with an aim of providing a product or service with consistent specifications does not want its processes behaving randomly and producing off-specification results. Out-of-specification products are a waste of money, time, and effort.

Large amounts of modern organizations' resources are devoted to controlling variability within their business and operating processes. The people involved in this duty carry titles such as executive, manager, supervisor, superintendent, or the like. Their role is to ensure that outputs are within prescribed limits. Anything outside those limits is urgently controlled. A business process with high variability means that outcomes range from good to mediocre to disastrous. Things are uncontrolled and volatile. This randomness is the exact opposite of what is required. It is much more profitable to get the right result every time.

Observing Variability

A simple tabletop game can help you understand why variability is a problem. It is a great introduction to controlling the variability of processes. In Figure 3.1, two lines cross at 90° with a 2 millimeter (mm) diameter circle drawn at their intersection. The next-larger circle is 10 mm in diameter, and the outer circle is 50 mm in diameter. Sit at a table, and with the nib of a pen at the height of a 300 mm ruler, try to drop the pen by hand into the smallest circle. A hit within the 2 mm circle is the "process" outcome that you require.

Repeat the targeting and dropping process at least 30 times. After each drop, measure the position of the new mark to an accuracy of half a millimeter. Record the horizontal distance from the vertical line (the X distance) and the vertical distance from the horizontal line (the Y distance) in a table like that shown in Table 3.1.

Observe the average and spread of the X and Y results. In Table 3.1, no hits are within the 2 mm circle; some are on the edge, or near, but most are far away. Even though great effort was made to control the "process," the results are spread across a wide range of outcomes; there is no repeatability. That is variability.

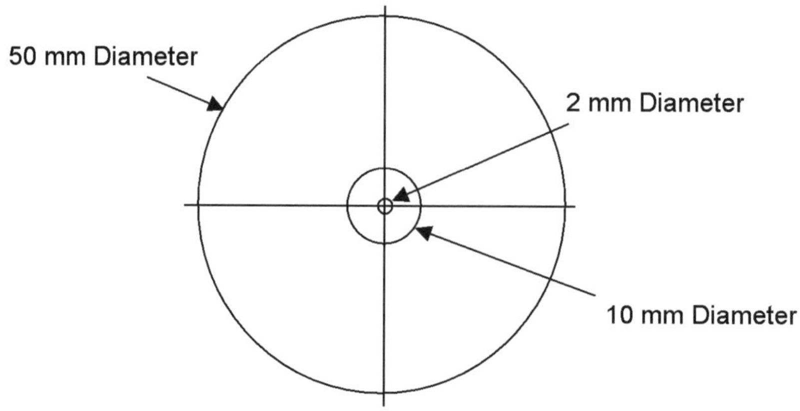

Figure 3.1—The Crosshair Game

Table 3.1—Record of Crosshair Game Hits

Hit No.	Distance X	Distance Y	Hit No.	Distance X	Distance Y	Hit No.	Distance X	Distance Y
1	8.5	16	11	1.5	5	21	1.5	5.5
2	7	9	12	1.5	20	22	3	3
3	4	16	13	3.5	3.5	23	3.5	0
4	3.5	2.5	14	2.5	12	24	2.5	6
5	5	24.5	15	3	24.5	25	0.5	2
6	5	16	16	4.5	6	26	1	2
7	7	10.5	17	4	12.5	27	3.5	10.5
8	5.5	9.5	18	5.5	5	28	1	9
9	2	3.5	19	1	9	29	4	14
10	3	2	20	6	4.5	30	0.5	3.5
			Average	X = 3.48	Y = 8.90			
			Spread	0.5–8.5	0–24.5			

If the aim of the game is to have the pen drop inside the 2 mm circle every time, then we have a very poor process for doing that. To get better results, you must change the process. To be successful at the game, you must invent a different process that puts the pen inside the 2 mm circle every time. The results reported in Table 3.2 were obtained from a process in which the pen was dropped after aiming at the circle from above, much like using a targeting sight to drop a bomb from an airplane.

The second attempt to play the Crosshair Game using a modified process got better results; the X and Y values are virtually the same. The averages indicate that the hits were closer to the intersection than they were for the first process. There is less spread. But the second process is still not suitable for meeting the requirements. It is unlikely that any process using human hands to drop a pen within a 2 mm circle from a height of 300 mm has sufficiently accurate control. Human hands cannot meet the required accuracy. You could tell the person dropping the pen to "try harder" or "improve the quality of their efforts," but that would be pointless because it is the process that cannot do what is required, not the person. To get the pen consistently within the circle requires a better process that removes the variability caused by the human hand.

Table 3.2—Record of Crosshair Game Hits Using a Sighting Process

Hit No.	Distance X	Distance Y	Hit No.	Distance X	Distance Y	Hit No.	Distance X	Distance Y
1	8	10	11	5.5	6	21	3.5	0
2	5	6	12	2	4.5	22	2	5
3	4	3.5	13	0	1	23	0.5	1
4	3	4	14	5	2	24	6.5	0
5	2.5	1	15	4	7	25	3.5	3
6	2	0.5	16	3	1	26	0	8.5
7	13.5	7.5	17	3.5	5	27	6	1.5
8	10.5	9.5	18	4	0	28	0	4
9	1.5	7	19	4	1	29	2	1.5
10	7.5	6.5	20	2	2.2	30	0	6.5
			Average	3.82	3.87			
			Spread	0–10.5	0–10			

Since the game was invented, it has been changed to enlarge the target zone to 10 mm and reduce the number of drops landing on the target to 10. The frequency distribution of results after playing the Crosshair Game some 400 times with the larger target is plotted in Figure 3.2. Many players got one or two drops inside the 10 mm circle. A lot of players got none.

Getting three or more drops within the 10 mm circle is difficult. The most successful player got nine drops on target. On his second attempt, he got five drops on target. Some people are gifted in a particular way that the rest of us are not. That is what makes gold-medal-winning Olympians so rare. To win Olympic gold, you must work at improving your ability, but first you must have a capability in your specialty that is well above average—you already have to be exceptional in order to

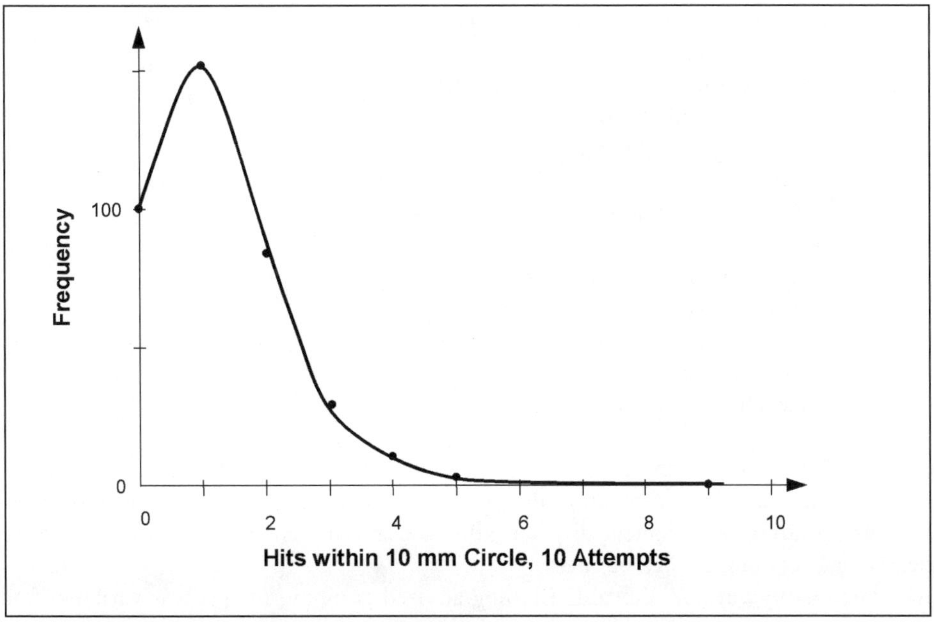

Figure 3.2—Frequency Distribution Plot of Crosshair Game Results

start training to win Olympic gold. But in business, you can't wait for extraordinary people to walk through your door—there just aren't that many of them. In business, you have to work with people like the rest of us.

Several Crosshair Game solutions have been proposed by past players. These include using a long, tapered funnel to guide the pen onto the target; using a tube in which the pen slides; using a V-shaped slide to direct the pen into the circle; using a guide rod with the pen fixed in a slider that moves up and down the guide; and using a robot with a steady manipulator to drop the pen. As good as these solutions are, they still involve human interaction in locating guides and maintaining equipment. When people are involved in a process, mistakes will be made at some point. The "human factor" causes variation and inconsistency. But if the solution were mistake-proofed, it would not matter where the pen dropped—it would always end up inside the circle.

There is one mistake-proof answer known to the author. It requires using the target in a different way. My thanks and respect go to the tradesman boilermaker who suggested it. Figure 3.3 depicts his solution: form the paper into a cone with the 2 mm circle at the bottom. No matter where the pen is dropped, it is always on target. Human error has no effect on the outcome. This mistake-proofed solution turns an impossible activity into one that is always done perfectly.

An answer jokingly suggested from time to time is to extend the acceptable distance to 50 mm diameter, and then everything will be on target. This suggestion defeats the purpose of having a process that delivers accurate results. Unfortunately, many businesses unwittingly select it as the solution to their problems. They choose to "widen the target" and accept any result—good, mediocre, or disastrous—rather than set high quality standards and improve their processes to meet them. A business that does not pursue excellence in the execution of its activities will not last.[2]

Examples of processes with inherent high variability are those that at some point

- Require decisions
- Require choices
- Permit human error
- Are done without exacting training
- Have no standards
- Have no or inadequate procedures
- Lack correct information
- Require a guess to be made
- Are ill defined
- Are based on opinion
- Involve emotion
- Can be done in multiple ways
- Are not measured
- Have high rates of equipment failure
- Involve interpretation of data
- Alter settings based on historical results

In these situations, randomness and uncertainty abound. This is particularly the case in sales and marketing, finance, human resources, administration, engineering, design, customer service, production, manufacturing, procurement, dispatch, after-sales service, and maintenance. In other words, it is the case in every process in a business.

The late quality guru W. Edwards Deming advised graphing the process variables and the process outputs across time on a run chart (a time-series plot) to identify uncertainty and variability.[3] When the run charts are used together, they identify the times and causes of poor results. If you want

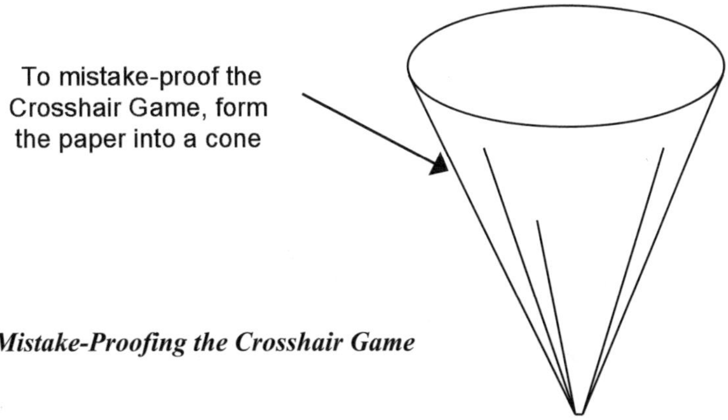

Figure 3.3—Mistake-Proofing the Crosshair Game

immediate control over a process, track the process variables—those factors that influence the result—so that they are observable as they change. If the change is bad, you have time to react and correct it before it does too much damage. If you want preemptive control of a process, trend the variables of the process inputs before they enter the process. By ensuring that the inputs to a process are correct, you can be more certain that the process they feed will behave right.

If you only want to know how well a process performed, monitor its final outputs after completion. Unfortunately, monitoring the final output puts you in the position of asking "what happened?" when something goes wrong, just like the company in Example 3.1, which had no idea what had changed to cause a spate of raw material stock-outs. By tracing the replenishment process on two run charts, the company exposed process fluctuations, and one phone call later, the underlying causes were revealed.

Example 3.1: Inventory Replenishment Mayhem

The stock replenishment process of a national company involved the ocean shipment of raw material from a manufacturer on one side of the country to the company's factory, which was located five shipping days away on the other side of the country. For some months prior to the investigation, the company had been running out of stock across a range of shipped materials. As a result, the company was unable to supply products to clients on time because the warehouse replenishment process could not maintain adequate raw material stocks. The company was using up safety stock and not getting resupply quickly enough to meet clients' orders. Annoyed clients reported the problems that this was causing in strongly worded correspondence and angry telephone calls. The company did not know why it was having the stock-outs.

The investigation began by collecting data on raw material stocked-out in the past. Table 3.3 shows the frequency of supplied material that had suffered stock-outs during the past two years. The company was suffering increased numbers of stock-outs, and the table confirmed the seriousness of the situation. The next step was to find what was causing the lack of supply.

It was necessary to look at the history of deliveries from the manufacturer. Historical records of delivery dates are plotted in Figure 3.4, which is a run chart of the ship departure dates. It shows a great deal of variability in the deliveries over the most recent months. Lately, deliveries have been up to two weeks overdue when they should have been arriving weekly.

Table 3.3—Frequency of Raw Material Stock-Out

Item	Total	Jun	May	Apr	Mar	Feb	Jan	Dec	Nov	Oct	Sep	Aug	Jul	Jun	May	Apr	Mar	Feb	Jan	Dec	Nov	Oct	Sep
T166	21	1	1	2	2	3	1	1				1	1		1		1	1		1	1	1	1
T129	14			2	1	2	1			1	1			1	1							1	
T209	13	1		2			1											1		2	2	1	2
T201	10	1	1	1		1	1		1											1	1		1
T281	10	2	1		2																1	2	
T126	9	1	1	1										1			1			1	1	1	
T169	8	1	2	1				1				1								1	1		
T241	5	1															2			1			
T321	4																					1	
T161	5			2									2	1									
T361	3			1														1					
160N	11	1			1	1	2	2	2	1	1												
120N	9	1	1	1		2											1			3			

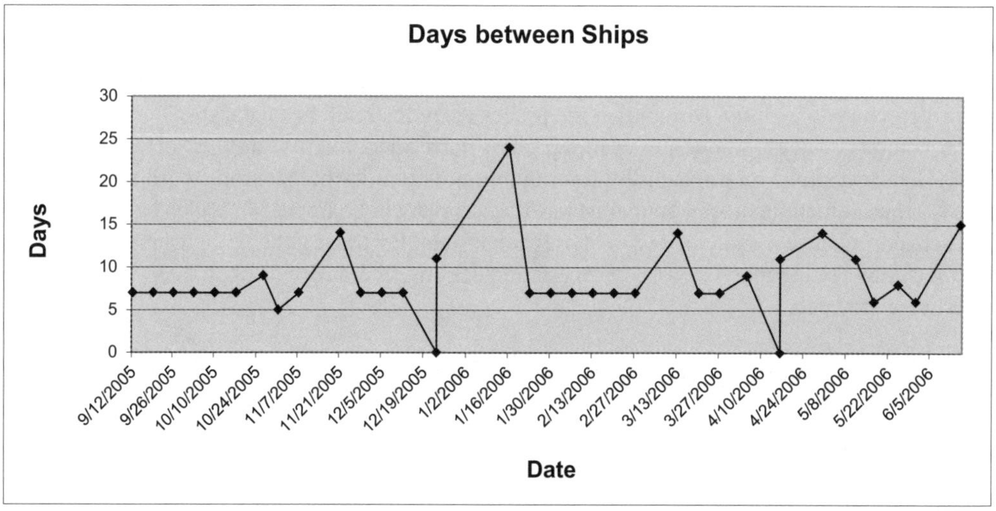

Figure 3.4—Ship Departure Dates

Figure 3.5 is a graph of the number of sea containers in each delivery. It shows variability in the amount of product sent in each shipment. Instead of having normal deliveries of 10 to 11 sea containers, the company was now receiving shipments ranging from 4 to 27 containers.

Inquiries with the freight forwarder revealed that the national shipping line regularly used for deliveries had a two-month maintenance outage for one of its two ships. Whereas once there had been regular weekly shipments, now the only ship left was running fortnightly. To get ordered material to the factory during the maintenance outage, the manufacturer started booking transport with international shipping companies. These ships had irregular departure schedules and only took the number of sea containers needed to fill the empty bays that were left after meeting prior commitments. Sometimes they took a few containers, and other times they took many. The irregular departure of the international carriers with either small or large amounts of raw material resulted in the stock-out mayhem.

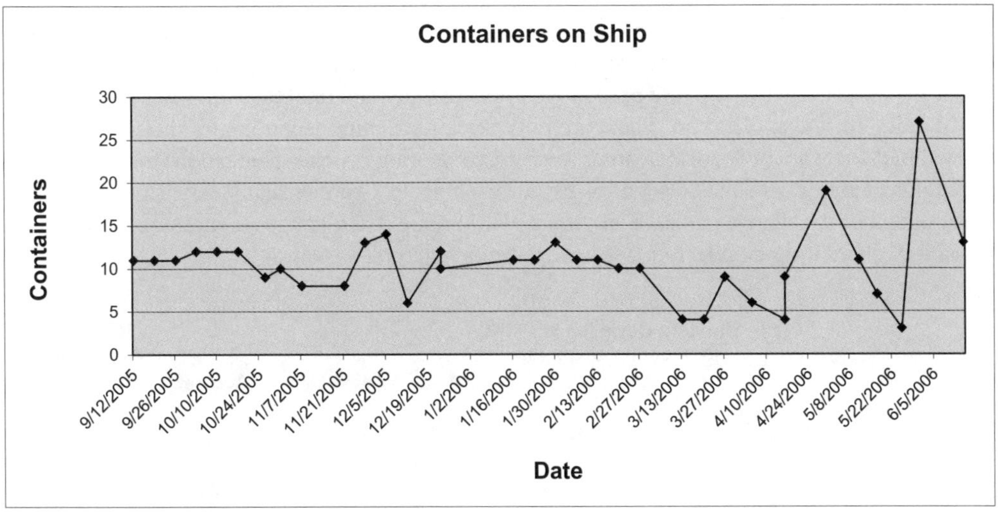

Figure 3.5—Numbers of Containers on Each Ship

The company's customers suffered because of the erratic supply of raw materials to the manufacturer. The irregularity was attributable to the high variability of international ocean shipping, which was further complicated by the feast-or-famine quantities of product on each ship. There were two possible responses to the temporary shipping problems until the return of the regular national carrier's serviced vessel. The company could increase the order size shipped fortnightly, which would effectively raise its inventory levels in transit, or it could book transnational rail delivery. Most importantly, to prevent future stock-outs, the manufacturer needed to change its replenishment process to regularly monitor the ocean shipper's plans and to check for possible delays in shipments.

The disruption of regular delivery to the company in the inventory replenishment example was the result of a "special cause" event—the ship repairs. A special cause event is an extraordinary occurrence in a process that is not attributable to the process itself. Had there been no ship repairs, the weekly deliveries most likely would have been normal. The ship repair was outside the control of the replenishment process, but it had a negative impact on it.

Fluctuation that is attributable to the natural variability of a process is called "common cause" variation. The Crosshair Game is an example of the effects of common cause variation. Where the pen lands depends on the behavior of the process variables affecting the drop, such as steadiness of the hand, accuracy over the target, evenness of the release, and so on. A ±25 mm spread of hit locations is normal for the Crosshair Game. To have a pen fall into a 2 mm circle, or even land within a 10 mm circle, using a process with ±25 mm variation is mostly the result of luck. Dropping a pen by human hand from a height of 300 mm and having it fall inside a 10 mm circle every time is impossible—the common cause variability of that process is too great for the accuracy required. In order for the pen to fall inside the circle every time, a process without the element of luck in needed, not an increase in the knowledge and skills of the person doing the job. You could spend a fortune teaching people how to drop pens into circles, and still everything would be scrap.

An example of a classic misunderstanding of variability that causes equipment to break down is the tightening of fasteners. This confusion is the root cause of many flange leaks, fastener looseness, and machine vibration problems. Figure 3.6 shows the variation in the typical methods used to tighten fasteners.[4] The method that produces the greatest variation, ranging ±35%, is "Feel—Operator Judgment," in which muscle tension is used to gauge fastener tightness. Even using a torque wrench has a variation of ±25% unless special practices are followed that reduce it to ±15%. Industrial experience is that fastener tensioning problems disappear when the final tension is within ±10% of the correct value for the application. There are only three tensioning technologies that, by design, can stay within that accuracy: load-indicating washers, measuring fastener elongation (for example, with hydraulic tensioning, in which a set pressure corresponds to a known shank extension, or ultrasonic measurement of the change in shank length), and mounting strain gauges to measure microstructure elongation. Other torqueing techniques have too much natural variation.

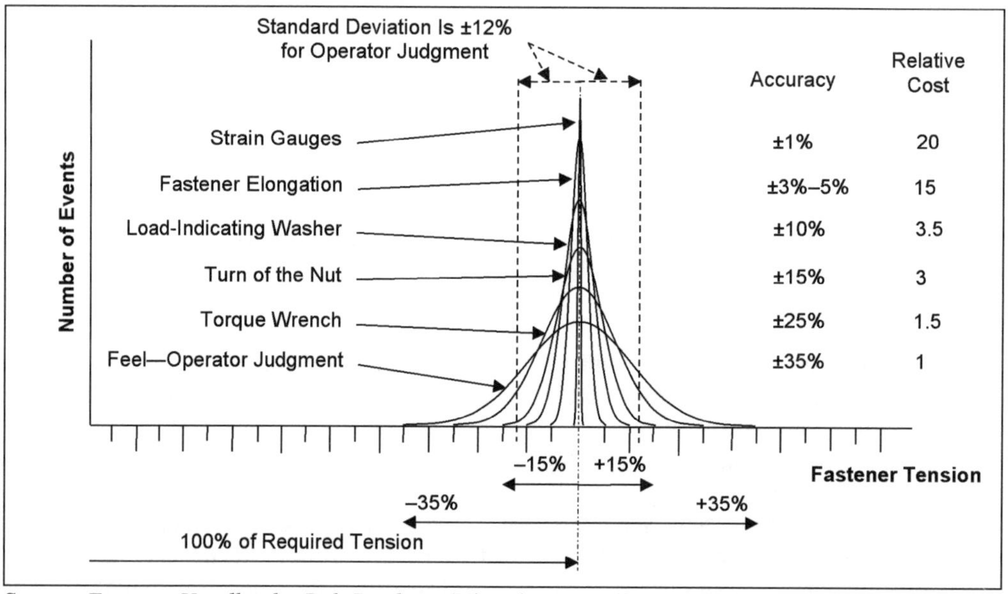

Source: *Fastener Handbook—Bolt Products* (Victoria, Australia: Ajax Fasteners, 1999).

Figure 3.6—Variability in Methods of Providing Correct Torque for Fasteners

It is impossible to guarantee tension accuracy when tightening fasteners by hand. Using a process that ranges ±35% to get within ±10% of a required value is like playing the Crosshair Game—it requires much luck. Joint and connection failure are inherent in a hand-tensioning process. Companies that approve the tensioning of fasteners by operator judgment must also accept that there will be many cases of loose or broken fasteners. It cannot be otherwise because processes that use hand torqueing to cause a required growth of fastener length have guesswork built into them. It would be a foolish manager or engineer who demanded that his or her people stop fastened joint failures but only allowed them to use operator feel or tension wrenches to control the accuracy of their work. Such a manager or engineer would come to believe that he or she has poorly skilled and error-prone people employees, when in reality, it is the method that the manager specified and approved that is causing the failures. In this case, the manager totally misunderstands that it is the process itself that is not accurate enough to ensure correct fastener tension, not the people in the process.

Stopping fasteners from failing necessitates a method that surely delivers a required fastener shank extension. The fastening process must be changed so that it guarantees the necessary shank stretch. Only after that management decision is made and followed through by purchasing the necessary technology, quality-controlling the new method to limit variation, and training the workforce until competent in the correct practice can the intended outcome always be expected.

There is one other phenomenon about variation that you need to grasp. Notice that one standard deviation about the target tension value is ±12% for "Feel—Operator Judgment." It is very close to the ±10% variation about the required tension after which fastener problems disappear. One standard deviation to either side of the mean on a normal distribution bell curve captures 68% of all results. This implies that about two-thirds of all tensioning done using "Feel—Operator Judgment" will not be problematic—it will be close to or within ±10% of the required tension. Because nearly two-thirds of all hand fastening done is "good enough," there will be many workforces in which people on the crews have hardly ever had a fastener failure. It would appear that these people have got hand tensioning down to a fine art. That might be true for some of them, but for many people, they have simply been lucky. The situation is much like golf—professional players are highly tuned to make par or better, and they consistently get those results. When nonprofessionals make par, it is luck, not their golfing expertise, that produces success.

Any operation using peoples' muscles to control fastener tension has failure built into its design—having some loose and broken fasteners is the nature of the process. This is why Deming gave his famous warning to managers: "Your system is perfectly designed to give you the results that you get." Poor equipment reliability is the result of choosing to use business and engineering processes with inherently wide variation. These processes are statistically incapable of delivering the required performance with certainty, and so equipment failure from their use must be expected. Occasional failure is a design outcome of these processes, and luck helps keep their users in business.

Another process that designs failure into equipment is the common maintenance practice of changing oil after it is dirty. When managers decide to replace lubricant only after it is chemically degraded or black with solid contaminants, they unwittingly agree to let their equipment fail. Table 3.4 lists some ISO 4406 solids-in-oil contamination ranges numbers.[5] Each value has twice the count of particles in a milliliter of lubricant as the previous range (a volume equal to about 20 drops of distilled water). The range number 21 (dirty lubricant) contains 128 times the number of particles in each milliliter than 14 (clean lubricant). The size and amount of solids in lubricant directly impact the likelihood of roller bearing failure.[6]

Table 3.4—ISO 4406 Particle Count for Lubricant

Range Number	Number of Particles per Milliliter		Increase in Particle Count from 10 Range	Visual Color
25	160,000	320,000	32,000	
24	80,000	160,000	16,000	Dark
23	40,000	80,000	8,000	
22	20,000	40,000	4,000	
21	10,000	20,000	2,000	Dirty
20	5,000	10,000	1,000	
19	2,500	5,000	500	
18	1,300	2,500	250	From new drum
17	640	1,300	130	
16	320	640	64	Clear
15	160	320	32	
14	80	160	16	Clean
13	40	80	8	
12	20	40	4	
11	10	20	2	
10	5	10		

Depending on the lubricant regime (hydrodynamic, elastohydrodynamic), viscosity, shaft speed, and contact pressure, roller bearing elements are separated from their raceways in the load zone by a lubricant thickness of 0.025 μm to 5 μm.[7] Eighty percent of lubricant contamination is composed of particles smaller than 5 μm.[8] This means that in the location of highest stress—the load zone—tiny solid particles can be jammed against the surfaces of the roller and the race.

The bottom diagram in Figure 3.7 shows particle contamination in the load zone of a bearing. A solid particle carried in the lubricant film is squashed between the outer raceway and a rolling element. Like a punch shearing a hole through sheet steel, the contaminant particle causes a high load concentration in the small contact areas on the race and roller. Depending on the size of the stress that develops, the surfaces may or may not be damaged by the particle. Low and average stresses are accommodated by the plastic deformation of the material of construction. However, an exceptionally high stress will punch into the microstructure, creating submicroscopic cracks in the surface and subsurface.[9] Once a crack is created, it becomes a stress raiser and grows under much lower stress levels than what is needed to initiate the crack.

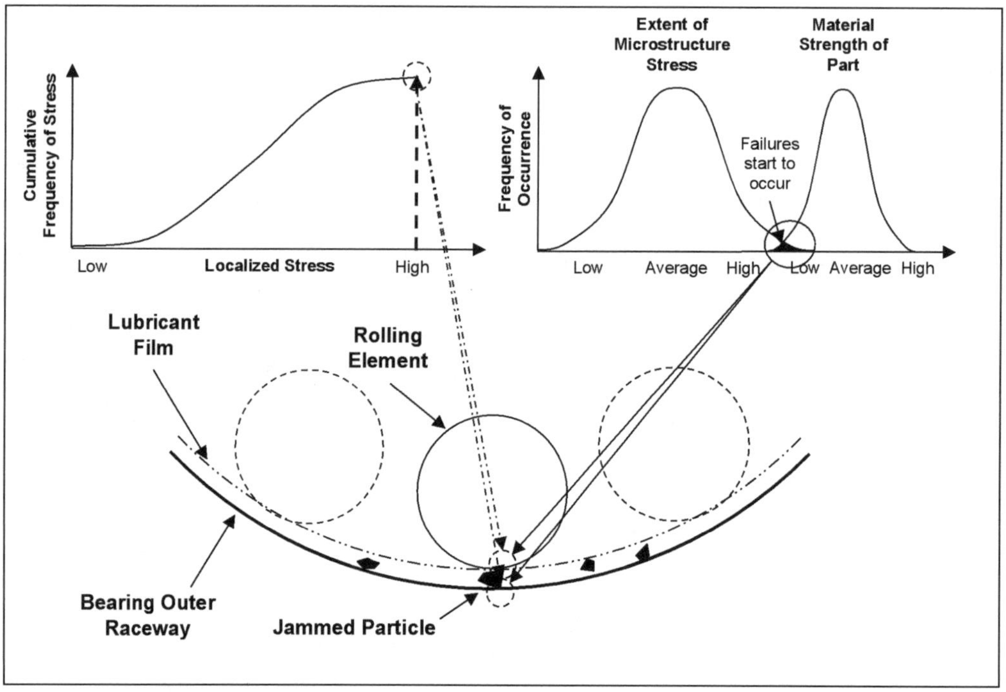

Figure 3.7—Solid Contaminant between Roller and Raceway Overloads the Microstructure

High surface stresses also result from cumulative forces in which loads, each of which is individually below the threshold that damages the atomic structure, act together. Such circumstances arise when a light load supported on a jammed particle combines with additional loads from other stress-raising incidents. These events include impact loads from misaligned shafts, tightened clearances from overheated bearings, forces from out-of-balance masses, and sudden operator-induced overload. All of these stress events are random. They might happen at the same time and place that a contaminant particle is jammed into the surface of a raceway, or they may not. Whether they combine to produce a sufficiently high stress to create new cracks, or they combine on already damaged locations where lesser loads will continue the damage, is a matter of probability.

When a roller bearing is in use, there is relative motion between the raceways and the rotating rolling elements. The odds that a damaged area on a roller will be repeatedly stressed are low because the roller moves to a different spot. However, a damaged area on a race remains exposed to all of the rolling elements that pass over it. The passage of each roller is an opportunity for solid contaminants to be squashed against the race surface. The size and frequency of stress caused to a bearing from solid particulate is random. You could have very clean lubricant, and although the odds are extremely low, you may be unlucky enough to jam the only solid particle in the neighborhood between roller and raceway at the same time that a rotating misalignment force spike passes through it. But as contamination by solid particles rises, more solids are in the lubricant. With each rolling element that turns over a raceway, the growing number of particles provides ever-increasing opportunities for one to be punched into the surface. As lubricant gets more contaminated, the odds of damaging bearings, blocking oil flow paths, or jamming sliding surfaces climbs.

To significantly reduce bearing failures, gear failures, and sticking hydraulic valve problems, the ISO 4406 particle count must be kept at clean levels, or below, so that the oil has hardly any solid contaminants in it. Companies mistakenly allow their gearboxes, drives, bearing housings, and hydraulic system oils to get dirty and blacken from wear particles before they replace them. Often they wait for an oil analysis to indicate that contamination is too high, or they replace dirty oil on time-based preventive maintenance schedules. Unfortunately, by the time the lubricant becomes dark from particle contamination, the probability of jamming a particle between two contact surfaces has increased markedly, and failure sites have likely already been initiated. The risk of failure to a company's plant and equipment from wear particle oil contamination is the direct result of the management processes applied (or not applied) to decide how much solid contamination will be sanctioned in the oil. When oil is changed after it is already darkened by particulates, it is far too late to greatly reduce the probability of failure. Greases and oils must never be blackened by particle contamination in the first place if you want to stop the chance of having lubricated and hydraulic equipment breakdowns.

The failure of machinery is directly related to the corporate maintenance strategy and the processes chosen to maintain and operate the plant. You could be destroying your equipment and think you have great operating and maintenance strategies! The managers and engineers in these companies are fervent that they do the right maintenance practices and have excellent preventive maintenance processes in place. They are wrong, of course, because the processes they use cannot guarantee the results they want. Many organizations try to achieve results using business, engineering, and operating processes with common cause variation that cannot reliably produce the performance they need—they are playing the Crosshair Game in what they do. Such businesses employ processes containing inherent volatility that naturally produce outcomes outside requirements. Trying to manage an organization with systems and processes that produce highly variable results is an exercise in futility that causes waste, distress for all involved, and emotional burnout for managers, engineers, and supervisors.

Controlling Process Variation

Controlling common cause problems requires design improvements to the way a process operates. In contrast, special cause variability is controlled by stopping the influence of the extraordinary event. Preventing the effect of the supply ship's repair that caused late raw material deliveries in Example 3.1, "Inventory Replenishment Mayhem," is done by using other reliable modes of transport to replace the unavailable ship. As soon as on-time delivery by ship is not possible, rail freight needs to be booked. You address special cause issues by stopping them from happening or by preventing them from impacting your business. But common cause issues are inherent in the process, and their elimination requires changing the process.

It is the nature of every process to produce variation. The challenge for business and operations processes is twofold. First, it is to have only "natural" variation and no special cause variation. Second, it is to select or develop processes with natural variation that is well within the required performance. This allows the organization to focus mainly on stopping special cause problems, sure in the knowledge that the process is inherently stable and produces good results. When a business or operating process no longer performs within its normal limits, look first for a special cause of the change. Only after all special causes are eliminated can you be sure that only natural common cause variation remains. If the common cause variations are still too volatile, you have justification for improving or changing the process. By following that sequence, you confirm whether any special cause variations are masking the natural process variability with effects that confuse the analysis. If a special cause is mistaken for a common cause, you will make the wrong decisions to address the problem.

So far, we have seen examples of variability in a game, in the supply chain of an organization, and in the maintenance strategies used on plant and equipment. Being able to get a picture of variability with run charts and tables brings a clearer appreciation of what is happening within the process. It allows us to ask relevant questions that lead to a more profound understanding of a situation's causes and resolutions. Great value is gained when an organization observes the irregularity of its business processes. Once a picture of process behavior is available, you can make focused efforts to control unacceptable variability.

The next case study focuses on an ore-processing plant where the consensus was to invest $250 million to expand production by 50%, when in fact that investment may have been unnecessary if production variability had first been addressed.

Example 3.2: The Hidden Factory

This is an example of the value of identifying the causes of variability in a business and removing them before spending new capital. The production from an ore-processing plant is trended in a simple bar graph in Figure 3.8. It shows the hourly production rates in a milling operation running 24 hours a day, seven days a week during eight consecutive weeks. It contains a lot of valuable information about the operation's capacity as well as a clear indication that the business is suffering from wild fluctuations in its production throughput. Examination of the graph provides insight into the facility's dilemmas.

The eight weeks of production shown on the graph represent 1,344 production hours. For 275 hours, there was no production, so for 20% of possible production time, the plant was standing still. The plant design capacity is 1,500 units per hour. For 615 of the remaining hours, or 57% of the time it was running, it delivered substantially less than the designed production rate. The actual average throughput for the eight weeks was 1,000 units per hour, which is two-thirds of the design duty. This facility is suffering from severe production problems and needs to investigate why it is not producing consistently at design capacity.

There is additional information in the graph. It is clear that for a significant number of hours, the plant ran above its design rate. There are two possibilities here. One is that in trying to make up for lost production, the plant was overloaded, which led to even more equipment failures and added downtime. The second is that the plant could be run at more than its design duty. Confirming each possibility requires an engineering design investigation. There is a good chance that with minimal engineering changes, the plant could be run consistently at 2,000 units per hour, which is one-third

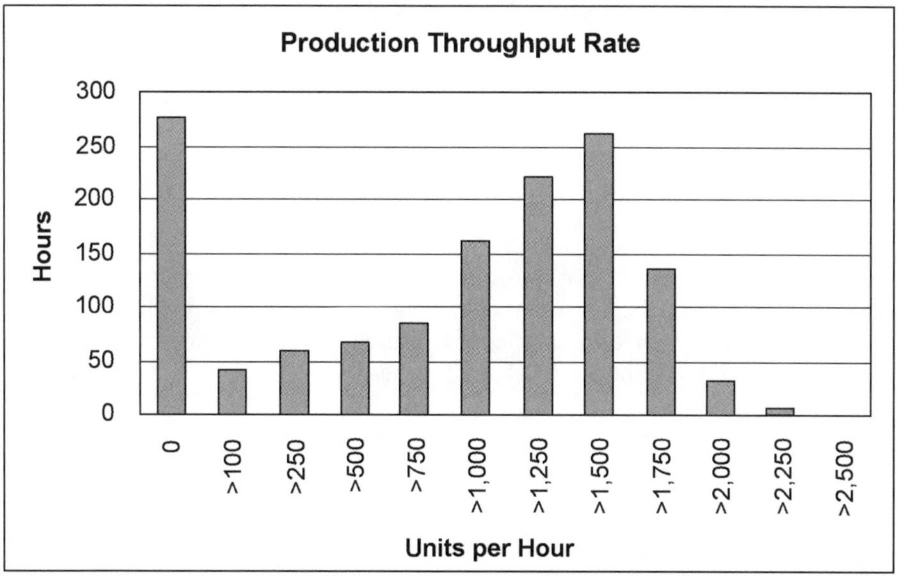

Figure 3.8—Production Rates Chart

greater than the current design capacity and twice the current average production. The overstressing of parts would be a major concern at the increased production rate and would need to be addressed by a full design review. An operating risk analysis based on Physics of Failure consequences would be conducted and problems designed out as part of the decision to increase production to a higher rate than the original design.

There are obvious questions to ask of a plant with this extent of variability in performance. What is causing the stoppages and below-design throughput so often? If the plant can produce at higher rates by accidents of circumstance, then what could be consistently produced if those circumstances were deliberate? It would be sensible to identify the causes of the disastrous production losses and solve them while making the fortuitous events of the past intentional. The total "lost" throughput represented by the work stoppage and slow running, plus the higher production rates available from reengineered capacity, means that this operation has plenty of opportunity to deliver a large production increase without significant capital investment.

This company's decision to spend $250 million on a major capital upgrade to boost production 50% may not have been necessary. By recovering the downtime and low production rates and reengineering bottlenecks for higher throughput, the extra capacity was probably achievable with the old plant. It was only necessary to conduct root cause investigations on why the production losses were occurring and how to solve them. The financial return on such an investment would be unbelievable. All of these options become clear simply by measuring production variability.

Constructing a graph like that in Figure 3.8 requires collecting the hourly production figures for a sufficiently long time to observe the full range of variability affecting the process. The figures show a range of performance around a mode value (the quantity in a data set that occurs most often). The extent of the spread below the mode indicates whether there are production problems hampering throughput. The range of the spread above the mode indicates whether there is spare capacity available. If the spread is tight around the modal production rate, then throughput is stable, although not necessarily optimal. But if the spread below the mode is wide, as in Figure 3.8, then the plant has "hidden" efficiency opportunities to improve its production performance.

When production throughput graphs have a wide spread of production rates, there is potential to increase plant capacity by removing the causes of operating losses with minor engineering upgrades or removing the variability by adopting improved procedures and useful training. Before you invest more capital to expand plant capacity, investigate the variability of current production, as there may already be a "hidden factory" within your plant.

Controlling Business Process Performance

When any process is run, it produces a range of outputs that are its characteristic signature—you get what the process does. Business process variability produces a frequency distribution of results. The repetitiveness of outcomes can form a normal bell distribution, as in Figure 3.9, although processes often have non-normal distributions with skewed shapes.[10]

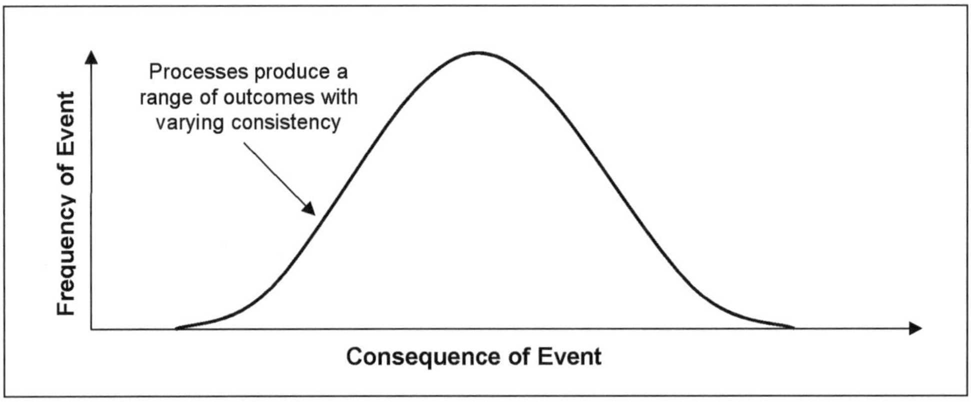

Figure 3.9—Processes Produce Their Own Results

The way to tackle variability is to put a limit on the acceptable range of variation and then build, or change, business processes to ensure that only those outcomes can occur. This approach drives process innovation until the goal is reached. Figure 3.10 shows a minimum specification of performance for a process producing wide variation. The acceptable range is categorized by zones of accuracy: "good" is tolerable, "better" is finer performance, and "best" represents world-class results. Only outcomes that are within the good standard are acceptable. Any lesser results are defects and rejects. Changing to a process that produces a performance curve that is always inside the good range requires designing and installing better methods that remove performance fluctuations, reduce volatility, and stabilize process variation inside the tolerable range.

With volatility controlled, the results tighten around a consistent mean, as shown in Figure 3.11. Variation still exists, but it is within the desired limits. A process that always produces stable, repeatable outcomes within its specification limits is in control and capable. It becomes highly predictable, and the results can be guaranteed. Business process performance is mostly within our control. You can improve your processes by choosing policies and practices that minimize the chance of bad events happening and maximize the chance of good outcomes occurring.

The purpose of controlling variability is to provide certainty of performance. Once variability is identified, it becomes necessary to make the decision to leave the situation alone and accept the range of outcomes that happen, or to address the underlying problems causing the fluctuations. When you make process improvements, you must find the causes of the problems and then identify ways

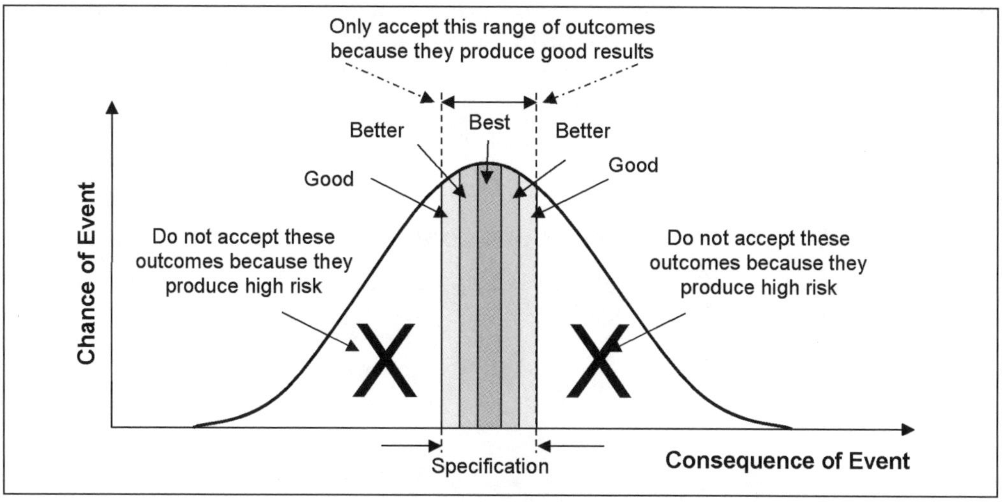

Figure 3.10—Controlling the Chance of a Process Event

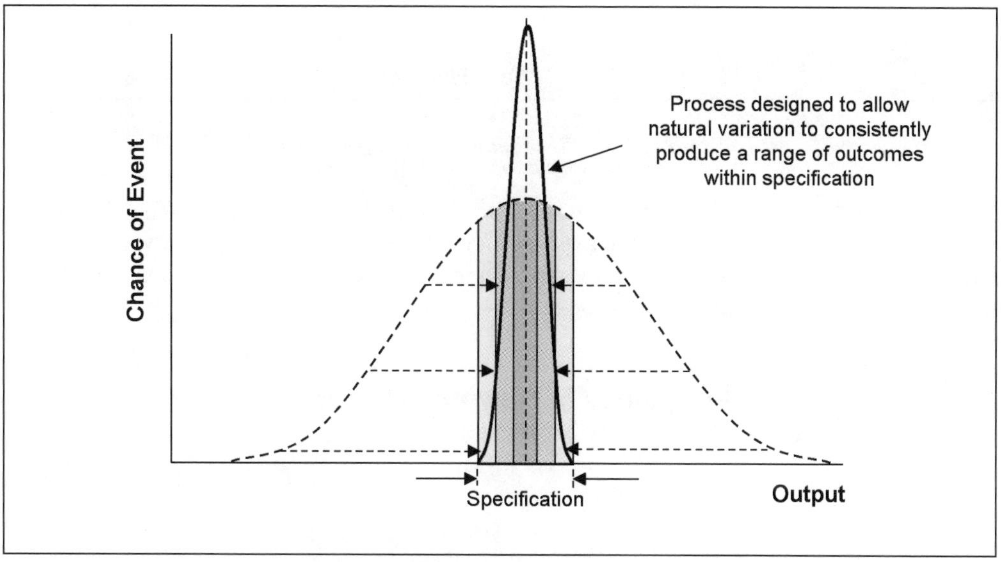

Figure 3.11—The Effect of Removing Volatility from Business Processes

to design them out of the process. Example 3.3 tells of a company struggling with constant maintenance problems despite doing the very best that it knows how to do.

Example 3.3: The Factory That Made Its Machines Break Down

Most industrial businesses make their equipment fail. Misunderstanding risk, variation, and probability leads managers and engineers to use processes that cause trouble. An analysis of a real business illustrates the effects of this all-too-common management error. Figure 3.12 is a run chart of a company's total breakdown hours per week over 16 weeks. If the graph is representative of normal operation, the time series can be taken as a sample of the company's typical business performance. Important information about the company's operations is exposed by using two basic statistical analyses of the data: a frequency distribution and a control chart.

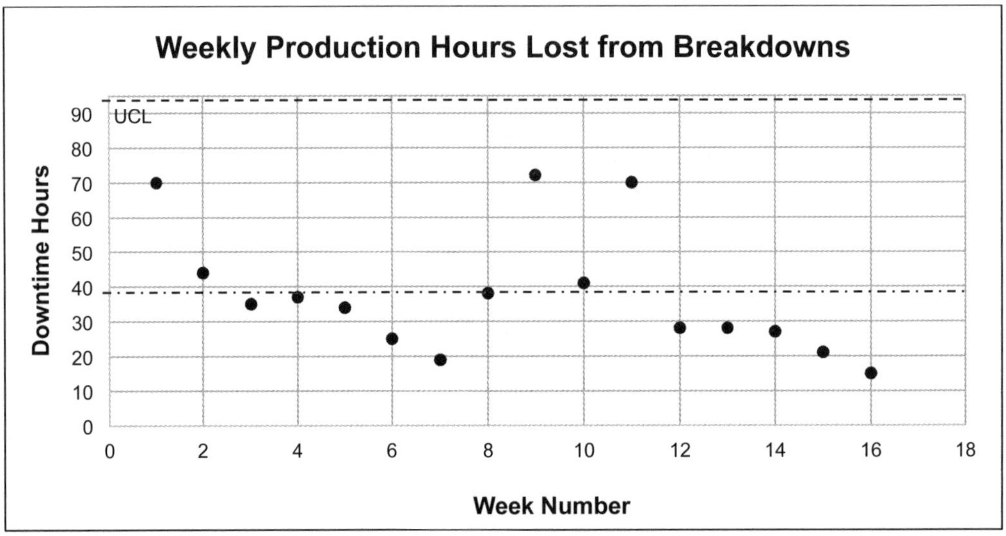

Figure 3.12—Breakdown Hours per Week

The frequency of total downtime hours each week due to breakdowns is plotted in Figure 3.13. The distribution is twin peaked, with the lower portion looking like a normal bell curve, but suddenly there is a discontinuity between it and the higher breakdown hour weeks. This company has two types of breakdowns: normal and catastrophic.

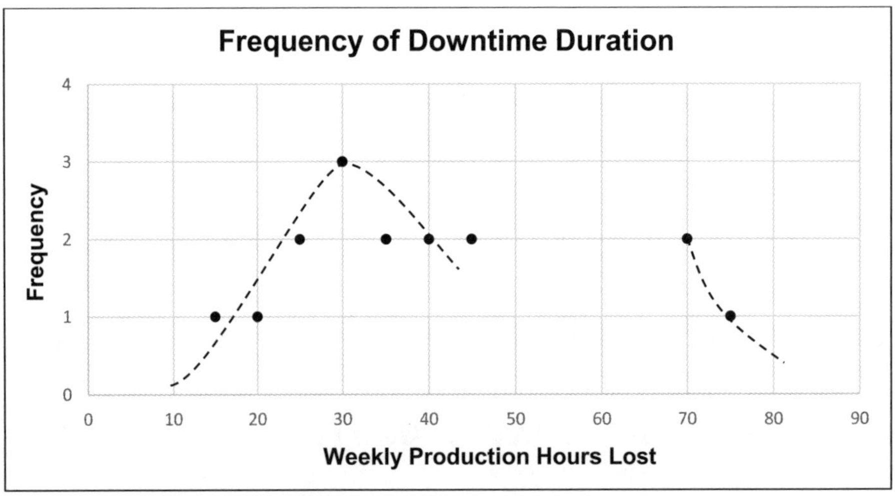

Figure 3.13—Frequency of Weekly Breakdown Hours

For an x-bar control chart, the average downtime due to breakdown is 38 hours per week. The standard deviation is 18 hours. The upper control limit is 92 hours. The lower control limit is 0. These control lines are shown in Figure 3.12. Because all data points are within the statistical boundaries, the analysis indicates that breakdowns are common to the business processes and not caused by outside influences. This company has a statistically stable system for making its equipment break down. Breakdowns are one of the "products" made every week.

Because the breakdown creation process is stable, the future generation of breakdowns is predictable and certain. If this time series is a true sample of normal operation, it can confidently be said that an average of 38 hours will always be lost to breakdowns every week in this business. In the three weeks following the 16-week period represented, the "normal" weekly breakdown hours were 25, 8, and 25 hours, respectively. This business has built breakdowns into the way it operates because the process of breakdown manufacture is part of the way the company works. The only way to stop the breakdowns is to change to processes that prevent breakdowns.

What Quality Is

In his book *Out of the Crisis,* Deming advised that "quality must be built in."[11] Quality, Deming tells us, is installed at the source. It is designed in and made part of the product or service; it is delivered by the business process design itself. Quality is a definite and "hard" measure that can be clearly identified. It is quantified with engineering values—the "numbers" that, when achieved, deliver customer satisfaction. In Deming's view, a product or service has the right quality when customers are so satisfied that they boast about it to the people they meet. The quality of the product or service is designed to satisfy the customer. Word of mouth markets it.

The same certainty over quality, but applied to equipment parts, is necessary to deliver the outstanding equipment reliability and plant availability that produces world-class production performance. What is important to know about quality is that it must be measurable. Quality is not left up to people to interpret—it is management's responsibility to define it. It needs to be quantifiable—a length, a thickness, a resultant force or pressure, a color, a smell, a viscosity, a period of time, a rate of change. You require a specific engineering value, even a collection of values, that defines a level of performance. Once the values are attained, performance is certain and the required quality is achieved.

To have quality, you need a target and a range of acceptable outcomes. It is impossible to know how to control quality until standards of allowable variability are set. Once a standard is specified, measures are made to identify whether the processes used to achieve it are statistically capable of meeting the standard. For the business in Example 3.3, "The Factory That Made Its Machines Break Down," the processes will never be able to deliver long periods of breakdown-free operation. The company's current production asset management system does not work. It is not designed to produce a breakdown-free week. In fact, the practices, methods, and processes used cause the failures. It is nearly impossible in this operation to expect more than a couple of days without breakdowns. This company needs to change its fundamental life-cycle asset management processes if it wants to improve equipment reliability. If the company set a target average of, say, 10 breakdown hours a week, a search for better methods and strategies to reach 10 hours breakdown per week would start. The great challenge for this company is to replace years of destructive practices in operations and maintenance with processes and methods that produce high reliability. This change can only occur when the company decides to create business processes that deliver more uptime.

It is necessary to design a new game plan when existing processes do not produce the required results. Figure 3.14 represents the strategic aim when changing processes to be capable. Deming said that it is the responsibility of management to improve a process—no one else can do it.

Figure 3.14—
Making a Process In Control and Capable

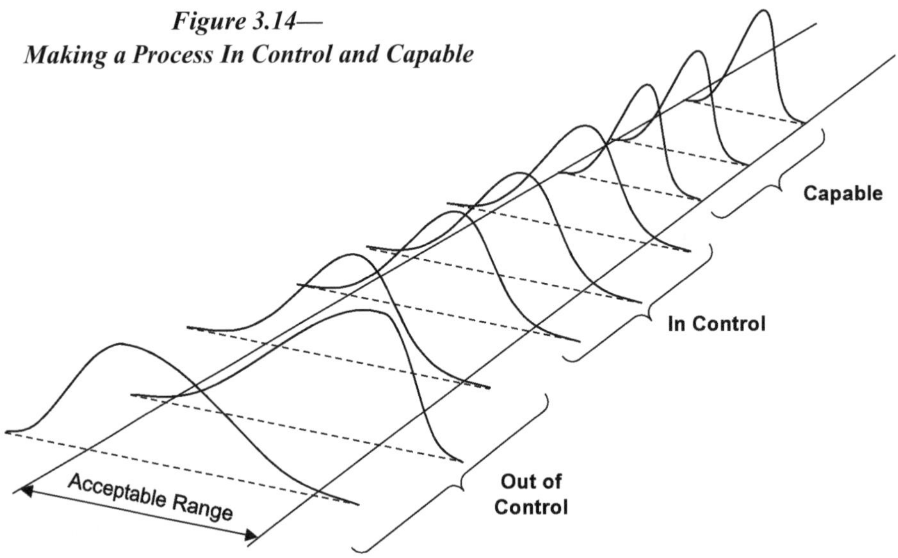

Setting Engineering and Maintenance Quality Control Standards

Reducing the influence of chance and luck on equipment parts starts by deciding what engineering and maintenance quality standards you will specify and achieve in your operation. Probabilistic plant reliability outcomes are controlled by delivering the conditions that produce excellent equipment dependability and performance.

For example, the degree of shaft misalignment that you tolerate directly impacts the likelihood of roller bearing failure.[12] If shaft misalignment is present on equipment, that does not mean that a bearing will be failed. Depending on the extent of misalignment, the operational abuse, clearance reduction from high temperatures, out-of-balance forces from unbalanced masses, and myriad other stress-raising possibilities, the size of the resulting stresses still may be lower than the strength of the bearing's material of construction. But it does mean that shaft misalignment increases the chances that its loads will combine with others and add up to produce a catastrophic failure. As more stress-causing opportunities become present in equipment, the chance of part failure grows ever greater.

If shaft misalignment between equipment is so critical to production uptime, then what is the shaft misalignment tolerance in your company? What number of contaminating particles do you permit in your lubricant? The lower the quantity of solids, the surer the likelihood that you will not have a breakdown. The frequency and scale of machine abuse permitted during operation directly affects the likelihood of machinery failure. What do your operating procedures say about running your plant and equipment for outstanding reliability? The rotating equipment balance quality that you achieve directly influences the likelihood of failing roller bearings.[13] The lower the residual out-of-balance forces, the smaller the possibility that out-of-balance loads will combine with other loads to initiate or propagate failures. What balance standard have you set for your rotors? How accurately will you specify fastener tension to prevent fasteners loosening or breaking? The more precisely the extension meets the needs of the working load, the less likely it is that a fastener will come loose or fail from overload. The temperatures at which bearings operate change their internal clearances, which directly influence the likelihood of bearing failure.[14] How well do you control bearing operating temperatures during machinery use so that clearances are always correct? Similar statements about the dependency of failure on the probability of failure-causing incidents can be said of every equipment part.

Chance and luck affect the lifetime reliability of all parts and, consequently, of all your plant items, machines, and rotating equipment. But the chance and luck that affect your equipment parts can be altered by your choices. Change your lubricant cleanliness limit so that far fewer wear particles are present, and you'll greatly reduce the number of contaminant particles in the lubricant film and, thereby, reduce chance and risk of failure. Combine that with ensuring that shafts are precisely aligned at operating temperature, that rotors are highly balanced, that bearing clearances are correctly set and sustained, and that operational abuse is banded and replaced with good operating practices to keep loads below design maximums, and you will greatly improve your "luck" with equipment reliability. You can have any equipment reliability you want by turning luck and chance in your favor through using and meeting ever-finer quality standards in your plant and machines.

Defect Elimination Strategy

Because variability exists in all processes, a range of outcomes will always result. The Crosshair Game and the operating problems discussed earlier in this chapter highlight some of the bad effects that process variability causes for organizations. When variability becomes excessive, defects occur and failures result. A defect is a "nonconformance to requirements or function." It is a deficiency. It means that bad quality went into service. Defects that escape correction lay hidden and may not become apparent until they cause a failure. A failure is "an event or circumstance which prevents the accomplishment of an intended purpose." A failure occurs any time a thing does not do its job. A failure happens whenever a system or component is unable to perform its designed role. Figure 3.15 is a modified version of the DuPont Chemicals defect and failure model.[15] It highlights some of the many processes by which failure-causing defects and errors come into an operation.

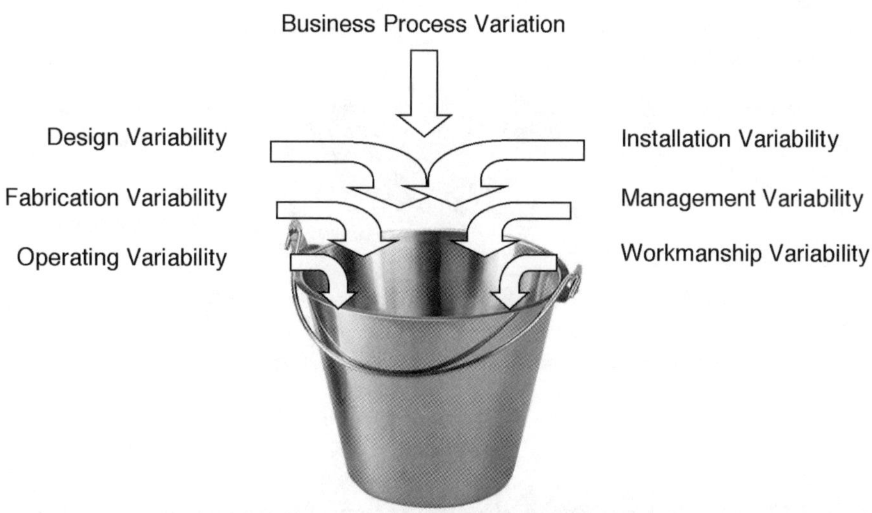

Your operation is a bucket for collecting defects

Figure 3.15—Defect Creation

Most businesses react to failure as shown in Figure 3.16. They introduce maintenance and repair systems to manage the presence of defects. They accept failure as normal. Consequently, they suffer production downtime and high maintenance costs by trying to limit the effects of the introduced problems and fixing failures caused by the problems they could not stop.

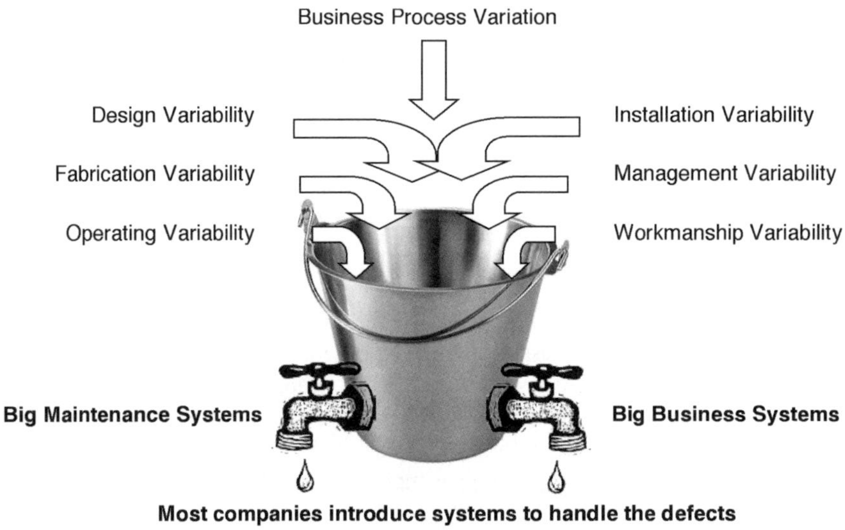

Figure 3.16—Defect Management

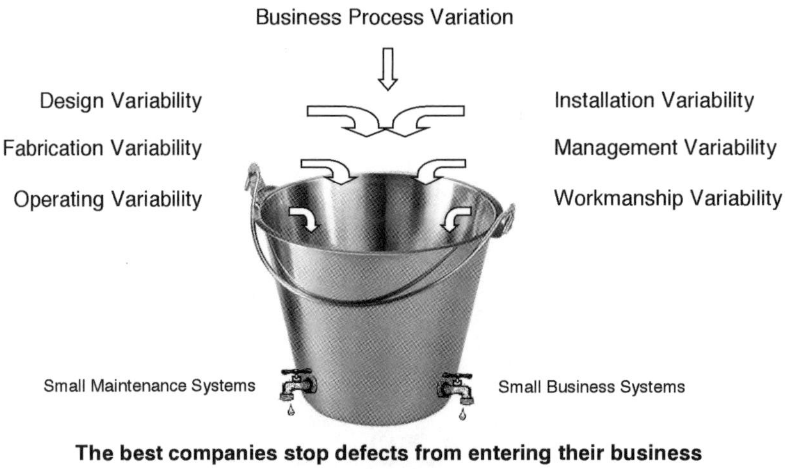

Figure 3.17—Defect Elimination

Figure 3.17 shows the best strategy: stop defects from entering your business in the first place. This forces your quality standards to improve, and, as a result, maintenance costs are reduced and production uptime is lifted. Because there are fewer failures, your equipment reliability, plant availability, and productivity rise. All of the money not spent on failure correction and repairs, plus the extra income from the added throughput in the recovered production time, can be banked as new operating profit.

Variability acts across distance and time. Variations in one process can reduce the effects of variability in an interacting process. Much like an ocean wave rebounding off a cliff, variability between interconnected processes may act to calm the waters, but usually the opposite happens, and variations combine to produce problems of greater magnitude—instead of calm, a surging wave of trouble is created. This was the case in Example 3.1, in which the international shipping line's failure to adhere to a fixed schedule and provide regular container slots compounded the replenishment problems of its users. Variability that worsens problems requires identification and redesign of the offending processes to remove the negative impacts.

Because every process in a business produces variable results, as more processes are used, there is greater opportunity for defects and failures. Organizations that try to do many things have many processes to manage and control. Each process step introduces opportunities for variation. A product or service is exposed to risks from the full range of variability in each activity employed during its creation—concept, feasibility, design, procurement, production, assembly, and delivery.

A common supply chain philosophy that compounds operational problems is when company purchasing policy requires the same item to be bought from a range of suppliers in the questionable hope of keeping costs low through competition. When external providers are used, it is necessary to protect against the worst excesses of their processes and ensure compliance with precise and agreed quality specifications. Companies with multiple suppliers end up suffering more problems and costs than a company using only one supplier. Working with a range of suppliers for the same item requires a great deal of extra effort and time in procurement, accounting and warehousing. Compounding the risk, each supplier brings its own process variability.

When a specified item is bought from many different suppliers, you increase the workload, and the range of variability within your business. This requires corrective measures to be added to your processes to fix the problems caused by slight differences between the goods from each supplier. Suddenly, the small amount of money saved at purchase is dwarfed by that wasted rectifying the troubles. But by staying with one supplier, you have smaller, simpler processes within your organization, you adapt your systems to the supplier's process variability, and the supplier will modify its processes to provide the product quality you want. Those companies that think having supplier competition reduces their costs increase variability problems throughout their business.

Variability of engineering component design and fabrication introduces two failure scenarios into plant and equipment. One is when the parts come from poorly controlled metallurgical or manufacturing processes and are at the weak end of their material capability. These outliers contain imperfections, defects, or flaws of one nature or another—cast, 3D-printed, and welded products can have internal faults. When these parts are put into machines and equipment, they suffer operational and environmental stresses. If the capacity of the part is not up to the difficulties of the situation, any inherent defectiveness can cause premature and unexpected failure. The second scenario is when part structure is well controlled in manufacture, but the part is wrong for the duty—it cannot take the stresses and degradation of service. In such circumstances, there is nothing wrong with the item, but it was selected for a situation that is beyond its ability, and unexpected failure again occurs. It is the responsibility of the design, reliability, procurement, and maintenance groups to prevent both scenarios.

Accepting process variability as inevitable is sensible, but accepting the accompanying failure consequences as inevitable is disastrous. Organizational processes need to be designed and built to hit all of their quality targets the first time. Proactive defect elimination and failure prevention removes process variation. The best way to fix a problem is not to have it at all. If you want to reduce the number of failures that happen in your business, build processes that always deliver the quality standards of excellence you want.

References

1. Leonard Mlodinow, *The Drunkard's Walk: How Randomness Rules Our Lives* (New York: Pantheon Books, 2009).

2. Chris Denove and James D. *Power IV, Satisfaction: How Every Great Company Listens to the Voice of the Customer* (New York: Portfolio, 2006).

3. W. Edwards Deming, *Out of the Crisis* (Cambridge, MA: MIT Press, 2000), 49.

4. *Fastener Handbook—Bolt Products* (Victoria, Australia: Ajax Fasteners, 1999), 48.

5. "ISO 4406:1999 Hydraulic Fluid Power—Fluids—Method for Coding the Level of Contamination by Solid Particles," accessed at http://www.iso.org/iso/catalogue_detail.htm?csnumber=21463, July 1, 2015.

6. "Contamination in Lubrication Systems for Bearings in Industrial Gearboxes," *SKF Ball Bearing Journal,* no. 242 (1993).

7. William R. Jones, Jr., and Mark J. Jansen, *Lubrication for Space Applications*, Publication NASA/CR-2005-213424 (Washington, DC: NASA, 2005).

8. Wayne Bisset, "Management of Particulate Contamination in Lubrication Systems" (presentation at the IMRt Lubrication and Condition Monitoring Forum, Melbourne, Australia, October 2008).

9. FAG OEM and Handel AG, "Rolling Bearing Damage—Recognition of Damage and Bearing Inspection," Publication WL82102/2EA/96/6/96.

10. Many real-world process outputs are normally distributed, but distributions can also be skewed or multipeaked.

11. Deming, *Out of the Crisis,* 49.

12. John Piotrowski, *Shaft Alignment Handbook,* 3rd ed. (Boca Raton, FL: CRC Press, 2007).

13. "ISO 1940-1:2003 Mechanical Vibration—Balance Quality Requirements for Rotors in a Constant (Rigid) State—Part 1: Specification and Verification of Balance Tolerances," accessed at http://www.iso.org/iso/catalogue_detail.htm?csnumber=27092, July 1, 2015.

14. FAG OEM and A. G. Handel, "Rolling Bearing Damage."

15. Winston J. Ledet, "Engaging the Entire Organization in Improving Reliability," accessed at http://manufacturinggame.com/resources/engagingtheentireorganization.pdf, July 1, 2015.

4

PREVENTING LIFE-CYCLE RISKS

The plain truth is, "first parts fail, then machines stop."[1] It follows that stopping parts from failing will stop equipment failures. Broken machines are the end of a chain of causes and effects that begins with a failure trigger introduced at some point in the life cycle. It could be a defect built into a part or a bad event that occurred during the machine's service life. To get outstanding equipment reliability, you need to eliminate defects and stop failure-initiating events from happening to parts. Prevent the first cause of the final cause of a part's failure, and repeatedly do that for all machines in your operation, and you are guaranteed to create world-class plant and equipment reliability.

Equipment is designed using drawings, images, and words. Designers turn imagination into blueprints, specified materials, and written instructions. Parts are made and put together in working assemblies, and the assemblies are brought together into machines and equipment. The designer expects the machine to be used as it was designed to be used: all parts stay within positional tolerance at operating conditions, lubricant is perfectly clean and in the required chemical composition, the stresses and strains in components stay a factor of safety below the capability of the selected materials of construction, and loads and forces act through the paths designed for them to follow. Few designers realize, unless they've worked in equipment maintenance for many years, that such demands can turn their vision into an owner's nightmare.

From the designer's perspective, there is no reason that the equipment should fail unexpectedly because it was designed to work properly, with each part working within the physical limits of its materials of construction. Yet plant and equipment fail often. When they do, the failure can cause unbelievably huge business-wide losses—and even many deaths. To understand how defects in equipment arise, you need to know the activities involved in design, manufacture, storage, installation, operation, and maintenance. Figure 4.1 shows summary steps for each life-cycle process that affects equipment reliability.

The simple flowcharts hide the great complexity in each process. Each step has its own processes and subprocesses and an accompanying suite of procedures. There are hundreds and even thousands of opportunities for failure-causing variations to arise during the life cycle. Every step involves numerous activities and actions requiring many decisions and choices performed to varying degrees of uncertainty. Throughout the life cycle, there are many opportunities for small errors, misunderstandings, and inconsistencies. Randomness and variability introduce defects and failure causes. Faults and discrepancies accumulate, leading to microstructure loads and conditions outside the equipment's design parameters. Machine parts get stressed and strained, and at some point an overly excessive load, or simply the accumulated fatigue from many loads (the proverbial straw that broke the camel's back), causes a part's failure. Each cause can be the start of a defect and all of the future business problems and losses it brings. To stop plant breakdowns, you need to stop defects and failure causes by removing the variations that create them. If you want to have extraordinarily reliable equipment, you need extraordinary certainty that there will be no flaws in the life-cycle processes that impact your equipment's reliability.

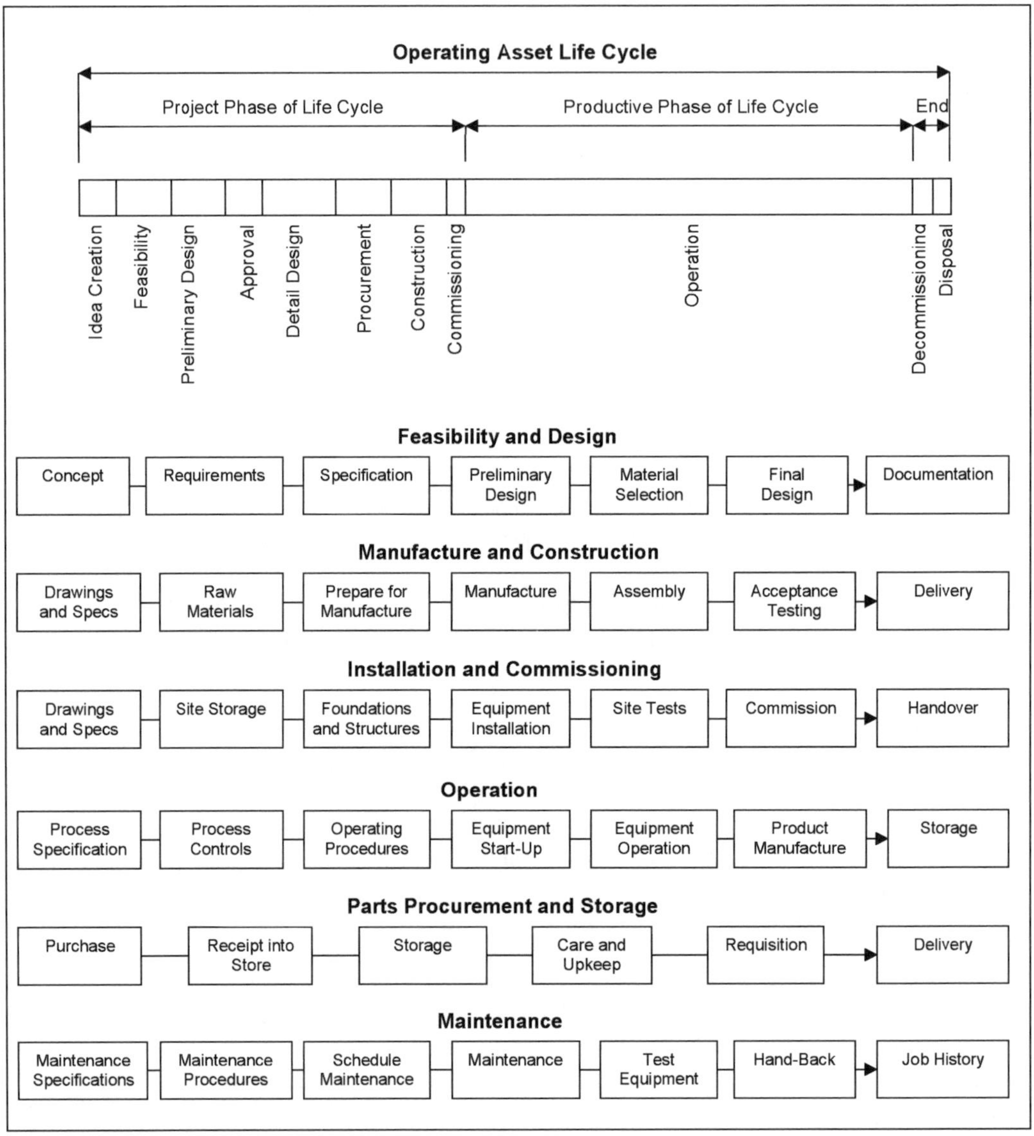

Figure 4.1—Life-Cycle Process Chains Are Complex

Raising the "R"

To get reliability and maintenance excellence, three things ensure success.

1. Prevent stress, fatigue, and degradation of your critical equipment parts' microstructures—your machines will only stop when their parts fail.
2. Control work quality and task accuracy throughout the life cycle of your equipment parts to protect against human knowledge and skill errors introducing defects. Mistakes, misunderstandings, and ignorance during design, manufacture, selection, storage, installation, operation, and maintenance cause most plant availability and equipment reliability problems.
3. Build life-cycle asset management, supply chain, operation, maintenance, and reliability processes that deliver risk prevention, defect elimination and zero-failure strategies and practices to equipment parts.

Prevent part failures, and you will create highly reliable machines. The health of your equipment parts has a fatal impact on your chance of Operational Excellence. Minimizing the stresses that happen to the assemblies and parts in your equipment requires using Physics of Failure knowledge to eradicate the causes of failure. Stress in a part is minimized by preventing microstructure damage (e.g., bending, twisting, shearing, impact, excess pressure, etc.) and material removal (e.g., corrosion, chemical attack, abrasion, etc.). You can derive the minimal reliability excellence strategy by considering an individual part's Physics of Failure mechanisms.

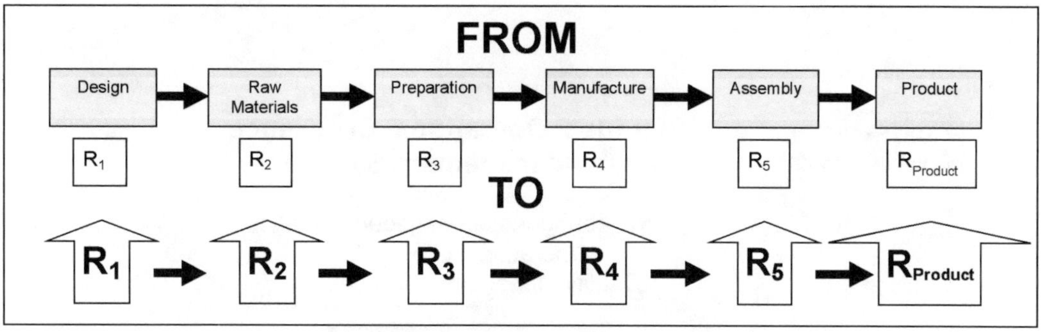

Figure 4.2—Raise the "R" with Higher Reliability in Every Process Step and Equipment Part

World-class reliability is not an accidental result; there is little luck involved in having an operation with outstanding equipment performance, low-cost production, and a healthy and safe workplace. You start the climb to world-class reliability by introducing the causes of reliability into your business processes. Then you teach your people how to do them expertly. Your company will get high equipment reliability when it uses the right processes, techniques, and methods that "raise the R"—reliability, the chance of success—of every step in every business process across the life cycle of your equipment. Your aim is to stop unwanted variation—the archenemy of reliability—so that only success remains.

Figure 4.2 symbolizes intentionally raising the reliability of every process step, activity, and part. You move reliability from its current boxed-in performance and make it grow. You remove the risks in a step, activity, or part so that the chance to start a failure is reduced by orders of magnitude. Once the chance of failure is eliminated or greatly reduced, what is left is a vastly improved chance of success. The math behind this logic is seen in Formula 4.1—when success is certain, failure cannot happen.

Formula 4.1

$$\text{Chance of Failure} = 1 - \text{Chance of Success}$$

The Plant Wellness Way

All that you have read so far needs to be put into a methodology for delivering the right project design, operating practices, and maintenance methods to produce lifetime reliability. World-class asset management, maintenance, and reliability need defect elimination processes, techniques, and methods used throughout the life cycle of plant and equipment. Operating plants and physical assets rely on us to get the working conditions right for them lifelong. The best strategies for improving reliability are those that extend the failure-free life of parts. When machine parts live and work in conditions that limit microstructure stresses to values that deliver long operating lives, they will return maximum reliability. You must kill all the "gremlins" in the life cycle.

Stress-to-Process Life-Cycle Asset Management Model

The Plant Wellness Way is life-cycle asset management of parts and components, not plant and equipment. Its driving principle is the elimination of microstructure failure by reduction and prevention of stress in components' materials of construction. It uses Physics of Failure knowledge to identify

Figure 4.3—Atomic Stress to Business Process Asset Management Model

how each working component can suffer stress or degradation, after which you develop defect elimination activities and zero-breakdown strategies to use during the phases of the life cycle that prevent all the failure-initiating events from happening. The analysis is based on risk elimination and reliability creation principles that ensure you use the right answers to get operational and maintenance excellence.

The enterprise asset management, maintenance, and reliability methodology used in the Plant Wellness Way is called the Stress-to-Process Model. Figure 4.3 introduces the Stress-to-Process Model for asset management success. With it, you engineer and install world-class reliability in your company. The Stress-to-Process life-cycle management methodology lets you discover exactly how to produce world-class reliability and embeds the best solutions in your organization's processes. It is a scientifically based approach for designing and building the least cost, least manpower, and most successful enterprise asset management system. It gets you to build and use the life-cycle processes and practices that create healthy, long-lived parts and thereby create outstandingly reliable equipment. It turns a company into an Accuracy-Controlled Enterprise (ACE) by making you "lock" the best solutions for outstanding equipment reliability into a life-cycle-long, company-wide quality assurance system that eliminates operating risks. Figure 4.4 on the following page, is a more detailed view of the bottom-up Plant Wellness Way Stress-to-Process Model. From the causes of the causes of component stress, you design the business processes you use to reach the pinnacle of world-class plant and equipment reliability.

Plant Wellness Way Methodology

All physical asset failures can be grouped into one of two categories: microstructure distortion or atomic degradation. Distortion causes parts to suffer such high stress or fatigue that their microstructure fails, whereas when a part degrades, the material of construction is attacked by elements in the contacting environment. Physics of Failure methods let us analyze equipment for situations that cause a part's atomic structure to disintegrate and/or its microstructure to suffer excessive stress. You identify potential causes of microstructure distress and then institute the fewest life-cycle activities needed to keep each part at its highest reliability so that your operating plant and equipment reaches the highest availability.

Using the Stress-to-Process framework, you derive the activities to put into your financial management, project engineering, supply chain, warehousing, maintenance, and operational management processes to prevent the deformation and degradation of each part in your machines and equipment. You engineer a life-cycle asset management system—a system of reliability—to deliver parts with long, failure-free lives. With the right processes from bottom to top in your company, you will naturally get the world-class asset maintenance and reliability results needed for Operational Excellence.

The five foundational business and reliability understandings used to improve equipment reliability for Industrial and Manufacturing Wellness (IMW) are as follows:

1. The costs of defects and failure are directly connected to the number and size of risks carried by your business—the more risks tolerated, the greater the opportunity for errors and the greater the costs, losses, and waste that eventually accrue.
2. Failure events do not have only localized consequences; rather, failure costs surge company-wide. Your business always pays every cent for all the costs of its failures.
3. All organizations, machines, and work are series processes, and the success of every series process depends on the success of each individual step.

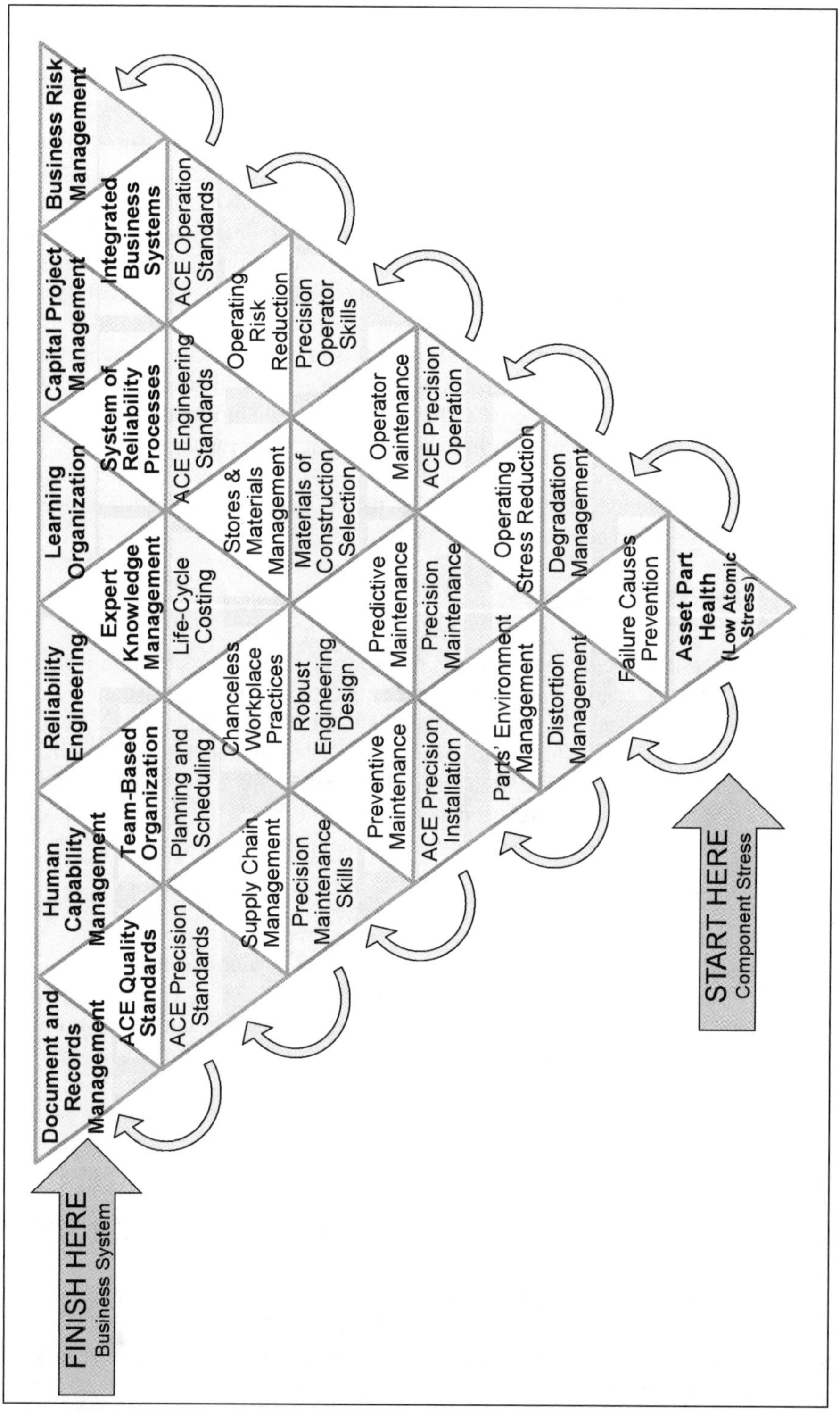

Figure 4.4—Plant Wellness Way: Up to the Top of World-Class Reliability

4. There are natural physical limitations in the materials used to make your plant and equipment. Throughout their microstructure, the stresses from imposed loads must always stay well within the elastic deformation range of the materials of construction.
5. Variation away from the standard for best results produces defects that create failures. For world-class reliability, use only processes throughout a component's life cycle that have natural variation within the outcomes that deliver excellence.

Figure 4.5 is an overview of the Plant Wellness Way methodology. It is the structured approach you will follow to arrive at the right design, operating, and maintenance strategies for maximizing equipment reliability. The methodology takes a life-cycle view of plant and equipment and recognizes that a lifetime of high equipment reliability depends on the reliability of the individual parts in a machine. It helps you develop the right engineering, project selection, plant construction, and operational and maintenance plans and practices for failure-free plant and equipment.

You are always trying to get the longest life out of your parts. If the parts do not fail, the equipment cannot fail. You improve your equipment reliability by eradicating, reducing, and controlling the risks to parts throughout their lifetimes. With fewer risks to parts, there will be fewer failures. This driving philosophy makes you continually reduce the chance of defects in critical working parts. These are the parts that stop your machines if they fail. By relentlessly reducing the likelihood of things going wrong in working parts, equipment reliability naturally improves because the parts carry lower and lower chances of failure. The methodology forces you to work out how to prevent risks to operating components from arising in the first place anywhere in the life cycle. It then requires you to implement risk elimination and prevention by embedding the solutions wherever in the life cycle they are most effective, including project feasibility, engineering design, plant construction, supply and procurement, and operations and maintenance.

From an IMW perspective, if you have to repair your equipment, the business processes used to deliver equipment health and well-being have failed. Your machines are not well if you have to continually correct problems and fix random failures. In the Plant Wellness Way, you do not maintain and repair your equipment in the traditional way. Instead, you create the right conditions for your operating plant and machines to stay well throughout their life cycles and permanently sustain those conditions. You produce lasting equipment health by causing components to have high reliability and removing the life-cycle causes of their failures. You instigate and forever continue the ideal health and wellness requirements and practices that your plant and equipment parts need for failure-free operation.

Plant and equipment wellness is not possible if your maintenance is failure focused. Companies using preventive (PM), predictive (PdM), and run-to-failure (RTF) maintenance strategies easily default to being failure focused. To achieve "wellness," you must create and endlessly sustain component health. Successful reliability creation solutions and methods that cause equipment parts to be healthy must be designed, installed and continually practiced in all business processes and activities. In this way, world-class reliability performance becomes a natural business habit and the organization's culture.

Asset Management Strategy from Physics of Failure Factors Analysis

You derive the minimal reliability excellence strategy for a machine by considering each part's Physics of Failure mechanisms. First the causes that can destroy a part's microstructure are established, and then the types of events that cause those factors are identified. Once the causes of the causes are found, you determine the most suitable and effective solutions to eliminate or prevent each one and install them in appropriate procedures used during the part's life cycle.

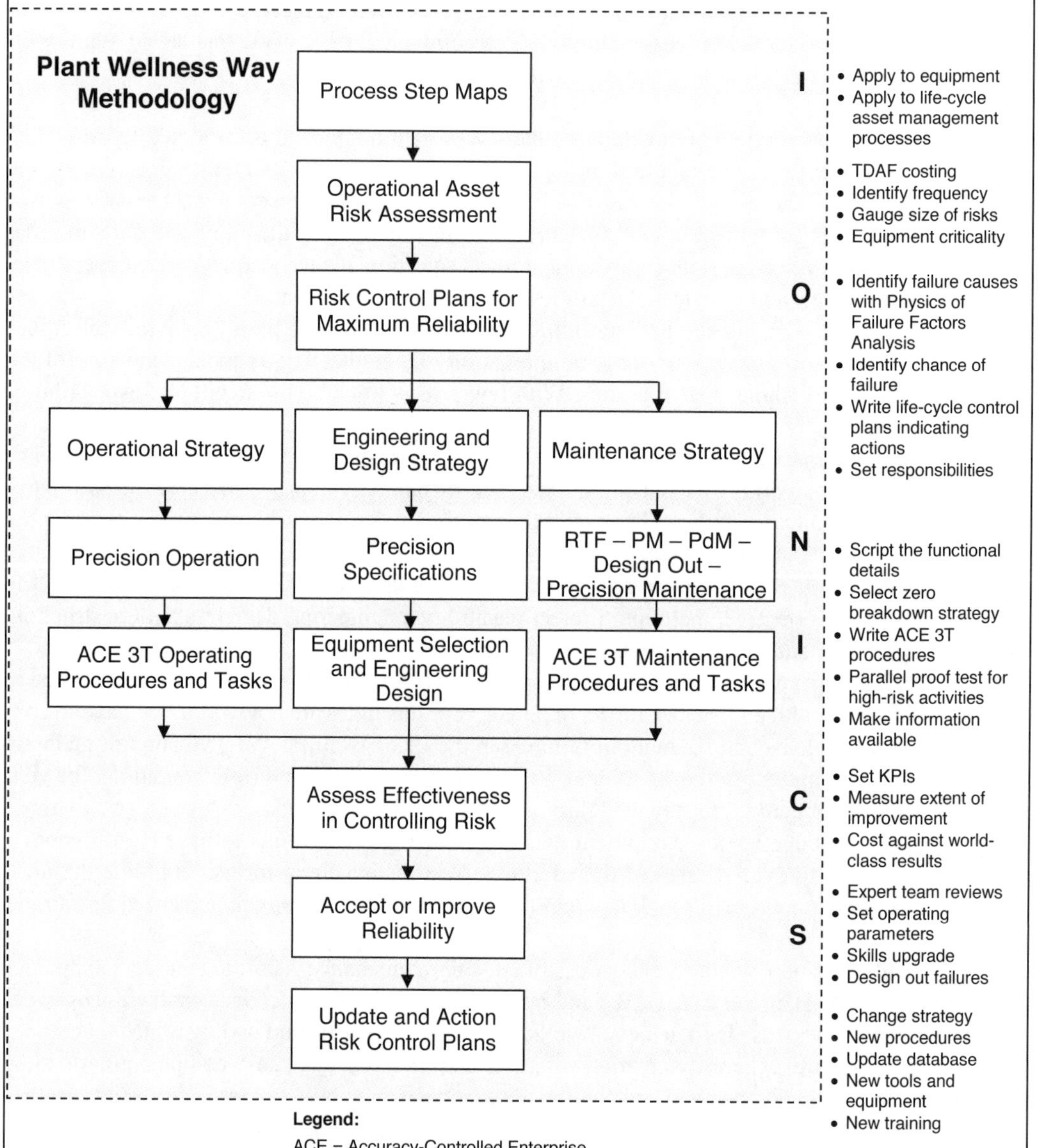

Legend:

ACE = Accuracy-Controlled Enterprise
IONICS = Six Plant Wellness Processes
KPI = Key Performance Indicator
PdM = Predictive Maintenance
PM = Preventive Maintenance
RTF = Run-to-Failure
TDAF = Total Defect and Failure
3T = Target, Tolerance, Test

Figure 4.5—Remove Operating Risks for Industrial and Manufacturing Wellness

The microstructures of metals and plastics are failed in the ways listed in Table 4.1. These principal mechanisms of material breakdown are called the Physics of Failure Factors. There are not that many ways to physically destroy atomic bonds and cause solids to come apart. Occasionally, peculiarities exist that are specific to the atomic structure of a group of solids. When you encounter such a unique factor, add it to the table for that group of materials.

Table 4.1—Physics of Failure Factors for Solid Materials

Metals	Plastics
Compressive force crushes the microstructure until it collapses	Compressive force crushes the microstructure until it collapses
Tensile force stretches the microstructure until it separates	Tensile force stretch the microstructure until it separates
Shear force tears the microstructure until it rips	Shear force tears the microstructure until it rips
Cyclic stress fatigue from compressive, tensile, and/or shear forces	Cyclic stress fatigue from compressive, tensile, and/or shear forces
Melting or overheating the microstructure	Melting or overheating the microstructure
Separation of the microstructure (e.g., dislocation)	Separation of the microstructure
Material missing from microstructure (e.g., cavities)	Material missing from microstructure (e.g., cavities)
Material mechanically ripped from the microstructure	Material mechanically ripped from the microstructure
Foreign inclusion in the microstructure	Foreign inclusion in the microstructure
Electromagnetic radiation destroys atomic bonds	Electromagnetic radiation destroys atomic bonds
Chemical reaction destroys atomic bonds	Chemical reaction destroys atomic bonds
Crystal lattice attack of microstructure grain boundaries	Depolymerization decomposition
Electrical discharge	

Table 4.2—Mechanisms Causing Microstructure Failure Factors

Tensile Force Stretches the Microstructure until It Separates	Material Mechanically Ripped from the Microstructure	Chemical Reaction Destroys Atomic Bonds
Pressure	Overloaded	Foreign inclusion in material of construction
Overloaded	Punch (impact load on small area)	Corrosion (pitting, galvanic, crevice, etc.)
Pressure hammer	Hammer impact, dent	Acidic atmosphere
Expansion	Gouge	Product ingress/egress
Unbalance	Abrasion (wear material away)	Chemical reaction
Gouge	Solid object impact (e.g., vehicle, lifting chains)	Oxidization
Hydraulic shock (water hammer)	Impingement (jet of fluid)	Dissimilar materials
Physical abuse	Detach-debond-delaminate	Hygromechanical (moisture absorption)
Acts of God/acts of nature	Physical abuse	Inclusions in contacting process
	Friction	Crystal lattice attack
	Physically deformed (bend, twist, squash)	Solar radiation: UV and thermal effects
	Erosion	Hydrogen attack/embrittlement
		Stress corrosion cracking
		Chemical attack
		Electrical discharge

There are numerous causes that can produce the mechanisms that fail microstructures. Table 4.2 lists a range of causes for three example microstructure failure factors. Each principal microstructure failure factor has multiple failure-causing mechanisms that can arise to produce the circumstances under which the atomic structure of the item is destroyed. Parts will fail any time their structure is not strong enough for the stresses inside them. The events that can produce those stresses number in the hundreds and even thousands of opportunities during a part's lifetime.

For example, free water in oil will lead to roller bearing failure by thinning the lubricant film between the raceway and rolling elements until there is metal-to-metal contact. Once moving metal parts hit against each other, their surfaces will be ripped away or crack. But free water can only be present in oil if it gets into the lubricant. Prevent water ingress, and there is no cause for the bearing to fail from watery oil or grease. When you look across the life cycle of lubricating oil, there are dozens of ways for water to get into a machine's lubricant. Water can get into oil if it is put in by the oil manufacturer; if the equipment maker puts water in the machine during cleaning or testing; if rain gets inside the equipment during storage; if rain or sea water gets inside the equipment during shipping or road transport; if it is flushed into the machine during commissioning or maintenance; or if it leaks into the sump from inside the equipment. Water could also come from rain falling on unsealed oil drums; from capillary action down the threads of bungs on sealed drums when water pools on the lid; from hosing down equipment so that water pressure pops open a shaft seal; from condensing humidity drawn through a breather; from leaking cooling system pipework; and from many other unintended events that can happen over an equipment's lifetime. When you adopt the Plant Wellness Way, the answers that prevent each of those risks become the reliability creation strategy you put into place throughout your business, along your supply chains, and across the life cycle.

The technique used to discover the causes of the causes of microstructure damage is called Physics of Failure Factors Analysis (POFFA). It is the starting point in the Stress-to-Process Model to discover the ideal life-cycle reliability strategy for each part. You work from the parts' microstructure properties up to the correct business processes needed to sustain outstanding component longevity. By finding all the ways that a component's microstructure can fail, you can proactively select the correct life-cycle strategies to eliminate the risk every time it might arise. You use degradation and deformation prevention practices to keep every part healthy, well, and safe throughout its life. You apply high-reliability methods and work quality assurance to install, maintain, and operate the part. You train people in the right reliability techniques and solutions. You stipulate the supply chain risk controls that your vendors must use to protect your physical assets' lives. Thus, you intentionally design outstanding and lasting plant and equipment reliability into your business.

Plant and Equipment Risk Identification

Because machines fail after their critical working parts fail, the Plant Wellness Way requires you identify the parts in a machine that will stop it from operating if they break. These parts are investigated using Physics of Failure Factors Analysis to find all of their life-cycle risks. Unacceptable risks are eliminated, and when that is not possible, the risks are controlled and managed with the least number of suitable engineering, operating, and maintenance strategies. Your intention is to reduce the chance of equipment failure to less than once in a span of three times the service life of the equipment in the operation. For example, for equipment expected to be in production for 20 years, the likelihood of a failure incident is to be less than a 1-in-60-year event. For an asset that is expected to be in operation for just 5 years, the target is no more than a 1-in-15-year event. Every critical component in the equipment must therefore have a far lower chance of failure in order for their combined odds to produce the required equipment likelihood of failure.

In all cases, the decision to apply component risk elimination or risk control strategies are decided by the economic value of their adoption and use. The total expense of providing risk mitigations to reduce equipment failure frequencies to once in three times the service life may be too costly to justify. Nonetheless, the philosophy of drastically reducing component risk to get outstanding equipment reliability is sound and sensible asset management.

A risk matrix is used to show risk levels. It is derived by using Formula 4.2:

Formula 4.2

Risk ($/yr) = Consequence ($/event) x Likelihood (events/yr)

On the risk matrix, the intersection of the "consequence" and "likelihood" values ranks the risk. The consequence is the severity of an event, and the likelihood is the probability that it will occur. Table 4.3 is a common 5 x 6 risk matrix used to gauge the risks in a business. It has five consequence columns and six likelihood rows. A 5 x 5 matrix is often used in occupational health and safety systems to assess job risks. Such risk matrices are developed using the recommendations of international risk management standards. The business-wide consequences for people, reputation, business processes and systems, and finance are explained and scaled to reflect the organization using the matrix. The methods and principles to apply in addressing risk can be advised in the "Risk Management Philosophy" box at the top-left side of the matrix.

The risk matrix is used to gauge whether an item or situation has an acceptable, low, medium, high, or extreme level of risk. As a general intent, risk mitigations need to drive risk levels lower by two levels or more. Extreme and high risks are reduced to medium and low, respectively, and a medium-level risk is reduced to low or less. Unknown to most managers, this is this yet another Crosshair Game practice because there is no certainty of the result. In the Plant Wellness Way, risk is driven down to levels that create the required reliability.

It is important to retain mathematical accuracy when developing a risk matrix. The design of the matrix corresponds to the use of log to the base-10 ($\log_{10}$) math. The numbers used in the consequence and likelihood scales are $\log_{10}$ values. For example, the $\log_{10}$ of 100 is 2; the $\log_{10}$ of 1,000 is 3; the $\log_{10}$ of 10,000 is 4; the $\log_{10}$ of 1,000,000,000 is 9; and so on. By using $\log_{10}$ values, we can keep the matrix small. If we used linear scales, this would require a huge piece of paper to show the graduation mark for a value of 1,000,000,000 because it would be far, far to the right on the scale. By using $\log_{10}$ scales, we can shrink the matrix to one page. Because the risk scales are $\log_{10}$, we add the indices. For example, 100 x 1,000 = 100,000 is $10^2 \times 10^3 = 10^5$; as a $\log_{10}$ calculation, it becomes 2 + 3 = 5. That is how the numbers in the cells of the sample Risk Matrix worksheet are determned.

The Risk Matrix worksheet can be found online in the downloadable spreadsheets accompanying this book.

The various business reputation, occupational health and safety, and business system consequences in a column need to correspond to the financial value of the column. Similarly, for likelihood, each row is representative of the intervals at which the risk occurs. The $\log_{10}$ numbers corresponding to each level of likelihood and consequence can be added together to provide a numerical indicator of the risk. This is useful for comparing dissimilar risks in order to set priorities or when a simple means (not involving quantitative risk calculations) is needed to give each risk a representative value. By ensuring mathematical and financial accuracy, the use of a matrix is defensible as sound and reasonable when making decisions.

Table 4.3—5 x 6 Risk Matrix for Determining the Risk Level

RISK MANAGEMENT PHILOSOPHY		Business-Wide Consequence					
E – Extreme risk: Detailed action plan approved by CEO **H – High risk: Specify responsibility to senior manager** **M – Medium risk: Specify responsibility to department manager** **L – Low risk: Manage by routine procedures** **Extreme** or **high** risk must be reported to senior management and requires detailed treatment plans to reduce the risk to **low** or **medium**		**People**	Injuries or ailments not requiring medical treatment	Minor injury or first aid treatment case	Serious injury causing hospitalization or multiple medical treatment cases	Life-threatening injury or multiple serious injuries causing hospitalization	Death or multiple life-threatening injuries
		Reputation	Internal review	Scrutiny required by internal committees or internal audit to prevent escalation	Scrutiny required by clients or third parties, etc.	Intense public, political, and media scrutiny (e.g., front-page headlines, TV, etc.)	Legal action or commission of inquiry or adverse national media
		Business Process & Systems	Minor errors in systems or processes requiring corrective action or minor delay without impact on overall schedule	Policy procedural rule occasionally not met or services do not fully meet needs	One or more key accountability requirements not met; inconvenient but not client welfare threatening	Strategies not consistent with business objectives; trends show service is degraded	Critical system failure, bad policy advice, or ongoing noncompliance; business severely affected
		Financial	**$500**	**$5K**	**$50K**	**$500K**	**$5,000K**
			Insignificant	**Minor**	**Moderate**	**Major**	**Catastrophic**
Historical Likelihood			**1**	**2**	**3**	**4**	**5**
Event occurs at this site annually or more often	**6**	**Certain**	**M**	**H**	**H**	**E**	**E**
Event regularly occurs at this site	**5**	**Likely**	**M**	**M**	**H**	**H**	**E**
Event is expected to occur on this site	**4**	**Possible**	**L**	**M**	**M**	**H**	**E**
Event occurs from time to time on this site	**3**	**Unlikely**	**L**	**M**	**M**	**H**	**H**
Event occurs in the industry, and could on this site, but doubtful	**2**	**Rare**	**L**	**L**	**M**	**M**	**H**
Event hardly heard of in the industry. May occur but in exceptional circumstances	**1**	**Very Rare**	**L**	**L**	**L**	**M**	**H**

The costs used to establish the risk boundary are the Total Defect and Failure (TDAF) costs that an organization is willing to accept. Figure 4.6 shows the risk boundary concept of investment to prevent failures. This company will not accept annualized TDAF costs of more than $20,000, and it is willing to invest money to reduce greater risks. If the risk is acceptable, nothing is done to stop it except to ensure that it does not change. Should the event happen, the business knowingly pays for its rectification. But if the cost of failure is unacceptable, mitigations are put into place to sufficiently reduce the risk. The mitigations to prevent the risk are seen as a better investment than paying to fix the consequences later.

A step-by-step example of the process for performing TDAF costing can be found in the document *The Instantaneous Cost of Failure,* included in the online materials accompanying this book.

Your risk matrix needs to be calibrated so that the bottom of the low risk level represents your business risk boundary. Calibrating a risk matrix means adjusting the consequences to match your company's circumstances. A risk that is significant to a small organization may be insignificant to a much larger business. The financial value of each consequence column on the risk matrix is set by

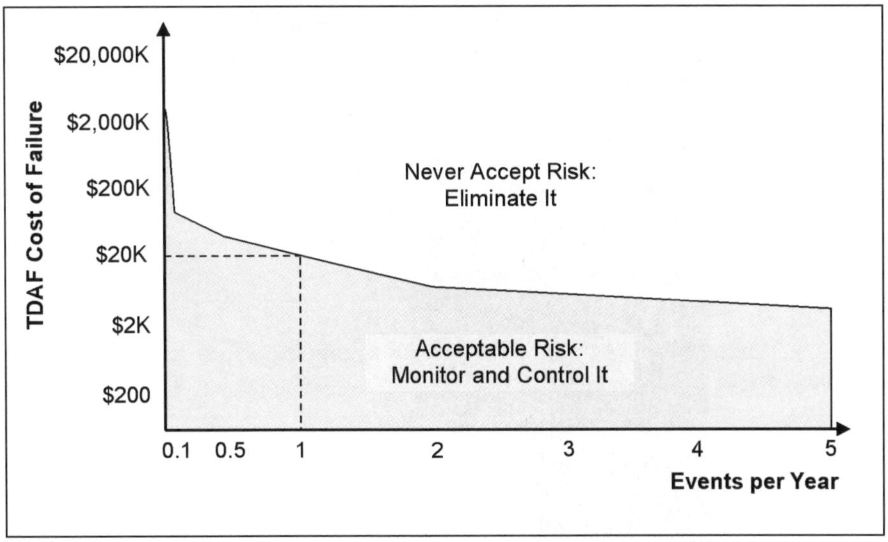

Figure 4.6—The Risk Boundary Concept

the amount that your senior management considers an insignificant cost. Each subsequent column then follows log to the base-10 rules to set its value higher by an order of magnitude.

In a standard 5 x 6 risk matrix, the cells furthest to the right in the table represent large amounts of money; it is necessary to alter the matrix to a finer scale for use in the Plant Wellness Way. A 16 x 13 risk matrix like the one in Table 4.4 is used. It is the previous 5 x 6 matrix with each column and row halved to create more cells covering smaller ranges. The table is also extended to lesser values to include small problems that happen often. This allows you to show smaller changes in risk on the 16 x 13 matrix that would not be noticeable on the 5 x 6 matrix. Even finer scales can be introduced if necessary.

Only financial values of consequences are used in the Plant Wellness Way. The business reputation, occupational health and safety, and business system consequences are not shown. Every situation's severity must be converted into the money lost by your business when it happens. This is a financial truth of doing business—in the end, everything bad that happens in your organization causes a financial loss. It is vital to know how much money is in jeopardy with every risk carried.

Color-coding in the matrix is used to indicate each risk level. Below the boundary, risks are acceptable but need to be monitored to ensure they do not change. Above the boundary, risks are acted on using strategies and actions that reduce the risk to below the low-risk boundary and, ideally, to the point that an equipment failure occurs less than once in three times the service life. By implication, risks less than low are the risks you live with. That does not mean you do nothing to mitigate the risk. For example, the risk of a planet-shattering meteorite hitting the Earth is extremely small, but it is not impossible. We monitor outer space looking for dangerous meteorites even though we can do nothing to prevent them from hitting the planet. In your business, you will have risks that are so small as to be unlikely to arise, but you still must watch out that situations do not change in ways that let those risks become potential realities. You cannot use the excuse that a risk is negligible on the risk matrix to not bother preventing it from happening.

The risk matrix in Figure 4.7 and Table 4.4 shows a low-risk boundary set at an annualized cost of $10,000 per event. The business will accept the equivalent of one failure per year if it costs less than $10,000, but it will act to reduce those situations in which failure events produce accumu-

Likelihood of Equipment Failure Event per Year				TDAF Cost per Event		$30	$100	$300	$1,000	$3,000	$10,000	$30,000	$100,000	$300,000	$1,000,000	$3,000,000	$10,000,000	$30,000,000	$100,000,000	$300,000,000	$1,000,000,000
Event Count per Year	Time Scale	Descriptor Scale	Historical Description			C1	C2	C3	C4	C5	C6	C7	C8	C9	C10	C11	C12	C13	C14	C15	C16
						1.5	2	2.5	3	3.5	4	4.5	5	5.5	6	6.5	7	7.5	8	8.5	9
100	Twice per week	Certain		L13	2	3.5	4	4.5	5	5.5	6	6.5	7	7.5	8	8.5	9	9.5	10	10.5	11
30	Once per fortnight	Certain		L12	1.5	3	3.5	4	4.5	5	5.5	6	6.5	7	7.5	8	8.5	9	9.5	10	10.5
10	Once per month	Certain		L11	1	2.5	3	3.5	4	4.5	5	5.5	6	6.5	7	7.5	8	8.5	9	9.5	10
3	Once per quarter	Certain		L10	0.5				3.5	4	4.5	5	5.5	6	6.5	7	7.5	8	8.5	9	9.5
1	Once per year	Almost Certain	Event will occur on an annual basis	L9	0					3.5	4	4.5	5	5.5	6	6.5	7	7.5	8	8.5	9
0.3	Once every 3 years	Likely	Event has occurred several times or more in a lifetime career	L8	-0.5						3.5	4	4.5	5	5.5	6	6.5	7	7.5	8	8.5
0.1	Once per 10 years	Possible	Event might occur once in a lifetime career	L7	-1							3.5	4	4.5	5	5.5	6	6.5	7	7.5	8
0.03	Once per 30 years	Unlikely	Event does occur somewhere from time to time	L6	-1.5								3.5	4	4.5	5	5.5	6	6.5	7	7.5
0.01	Once per 100 years	Rare	Heard of something like it occurring elsewhere	L5	-2									3.5	4	4.5	5	5.5	6	6.5	7
0.003	Once every 300 years			L4	-2.5										3.5	4	4.5	5	5.5	6	6.5
0.001	Once every 1,000 years	Very Rare	Never heard of this happening	L3	-3											3.5	4	4.5	5	5.5	6
0.0003	Once every 3,000 years			L2	-3.5												3.5	4	4.5	5	5.5
0.0001	Once every 10,000 years	Almost Incredible	Theoretically possible but not expected to occur	L1	-4													3.5	4	4.5	5

Equipment Risk Envelope

$100,000 event every two years

Figure 4.7—Risk Matrix Showing the Level of Business Risk

Table 4.4—Calibrated 16 x 13 Risk Matrix to Observe Changing Risks

Likelihood of Equipment Failure Event per Year				TDAF Cost per Event		$30	$100	$300	$1,000	$3,000	$10,000	$30,000	$100,000	$300,000	$1,000,000	$3,000,000	$10,000,000	$30,000,000	$100,000,000	$300,000,000	$1,000,000,000
Event Count per Year	Time Scale	Descriptor Scale	Historical Description			C1	C2	C3	C4	C5	C6	C7	C8	C9	C10	C11	C12	C13	C14	C15	C16
						1.5	2	2.5	3	3.5	4	4.5	5	5.5	6	6.5	7	7.5	8	8.5	9
100	Twice per week	Certain		L13	2	3.5	4	4.5	5	5.5	6	6.5	7	7.5	8	8.5	9	9.5	10	10.5	11
30	Once per fortnight	Certain		L12	1.5	3	3.5	4	4.5	5	5.5	6	6.5	7	7.5	8	8.5	9	9.5	10	10.5
10	Once per month	Certain		L11	1	2.5	3	3.5	4	4.5	5	5.5	6	6.5	7	7.5	8	8.5	9	9.5	10
3	Once per quarter	Certain		L10	0.5				3.5	4	4.5	5	5.5	6	6.5	7	7.5	8	8.5	9	9.5
1	Once per year	Almost Certain	Event will occur on an annual basis	L9	0					3.5	4	4.5	5	5.5	6	6.5	7	7.5	8	8.5	9
0.3	Once every 3 years	Likely	Event has occurred several times or more in a lifetime career	L8	-0.5						3.5	4	4.5	5	5.5	6	6.5	7	7.5	8	8.5
0.1	Once per 10 years	Possible	Event might occur once in a lifetime career	L7	-1							3.5	4	4.5	5	5.5	6	6.5	7	7.5	8
0.03	Once per 30 years	Unlikely	Event does occur somewhere from time to time	L6	-1.5								3.5	4	4.5	5	5.5	6	6.5	7	7.5
0.01	Once per 100 years	Rare	Heard of something like it occurring elsewhere	L5	-2									3.5	4	4.5	5	5.5	6	6.5	7
0.003	Once every 300 years			L4	-2.5										3.5	4	4.5	5	5.5	6	6.5
0.001	Once every 1,000 years	Very Rare	Never heard of this happening	L3	-3											3.5	4	4.5	5	5.5	6
0.0003	Once every 3,000 years			L2	-3.5												3.5	4	4.5	5	5.5
0.0001	Once every 10,000 years	Almost Incredible	Theoretically possible but not expected to occur	L1	-4													3.5	4	4.5	5

Notes:

Risk Level
Red = **Extreme**
Amber = **High**
Yellow = **Medium**
Green = **Low**
Blue = **Accept**

1) Risk boundary is set at low level
2) Based on HB436:2004-Risk Management
3) Identify "black swan" events as B-S (A "black swan" event is one that people say "will not happen" because it has not yet happened)
4) TDAF cost includes all business-wide costs, losses, waste, and opportunities missed because of a failure event

For a fully enlarged version of this table, see the Risk Matrix worksheet, which is included in the downloadable materials accompanying this book.

lated costs of more than $10,000 annually. Note that this also means the operation will accept a $100,000 loss event every 10 years and will do nothing else to prevent it beyond ensuring that the event happens no more than once every 10 years. If you can't afford even one $100,000 failure, it would be shrewd to get an appropriate insurance policy that moves the risk to the insurance company, while you pay only the premium.

The risk levels intentionally skew at the top-left side of the matrix. Frequent small, low-cost risks cannot be allowed to remain in your business, for two reasons. The first is that every small event takes resources and time to address. That time and effort could be better used to do far more valuable work. Small problems have massive opportunity costs, and so all small problems must be eliminated. Second, negligible risks can turn into large operational risks such as breakdowns, waste, losses, and unhappy customers. All small hazards become the doorways to large safety risks, such as injury, dismemberment, or death. Living with frequent small problems requires frequent intervention, which puts operators and maintainers at risk of harm many times. When you do not stop frequent small problems, it is only a matter of time before someone gets hurt or you have a serious production disruption and suffer another "missed-the-budget month." It is vital that small repetitive risks are escalated to a higher concern and importance so you justify getting rid of them before they become your future injuries and disastrous TDAF cost losses.

Your risk matrix is a financial model of your business. The consequential value you lose when a risk event happens is real money lost to your company. The likelihood of events comes from your own equipment failure history. Your risk matrix shows what happens in your business, and it exactly reflects the outcomes of the way your organization behaves.

Using the risk matrix lets you see the scale of a loss event and immediately spot the savings you would get if the risk were reduced to an acceptable level. The savings made from lowering a risk so that a failure event does not occur become new operating profits. When the potential profits are large, you have a strong business case for new expenditure to reduce the risk. For example, the $100,000 failure event occurring every two years represented in Figure 4.7 rates as a high risk. The annual risk calculated with the risk formula is $50,000 per year. If the likelihood of the event could be reduced to once every 10 years with suitable mitigating actions costing $10,000 annually, the business would get an additional $40,000 operational profit each year. The money to pay for the risk mitigations to gain the resulting profit becomes available when the risk goes down. With the risk reduced to a low-level occurrence of once a decade for a cost of $10,000 per year, the average annual operating profit rise is a recurring 400% annual return on investment. A project having a 400% annual return on investment presents a very strong business case to do it urgently.

Start with a Business Risk Impact Review

The business risk from operating equipment is identified by gauging the combined severity of a part's failure on the operation, safety, environment, and company reputation. You begin by asking two economic questions for each asset to identify the business-wide impact of the worst failure events.

1. Are the total business-wide consequences of any failure of the equipment acceptable?
2. Where failure is acceptable, how frequently can it occur before it becomes unacceptable?

Events that can cause human death are treated differently from those that only destroy or ruin operating assets. When people can die from an incident, you must include this in the consideration of business risk. Modern SFAIRP[2] safety philosophy dictates that you do all that is possible to prevent a fatality so that the likelihood of death-causing events is incredibly low.

Where an event's consequence and frequency plot on the company risk matrix determines whether the risk analysis is taken further. The operations group will know whether the failure of an equipment item will cause a production stoppage that has adverse business impacts. If the combined severity of equipment failure is economically acceptable, it is allowed to fail, and the analysis goes no further for that item. The default maintenance and operating strategy for such equipment is *run to failure:* no maintenance activity is performed on it, and no spare parts are carried in store for it. When the equipment breaks, the decision is made to rectify it or address the failure in an appropriate way. Corrective actions are instigated after failure, and the accompanying costs and time delays are accepted without concern because a failure of the equipment does not matter to business success.

When an equipment failure is considered financially unacceptable or minor failures are too frequent, the business-wide economic impact of a part's failure is determined using TDAF costing. The estimate needs to be acceptably accurate to be believed by managers and defendable when challenged by others. Aim to be better than ±20% of true business costs, losses, and waste in your TDAF cost estimates. The consequence is the worst TDAF cost if a critical part fails in service. The likelihood is the historical annual frequency for the failure event in the business (or the frequency from other comparable, similarly operated businesses if the imagined failure has not yet occurred). The business risk from each part is marked on a calibrated business risk matrix. This component risk assessment process is repeated for each critical part in the equipment. The greatest risk is the component failure event with the highest annualized cost. In this way, you quantify the worst business risks and make them visible to everyone. When component risks are above the risk boundary, action must be taken to reduce those risks to an acceptable level. Such components are then individually analyzed using Physics of Failure Factors Analysis to identify useful risk mitigations.

The risk to your operation varies with the equipment component concerned, the type of failure it suffers, and the risk mitigations your company has in place for each failure. This produces a spread of risk values. If you plot a point for every risk from the critical parts in an item of equipment onto your company risk matrix, there will be an area on the matrix covered in dots. The range of risks will have a shape. That patch is the *risk envelope* for the equipment. Figure 4.7 shows a risk envelope encompassing all of the points representing one equipment's range of risks for the operation using it.

Risk Reduction Decisions

When the business impact of an equipment failure is greater than what the organization will tolerate, you put into place appropriate actions that reduce risk below the acceptable boundary and keep it there. The mitigation methods and actions chosen must be effective in reducing the risk by either lowering the consequential business-wide impact of the event (its TDAF cost) or lowering the likelihood of the event occurring (its degree of uncertainty). You check the effectiveness of your chosen mitigations by plotting the risk before and after on a risk matrix to ensure the proposed risk controls can bring the desired results. You and your management must be convinced that what is done to reduce risk will work and that the risk will remain under the low level. If a mitigation action does not clearly deliver substantially reduced consequence and/or lower frequency, it is a waste of resources, and it is discarded and a better mitigation chosen.

Figure 4.8 shows how the effectiveness of mitigations is checked and tested for their potential economic value. In the example, the consequence of the $100,000 failure event is unchanged, but the likelihood has been substantially reduced from annually to once a decade because of the mitigations used.

When a number of risk mitigations are required to operate collectively in order to achieve the required risk reduction, it may be necessary to accept a less than optimal individual mitigation choice because it is an integral part of the total risk reduction solution. A common example is to carry critical spares in your warehouse and pay all the ongoing costs to keep them reliable and available even if the parts in service rarely fail. The spares are integral to your operating risk reduction solution, and

Likelihood of Equipment Failure Event per Year				TDAF Cost per Event		$30	$100	$300	$1,000	$3,000	$10,000	$30,000	$100,000	$300,000	$1,000,000	$3,000,000	$10,000,000	$30,000,000	$100,000,000	$300,000,000	$1,000,000,000
Event Count per Year	Time Scale	Descriptor Scale	Historical Description			C1	C2	C3	C4	C5	C6	C7	C8	C9	C10	C11	C12	C13	C14	C15	C16
						1.5	2	2.5	3	3.5	4	4.5	5	5.5	6	6.5	7	7.5	8	8.5	9
100	Twice per week	Certain		L13	2	3.5	4	4.5	5	5.5	6	6.5	7	7.5	8	8.5	9	9.5	10	10.5	11
30	Once per fortnight	Certain		L12	1.5	3	3.5	4	4.5	5	5.5	6	6.5	7	7.5	8	8.5	9	9.5	10	10.5
10	Once per month	Certain		L11	1	2.5	3	3.5	4	4.5	5	5.5	[illegible]	[illegible]	[illegible]	[illegible]	8	8.5	9	9.5	10
3	Once per quarter	Certain		L10	0.5				3.5	4	4.5	5	[illegible]	[illegible]	6.5	[illegible]	7.5	8	8.5	9	9.5
1	Once per year	Almost Certain	Event will occur on an annual basis	L9	0					3.5	4	4.5	[illegible]	[illegible]	[illegible]	[illegible]	7	7.5	8	8.5	9
0.3	Once every 3 years	Likely	Event has occurred several times or more in a lifetime career	L8	-0.5						3.5	4	[illegible]	[illegible]	[illegible]	6	6.5	7	7.5	8	8.5
0.1	Once per 10 years	Possible	Event might occur once in a lifetime career	L7	-1							3.5	4	[illegible]	5	5.5	6	6.5	7	7.5	8
0.03	Once per 30 years	Unlikely	Event does occur somewhere from time to time	L6	-1.5								3.5	4	4.5	5	5.5	6	6.5	7	7.5
0.01	Once per 100 years	Rare	Heard of something like it occurring elsewhere	L5	-2									3.5	4	4.5	5	5.5	6	6.5	7
0.003	Once every 300 years			L4	-2.5										3.5	4	4.5	5	5.5	6	6.5
0.001	Once every 1,000 years	Very Rare	Never heard of this happening	L3	-3											3.5	4	4.5	5	5.5	6
0.0003	Once every 3,000 years			L2	-3.5												3.5	4	4.5	5	5.5
0.0001	Once every 10,000 years	Almost Incredible	Theoretically possible but not expected to occur	L1	-4													3.5	4	4.5	5

Consequence Reduction
Likelihood Reduction
Money Saved by Mitigations
Risk = $10,000/year event

Figure 4.8—Business Risk Matrix Showing the Effectiveness of Risk Mitigations

all of them must be on site, even if they are dead money. It is only because the spares are handy as part of a larger strategy that your operation can operate at a very low risk of disaster.

Another seeming wasteful risk mitigation is to limit the number of times routable assemblies are rebuilt before being replaced with totally new units. Although a routable, by definition, is to be refurbished, its frequency of failure rises as the unchanged components age and fail. At some point, the rising failure rate of the rebuilt routable will cause high corrective maintenance costs from its frequent replacement. Even though the cost of a new unit will be more than the cost to rebuild the routable again, you would be wise to pay for the new unit. You will substantially reduce the operating risk to your plant by using new equipment with long times between failures rather than paying a lesser amount to fix old, tired assemblies that will not last for very long.

A business makes money if a risk can be prevented for less than the risk's equivalent annualized cost. The greatest opportunity for a company to manage risk for much less cost is to identify those methods, systems, and practices that reduce the chance of a risk arising and then implement them with great energy and vigor across the organization. Maintenance is one of the methodologies available to reduce the risk of equipment failure, but if it is used incorrectly, it becomes a consequence reduction strategy done after failure has started. In IMW, maintenance is used for proactive failure prevention. It combines with cost-effective engineering and operational choices to deliver chance reduction strategies that stop all failures from starting during the equipment lifetime.

Any risk mitigation or combination of risk controls you choose will have to meet the criteria of substantially reducing the original risk with much lower event frequency and/or a large reduction in consequences. The word "substantial" is defined in the *Collins English Dictionary* as "worthwhile; important; of telling effect." In the context of risk management, any improvement you select must clearly deliver better outcomes by a proportion that is indisputably valuable to the organization. On the risk matrix, you want to see a big distance between the original risk point and the final risk.

There are three decision rules used in the IMW for deciding when you should accept a risk mitigation for a physical asset.

1. A solution that reduces opportunity for failure is acceptable if it significantly extends the time between situations when component failure is possible. It is satisfied by mitigations that reduce the frequency of microstructure stress excursions and/or of contact environmental degradation events. Examples are engineering solutions that lessen the frequency of causes of failure factor mechanisms, and the application of world-class practices, such as precision maintenance and precision operation, throughout service life.
2. A solution that reduces the chance of failure or increases reliability must significantly reduce the stress in a part. It is satisfied by solutions that indisputably lower the size of component microstructure stress or improves the microstructure's capability to comfortably handle all imposed stresses. Examples include de-rating the service duty, reengineering the component by using parts made of greatly stronger materials or with quality properties that mitigate the effect of stress, changing part shape to reduce microstructure stresses, and practicing precision maintenance and operation all the time so stresses are always minimized.
3. A solution that reduces consequence is acceptable if it significantly reduces the TDAF cost. It is achieved by using mitigations that prevent large financial losses if a failure initiation event

Opportunity Cost of Run-to-Failure Decisions

If the business-wide risk of a failure is already below the risk boundary, do not spend time conducting an analysis. You will get greater value from your time by preventing higher-risk situations. For example, if a small 5 kilowatt conveyor gear box drive fails, it will cost $1,500 to $2,500 for a new one, depending on the gearbox model. It will take about four hours to change out the unit at a direct maintenance labor and services cost of about $1,000. If a new unit can be sourced within an hour after a breakdown, the business-wide impact of costs, waste, and losses in a continuous process operation could reach $35,000. If a breakdown happens no more than every five years, the annualized risk is $7,000. This risk is well below the $10,000 per year risk boundary of our imaginary manufacturer, so our maintenance strategy defaults to accepting that failure. However, if the same gearbox fails every two years and each failure costs $35,000, you will have to address the reasons for the failure, as the annualized cost of $17,500 is above the $10,000 per year risk boundary. In this case, you cannot accept a run-to-failure strategy and need to develop mitigations to reduce the frequency of failure.

A trap you fall into with a run-to-failure decision is that, by default, you design into your operation regular breakdowns of unimportant machines and equipment. These require you to use maintenance time and budget repairing equipment of minor importance. The value of the opportunities lost because limited maintenance resources are used to fix inconsequential equipment when they could be doing far more valuable work must be considered in selecting a maintenance strategy. That time and money could be better spent creating plant reliability or eradicating larger risks rather than repairing broken things of trivial importance. Run to failure might seem to be an acceptable risk reduction strategy, but not if it costs you opportunities to get far greater successes.

You remove the opportunity trap from the run-to-failure strategy by proactively replacing unimportant equipment shortly before it is likely to fail. In the case of the $2,500 gearbox in the continuous process plant, we originally devised a strategy to replace it new upon breakdown about once every five years, for a TDAF cost of $35,000. But when fixing a breakdown imposes such high cost, you need to check the value of doing an old-for-new replacement during a planned outage.

The gearbox will cost $2,500 to buy regardless of whether it's a breakdown or a scheduled production shutdown. The time needed to swap the old gearbox with a new one during a shutdown will be less because the plant is already handed over for maintenance. The same job done in a shutdown will have a total business-wide cost of about $5,000. It is clearly more economical to install a new gearbox as planned preventive maintenance for $5,000 than to have a TDAF cost breakdown of $35,000. The best strategy for this gearbox is not the default run-to-failure strategy but rather a zero-breakdown preventive maintenance strategy to replace it brand-new every four years in a planned shutdown and never let the equipment age so long that you risk failure in service. In fact, you could fit a new gearbox every two years so that failures are impossible, and it would still be far less costly than having a breakdown after five years. The zero-breakdown strategy will bring you modern, new equipment, lower maintenance costs, lower risk, and higher operating profits.

This simple maintenance strategy selection case study has an important message: when making maintenance decisions, you can only determine the best choice after you do economic modeling. Maintenance is an economic decision. Companies that are not modeling all of their maintenance choices are wasting hugh amounts of money. The right maintenance strategy is the one that brings the most money to the company—the least life-cycle cost choice is always the right one. Only do the maintenance that brings the most operating profit over the lifetime of the asset. The best financial choice for your company is also the best maintenance choice for your company. In the IMW Way, you can see on the risk matrix which maintenance strategy will bring you most operating profit.

References

1. First heard from retired professor David Sherwin in his three-day "Introduction to Reliability Engineering" course, Perth, Western Australia, June 2007.

2. So Far As Is Reasonably Practicable (SFAIRP) is a risk management framework used for industrial situations.

5

Process 1–Identify Business and Operational Risks

Summary Description of Process 1: Risk Identification

The climb to the peak of reliability excellence needs a sure path to the top. The first step you take is to choose to make the ascent. At the base camp level, you develop a believable, properly resourced plan to climb to the heights of world-class reliability. That is what you will have once you complete the six IONICS processes.

Your ascent up the Plant Wellness Way begins with Process 1, which is depicted in Figure 5.1. In Process 1, you gauge the total business-wide costs of the harm, waste, and losses incurred when your operating plant and equipment fail. If the costs are unacceptable, you must eliminate the risks or, at the very least, put in place effective risk preventions.

Develop Process Maps

Start by making a process map to illustrate what you are analyzing. For example, to analyze an equipment item for reliability improvement, lay out its assemblies in logical operating order. If you are analyzing a job with the intent to improve it, draw a flowchart of all its sequential tasks. When investigating a production line, the process map shows the individual production steps from beginning to end in order of product flow, starting with the raw materials supplied and progressing through the production process to the outputs that the business provides.

The process map is the foundation document for removing operational risk. It shows the design logic of the process, plant, or work flow. Simple rectangles joined together by arrows across the page provide a powerful visual tool for understanding how a system, process, or work flow operates. Using a process map helps you do a more complete, more detailed job of analyzing weaknesses and identifying business and operating risks. It allows you to see interconnectivity within and across processes and the impact of a step's reliability on process performance. Later, you will use the process map to design a better solution and to create key performance indicators to monitor and measure results. A process map helps you explain to others the improvements that are needed, why they will be effective, and how they will be implemented. Examples of process maps can be found in the preceding chapters as well as in the following pages.

An example of a process map for a maintenance job is shown in the sample Job Process Map worksheet, which is included in the downloadable spreadsheets accompanying this book.

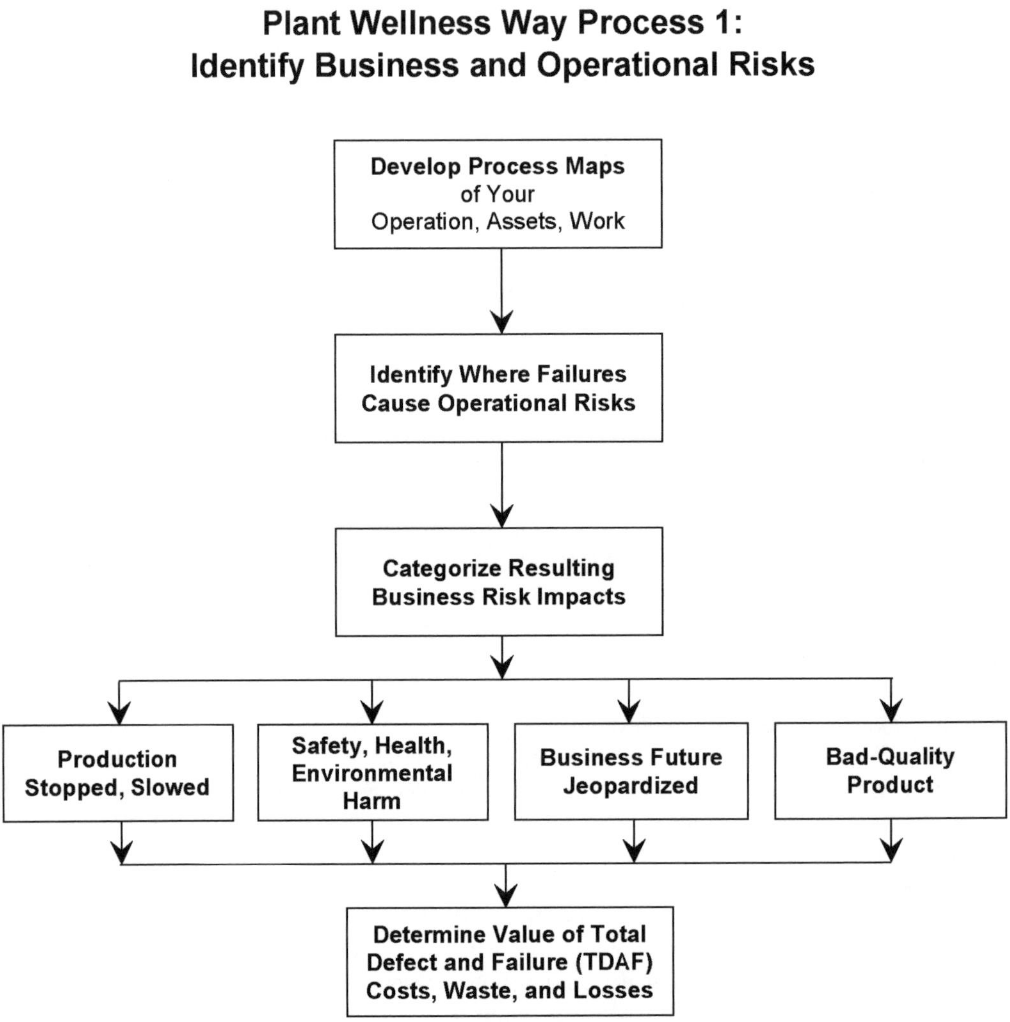

Figure 5.1—Process 1 Steps

Identify At-Risk Process Steps and Assets

Using the process map, create a spreadsheet in which to record all of the items shown in the flowchart. For example, in the case of a production line, enter each production asset, in order of use, into the spreadsheet. When analyzing an asset in order to improve its performance, list its subassemblies. For analysis of a subassembly, list all of its parts in sequence. For a job, list the individual work tasks in the order in which they are done. Leave nothing out of the list. You will not get protection from failure if items are not identified for analysis. Give each item in the list its own row in the spreadsheet and a unique number. The spreadsheet will later be expanded for other uses.

An example of such a layout for equipment items is in the Operating Criticality worksheet, which is included in the downloadable materials accompanying this book.

Categorize the Effects of Risk for Each Step

When analyzing a process or work flow, take each step listed in the worksheet and identify how it could fail in service. Write down failure or error events that could produce a loss or cost to your business. A failure is any incident or problem that affects quality, production rate, health, safety, or the environment or causes downtime. Record the known failures and the possible causes (i.e., those that might happen during the operation's lifetime) for each item. You can start creating this list from your memory of past events and knowledge of the situation. Finally, review the Physics of Failure Factors (see Chapter 4) to check for other possible risks to address. Next to each cause, indicate its effect on the operation's performance, people, and environment. If there are no operational risks whatsoever across the life cycle for a flowchart step, the analysis is taken no further because there are no risks to be addressed.

Analyzing equipment operating risk necessitates finding all of the critical working parts on the bill of materials and identifying the many ways that each one could fail in your operation. Known failure events will be recorded in the maintenance or operating history and recollected from peoples' memories. Finally, review each part using the Physics of Failure Factors mechanism guidewords (see Chapter 9, Table 9.2) to check whether there are other possible risks to the equipment.

Estimate the Total Defect and Failure Costs

For each failure cause, record the estimated TDAF cost for the worst-case scenario. The TDAF cost is the company-wide cost surge that a failure causes the business. The final value can total far more than the cost of maintenance and lost production. If you cannot calculate the full TDAF costs using the method described here, then calculate the direct costs of maintenance for the repair (maintenance labor, parts, outside services, and overhead) and multiply that value by 10 for continuous process operations and by 5 for batch processes. This factored cost gives a conservative lower value of the total surge costs and losses that each plant and equipment failure causes your business.

The TDAF cost is the loss consequence value used in the risk formula. As you work through the Plant Wellness Way life-cycle asset management strategy analysis, you select risk controls to reduce total operating risk to an acceptable level on the risk matrix—ideally, to less than one failure in three service lifetimes. It is presumed that all failure events will happen until you select mitigations to prevent them. Given enough time and opportunity, the severest possible event can happen in your operation. When a person could die in a failure event, indicate that possibility and the cost of their death to the business in the Comments field of the equipment item being analyzed. An important outcome from the analysis is to deliver strategies and actions that will keep your people safe.

Start with a Process Map of the Situation

Whether you are improving a work process, a production line, or an equipment item, the process map is a "picture" of the logic of how a thing works. Drawing a process map of a situation lets you more fully understand its operation and find its weaknesses. A process map lets you ask good questions that allow you to understand problems and make useful risk reductions. Mapping the logic of a situation is the way to start all equipment reliability and business process improvement. This is necessary so that you can design more successful solutions.

A process map of an industrial operation, a job, or a plant item allows reliability improvement principles to be used to reduce the chance of its failure and to lift the chance of its success. With a process map, Series Reliability Property 1 (see Chapter 1) is applied to find new solutions to mistake-proof each step so that errors do not lead to defects and failures. With a map of your process steps, you can investigate the best way to include parallel activities to lift each step's reliability. A process map also lets you use Series Reliability Property 2 to find ways to remove steps and simplify a process. Take the time and make a good effort to draw your process maps in detail so that you will be more successful.

Business Process Maps

Business process flowcharts start with a high-level overview of the key functions in the business and how they are interconnected. Each box in the high-level map will have its own process and subprocesses to be drawn. Process maps of the lower process levels are also needed in order to see the full details and complexity of how a business operates. You'll need to create a process map at the location and level of the business that you are investigating.

Figure 5.2 is an example of a high-level process map of the steps in an operating asset life cycle. Within each box are other processes that collectively achieve the high-level step's purpose. The life-cycle process map immediately shows two types of risk: it's a series arrangement across time with an immense number of interactions and choices, and it's got people working in it, so human error will be a major cause of failures. Without a flowchart, it would be difficult to imagine the whole operating life cycle and hopeless to decide where to focus in order to find the weaknesses in the processes. Next you make detailed process maps for the step of interest down to the level that you need to investigate. For example, in the case of the "operate" phase, you would develop flowcharts of the existing engineering, operations, procurement, storage, and maintenance processes and their interactions.

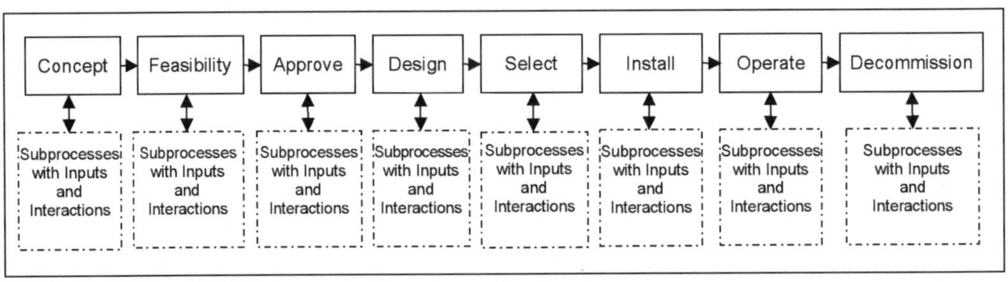

Figure 5.2—Process Map of an Operating Asset Life Cycle

Equipment Process Maps

The operating equipment process map is used to identify the parts and assemblies that need to be highly reliable, and it begins the development of strategies to get that outcome. Maximizing equipment reliability requires identifying and controlling component risks—by removing them if at all possible and unrelentingly reducing them if not. Figure 5.3 is a series of process maps for a centrifugal pump-set representing the equipment's construction and operation. It helps you identify where failures will stop the equipment from working. With it, you can spot the risks to the operation by asking at each step in the process, how does failure at this step affect the whole process outcome? Once you find problems, you can identify effective plans and actions to eliminate or reduce the risks.

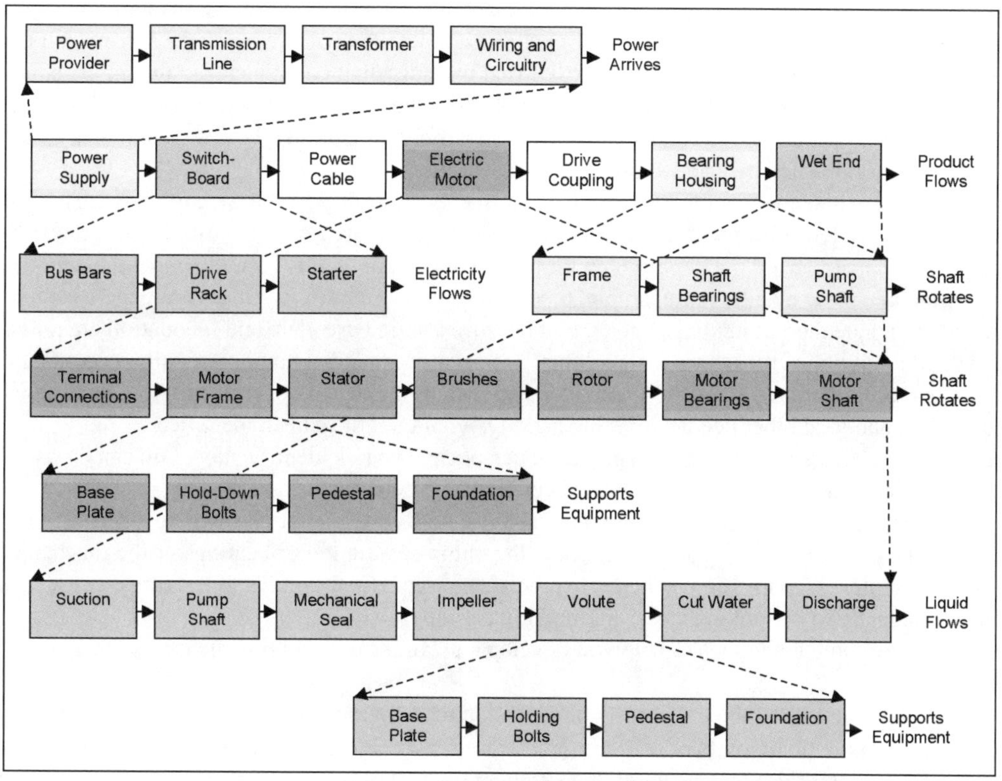

Figure 5.3—Process Map of a Centrifugal Pump-Set Delivering Product

You will be tempted to think it is sufficient to consult the equipment assembly drawing, exploded parts drawing, and bills of material to identify which components' failure will cause a breakdown. You will not draw a process map because it seems as if all you need to know is in the manufacturer's parts manual sketches and engineering drawings. Those drawings show you how a machine goes together, but they do not show you the logic of the equipment's design. They do not let you identify the interconnectivity between one part and another. The way a machine and its parts work is best explained in a logical flowchart.

Make the maps detailed enough to use them to identify the operational risks to the equipment being examined. For example, the mechanical seal in the wet end does not have a process map. When the working parts of a mechanical seal fail, the whole seal becomes unusable and the pump must be stopped. To identify the consequent impact of seal failure on the pump, we do not need to know every way in which a mechanical seal can be failed. At this stage, we only need to know that when the mechanical seal fails, so does the pump. Similarly, the shaft drive coupling is a single assembly that does not have its own process map because the box on the diagram sufficiently identifies the risks to the parent equipment if that part should fail.

Typically, your process mapping will encompass the major assembly level down to the sub-assembly level. In some cases, you may want to map an assembly right down to its individual parts and investigate the risks that each part carries. For example, you could expand the Wiring and Circuitry box in Figure 5.3 to identify the risks carried by individual components in the power supply system. If it is necessary to understand what might cause the mechanical seal or the coupling or the electrical wiring to fail, the process map for that specific item can then be drawn and analyzed to identify the risks carried by each component.

Expanding an item on a process map lets you see the specific parts causing the risks that it carries. If items are left off a process map, no purposeful risk controls will be installed to protect the equipment. When you are not certain how far to take the mapping, it is better to expand the process map to include more details. For example, the electric motor frame and pump volute were each expanded to include their mounting arrangement because a solid base plate and foundation are critical to the operating life of the pump-set. Had they not been expanded, a number of common causes of centrifugal pump failures associated with its base, such as a cracked concrete plinth, a bad support foundation, and vital other needs providing rigidity, would not have been identified.

Using a process map gives you a powerful perspective on risk identification. You can do "what-if" sensitivity analysis and visualize the effects of multiple causes of failure acting together. It lets you ask questions such as, if the motor frame is loose on its base, what else will be affected? Or, if the concrete pedestal cracks, how will it affect the pump-set and motor bearings or the mechanical seal? With a process map, you can better appreciate consequential failures from remote causes that cannot be identified by looking at the engineering drawings.

Here are some guidelines to help you develop a useful process map sequence:

- Follow the path of the force. From the location where a force is applied, follow the force and loads to the final points of restraint. For example, in Figure 5.3, the torque generated by the electric motor driving the pump is restrained by the hold-down bolts in the pedestal, which is itself embedded in the foundation supporting the whole machine. Ultimately, it is the foundation that restrains all operating forces and loads.
- Follow the energy flow. Draw maps starting from the energy source and follow the process to the end. For example, the rotation for the pump-set travels from the motor's rotating magnetic field through the electric motor shaft, through the connecting coupling, and into the pump shaft.
- Follow the product flow. Start mapping at the point at which the material enters and follow the process to the point at which it leaves. For example, the liquid moving though the pump enters at the suction nozzle and leaves at the discharge nozzle.

Most equipment types are used repeatedly in industry. Therefore, once you have the first process map for a type of equipment, you can copy it again and again. Alternating current (AC) electric motors are an example. You can reuse the process map for AC electric motors over a large range of sizes. A 5 kilowatt (kW) AC electric motor would have the same process map as an 11 kWh electric motor. This saves time analyzing all AC electric motors in an operation. You would not use an AC electric motor process map for a hydraulic motor. The hydraulic motor works in a totally different way from an AC electric motor. Therefor, a hydraulic motor needs its own process map. But once it is drawn, the process map can be used again for similar hydraulic motors and adjusted for peculiarities.

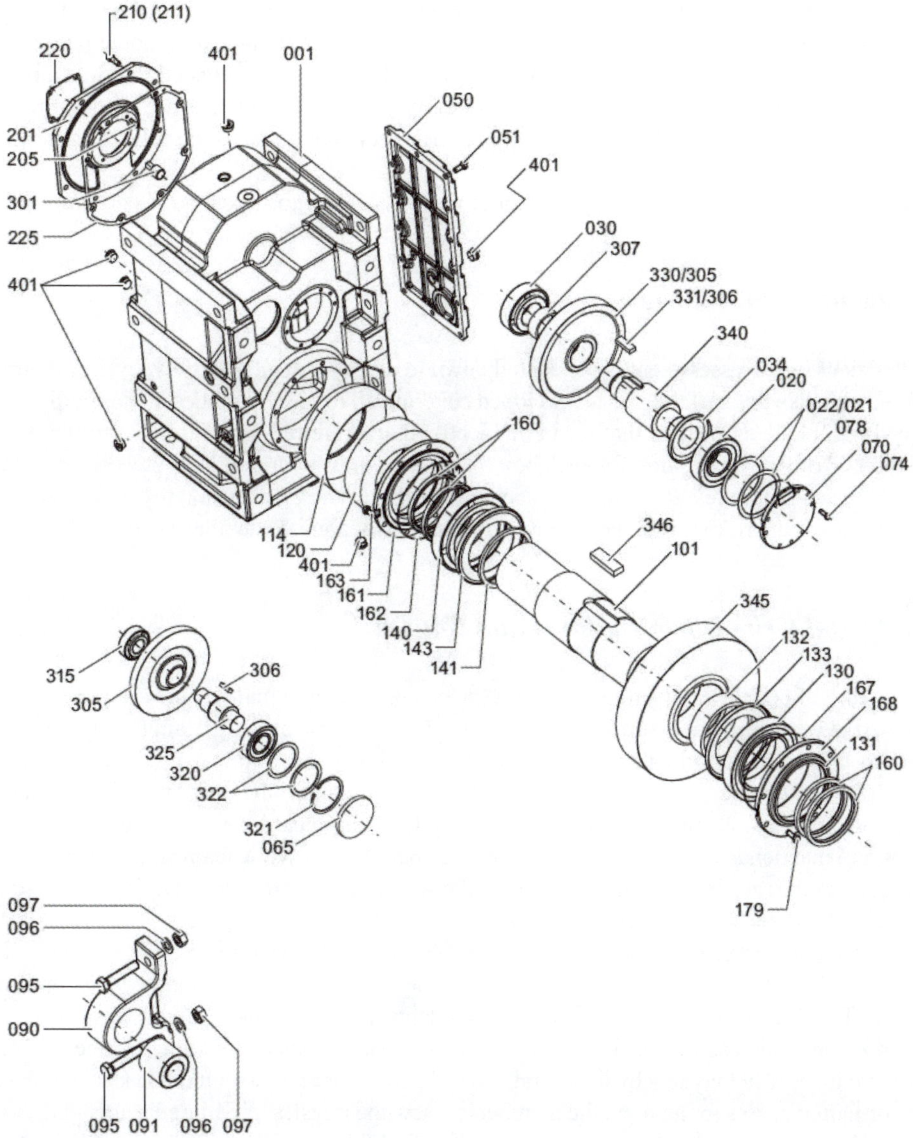

Figure 5.4—Exploded Assembly Drawing of a Gearbox

Collect All Technical Information

Up-to-date process drawings, equipment drawings, general assembly drawings, manufacturer manuals, parts assembly drawings, and bills of material are collected so that they can be reviewed to identify critical equipment in a production process. For critical equipment, critical parts are identified so that appropriate activities can be selected to keep each part at exceptional reliability. Figure 5.4 is an example of an "exploded" general assembly for a small industrial gearbox; it shows how the individual parts are assembled to make the equipment. The same manual will include a parts list that describes each part number shown on the parts assembly drawing.

When a manufacturer uses a specific part numbering sequence, it is adopted in the analysis as well for consistency and easy reference. Throughout the analysis, the same number and description of each part are used in combination with the manufacturer's part name and model number. When the manufacturer does not have a part numbering classification, you will need to create a part numbering and naming convention. If necessary, "dummy" numbers and codes are used to create individual numbers for all parts. A suitable part numbering convention takes the following form:

Manufacturer code—equipment type—model no.—assembly no.—part no.—part description

Equipment Failure Review

An equipment failure assessment is conducted down to at least the subassembly level. A failure identification table like that in Table 5.1 is developed covering all of the assemblies in the equipment. Give each entry a unique number so that it can be identified and referred to in future correspondence and discussion. Write the description for each box on the map in its own row in a spreadsheet so that you can record each failure related to the box as you identify it. The table eventually will encompass the risk analysis and the risk management strategy, plans, and actions you will use to control each risk.

Identify and Write Down Failures and Causes

A small team of people knowledgeable about the design, use, and maintenance of the equipment is assembled to work through the maps. They ask what could cause each operating equipment item assembly to fail, including identifying failures from possible combined causes and/or consequential effects. The composition of the team is critical and must contain the necessary design, operation, and maintenance knowledge and expertise covering the equipment being reviewed. Ideally, the operations and maintenance shop floor supervisors are part of the review team so that they understand the purpose of the review and can later support the workplace efforts needed to instigate and implement the risk control activities that will arise.

The team members brainstorm a list of possible failures for each item in the spreadsheet. They record likely equipment failure causes and their impact on the operation, along with estimated business-wide TDAF costs and useful explanatory comments. Places to look for failure history are in documented records, such as the breakdown and corrective maintenance work noted in the maintenance history, and in the shift logs kept by the operations and maintenance crews. Once the team has exhausted the list of failure causes for an item, the members review the Physics of Failure Factors guidewords to identify other possible risks. Add previously unidentified failures to the analysis spreadsheet as they are found. When the team consensus is that a failure cause cannot happen, indicate that in the spreadsheet's Comments field. But if a team member wants the cause recorded, then do so.

Work Activity Process Maps

Work tasks and activities can also be mapped. If job procedures are available, convert them into process maps. An example of a process map for a clerical task recording important cost information is shown in Figure 5.5. When job procedures are not available, ask people what they do and record the steps they follow. From the description, develop a work activity process map.

Work Process Risk Review

Work done by human beings can be done wrong. Mistakes cause risks ranging from no significance to disaster, and adverse risks need to be proactively identified, prevented, and controlled. For human error that cannot be eliminated, you must reduce the consequences of the error. The risk identification method used for equipment can also be used to identify human error and job quality risks in work processes. The tasks and actions noted on the work process map are recorded in a spreadsheet. Each step is analyzed to find the risks. Table 5.2 lists the work process for Figure 5.5, the Monthly Cost Report, as an example of identifying human error risks in work processes. Risk mitigations typically involve adding parallel test activities or independent double checks to create error-proofed tasks. Where risk control actions or parallel proof tests are self-evident, they are added to the table as the risk analysis is done.

Introduction to Total Defect and Failure Costing

Conducting a rigorous risk analysis means compiling the complete business-wide financial costs of a failure incident. The process of collecting, analyzing, and reporting all of the costs attributable to a failure is known as total defect and failure costing. It is a formalized accounting process that shows the vast amounts of money wasted throughout an organization from equipment failures. To assist in compiling the TDAF cost list, it is useful to use the company's chart of accounts, as it contains all of the accounting cost codes for allocating costs and charging payments in the organization. New cost centers usually need to be developed to capture all TDAF costs. The methodology brings together the financial, production, engineering, and maintenance groups. It provides a means for these normally separate groups to work together to solve problems of serious financial waste.

The TDAF cost methodology is activity-based costing (ABC) applied to a single failure incident. The goal is to identify the total costs of failure and to either accept such failures in the future or put in place mechanisms and solutions to prevent them. Activity-based costing is an accounting method that identifies the total costs of the deeds undertaken to perform a function or produce a product. Activity-based costing applied to TDAF costing allows an organization to determine the actual costs of all resources and services wasted by a failure. It is a powerful technique for measuring failure costs because it itemizes every expense and identifies its makeup. With TDAF costing, you trace failure cost surges throughout your organization and quantify all lost profit.

A step-by-step example of the process for performing TDAF costing can be found in the document *The Instantaneous Cost of Failure,* which is in the online materials accompanying this book.

Table 5.1—Risks Identification Table for Pump-Set Parts and Assemblies

Item No.	Equipment	Assembly	Subassembly	Components	Possible Causes of Failure
P01	**Supply Pump**				
1		Power Supply	1. Transmission Line		1. Power provider failure
					2. Lightning strike
			2. Transformer		1. Poor termination
			3. Power Cable		1. Poor termination
2		Switchboard	1. Panel		2. Fire
					2. Liquid ingress
					3. Impact
			2. Bus Bar		1. Loose clamp bolts
					2. Poor cable crimping
			3. Drive Rack		1. Dust from product
					2. Poor assembly
					3. Rack rusted in place
			4. Motor Starter		1. Overload
					2. Short circuit
3		Electric Motor	1. Termination		
			2. Motor	1. Frame	
				2. Base Plate	
				3. Holding Bolts	
				4. Pedestal	
				5. Foundation	
			3. Stator		
				1. Brushes	
				2. Windings	
			4. Rotor		
				1. Windings	
				2. Bearings	
				3. Shaft	
4		Drive Coupling			
5		Bearing Housing	1. Frame		
				1. Bearings	
				2. Shaft	

Table 5.1, continued

Effects of Failure	TDAF Costs of Failure	Comments
1. Downtime	$100,000	$25,000 per hour, minimum 4 hours if power is turned off
1. Downtime	$200,000	Minimum 8 hours if power is lost due to lightning
1. Downtime	$200,000	Minimum 8 hours if power is lost due to failure
1. Downtime	$200,000	Minimum 8 hours if power is lost due to failure
2. Downtime	$2,000,000	If a person dies, add $5,000,000 to the TDAF cost
1. Explosion	$2,000,000	Complete rebuild of switchboard takes two weeks; if a person dies, add $5,000,000 to the TDAF cost
1. Downtime	$200,000	
1. Fire in switchboard	$2,000,000	Complete rebuild of switchboard takes two weeks. If a person dies, add $5,000,000 to the TDAF cost
1. Fire in switchboard	$2,000,000	Complete rebuild of switchboard takes two weeks; if a person dies, add $5,000,000 to the TDAF cost
1. Arcing in switchboard	$200,000	
1. Arcing in switchboard	$200,000	
1. Downtime	$100,000	
1. Downtime	$100,000	
1. Major electrical fire	$2,000,000	Complete rebuild of switchboard takes two weeks; if a person dies, add $5,000,000 to the TDAF cost

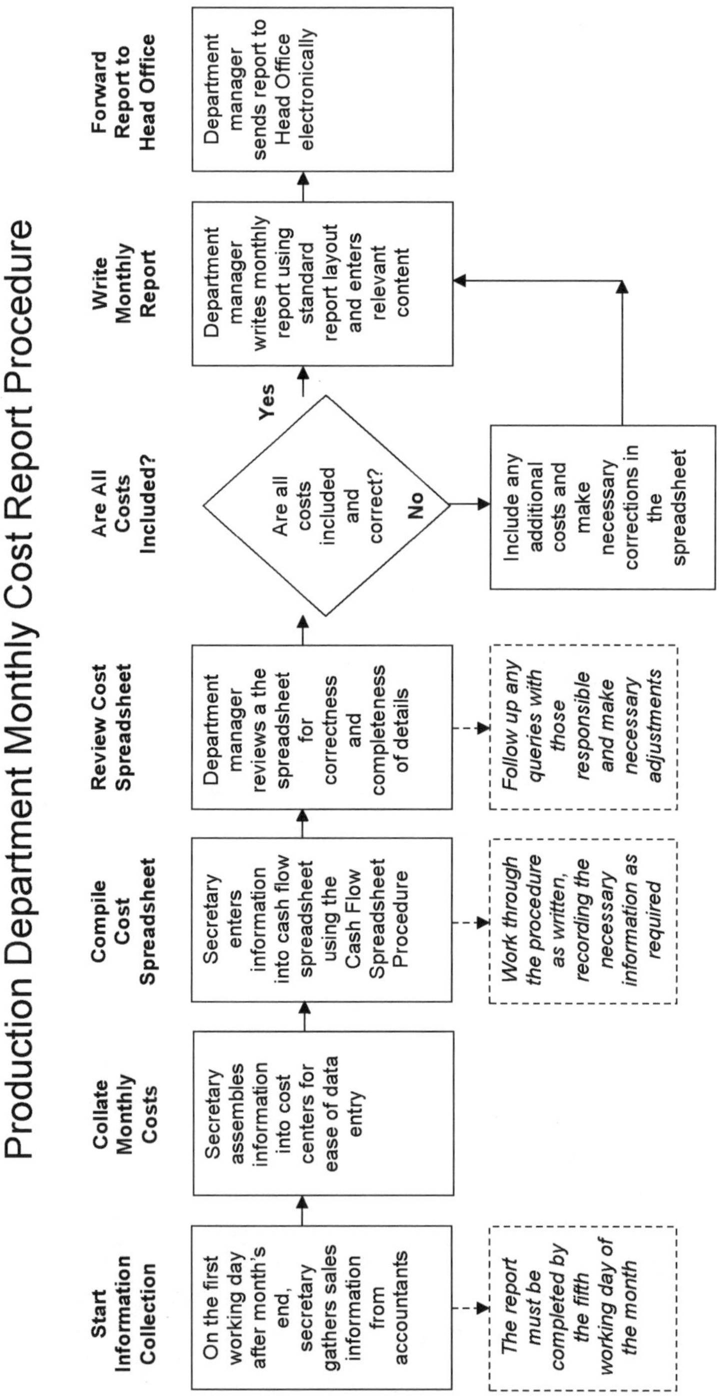

Figure 5.5—Job Procedure Converted to a Work Process Map

Table 5.2—Risks Identification Table and Risk Management Plan for a Work Activity Process

Department	Process	Job	Task	Possible Causes of Failure	Effects of Worst Likely Failure	TDAF Costs of Worst Failure	Risk Control Plans	Actions to Be Taken	Proof That Actions Have Been Completed
Production									
1	Monthly Cost Report								
2		Start information collection	Gather sales information from Accounts	1. Information not available	Report not completed on time	$500	Warn Accounts of impending report date	Set up an electronic schedule entry to automatically warn Accounts manager one week prior report due date	Department manager to check schedule entered
				2. Wrong information provided	Bad management decision	$10,000	Get Accounts to double-check cost information is correct	Accounts to include double-check actions into their work procedure	Accounts to send copy of revised procedure to department secretary for review
				3. Incomplete information presented	Bad management decision	$10,000	Get Accounts to double-check cost information is complete	Accounts to include double-check actions into their work procedure	Accounts to send copy of revised procedure to department secretary for review
3		Collate monthly costs	Put costs into cost centers						
4		Compile spreadsheet	Enter cash flow details using data entry procedure						
5		Review cost spreadsheet	Department manager checks spreadsheet						
6			Confirm all costs are recorded						
7		Write Monthly Cost Report	Department manager writes report						
8			Report forwarded to Head Office						

6

ASSET MANAGEMENT FOR PLANT WELLNESS

Enterprise asset management is the system of organization, processes, practices, knowledge, and information management used to optimize the value gained from your physical assets. It is also known as life-cycle asset management, engineering asset management, or physical asset management. Enterprise asset management, initially called *terotechnology,* began in Europe during the 1980s. Terotechnology is defined in British Standard 3811 as "a combination of management, financial, engineering, and other practices applied to physical assets in pursuit of economic life-cycle costs." Today, the word *costs* would be replaced by *profits.*

The standard continues, "Its practice is concerned with the specification and design for reliability and maintainability of plant, machinery, equipment, buildings and structures, with their installation, commissioning, maintenance, modification and renewal, and with feedback of information on design, performance and costs." Terotechnology adopted practices based primarily on maximizing the profits from all life-cycle functions through equipment failure data collection, failure data analysis, profit optimization, and feedback to original equipment manufacturers to improve machinery designs.

By the 1990s, the global interest in achieving high reliability of operational assets had prompted a change in terminology, from *terotechnology* to the more acceptable term *enterprise asset management.* The clear implication of the term is that the entire company is involved in its achievement. To fulfill an organization's strategic plan, its human assets, knowledge assets, intangible assets, financial assets, and physical assets must be integrated holistically across time. From this realization, important interactions and relationships emerge throughout the asset life cycle that must always be borne in mind. The groups that affect physical asset management success are those that design the asset, make the asset, select the asset, install the asset, operate the asset, and ensure the asset's ongoing availability. Fostering good decision making in each of these groups to optimize the success of the business is what you want to achieve from a life-cycle enterprise asset management system. The combined achievement of world-class plant reliability and optimal operational performance is known as Operational Excellence.

In January 2014, the international physical asset management standard ISO 55001, Asset Management—Management Systems—Requirements, was released. It indicates what an enterprise asset management system ought to contain but not how to deliver outstanding equipment reliability and operational performance. The drive for a universal asset management standard arose because the international quality standard, ISO 9001, Quality Management System—Requirements, did not focus specifically on the performance of physical assets.

Standards such as ISO 55001 require setting corporate policies and asset management plans specifying what to do, but they cannot themselves cause asset management, maintenance, and operational excellence. It is only by making the right reliability choices and using the right reliability practices in all of your business processes that you can get lasting operational success. Achieving world-class reliability and maintenance excellence requires failure-free parts, zero breakdowns, least maintenance costs, and utmost operating profit. That requires a life-cycle system of reliability; an enterprise asset management system alone will make little difference in the quest for Operational Excellence.

An effective asset management system as part of a life-cycle system of reliability will endlessly produce optimal life-cycle profits from your plant. The principal benefits of useful enterprise asset management, in no particular order, include the following:

- Utmost return on investment and profit growth through effective stewardship of assets
- The ability to demonstrate best value for money within a constrained funding regime
- Greater customer satisfaction from improved product and service quality and reliability
- On-time order delivery to the required specification and performance standards
- Evidence demonstrating legal, regulatory, and statutory compliance with controlled and systematic processes
- Improved risk management and corporate governance that provide a clear audit trail for decisions taken and the control of their associated risks
- Enhanced corporate status through superior shareholder value, improved marketability of product or service, greater staff satisfaction, and more efficient and effective procurement from the supply chain
- Proving that you are meeting community obligations, complying with corporate governance requirements, providing workplace health and safety, and achieving environmental sustainability with your physical asset selections and use

No matter what you call it—terotechnology, Operational Excellence, enterprise asset management, or the Plant Wellness Way—it requires a significant change in attitude and thinking at all levels in the organization.[1] It will take the following:

- A change in philosophy from detecting and fixing problems to not letting problems start
- Cooperative teamwork between finance, operations, engineering, and maintenance
- The realization that "good" is not good enough—only excellence is suitable
- Everyone in the organization proactively reduces risks of future failures in what they do
- All processes in the life cycle eliminate the causes of the causes of component failure

Attaining those goals requires developing and embedding the right design, project management, procurement, operational, and maintenance life-cycle strategies and actions into your company, then sustaining them as standard practices that are continuously improved and made more efficient.

Successful Asset Management by Design

A successful enterprise asset management system will deliver world-class physical asset performance and reliability. It has integrated business processes that maximize lifetime operating profits from your production assets. A truly effective physical asset management system delivers optimal plant operation, ensures failure-free equipment, and maximizes productivity by:

- Doing proactive operational risk elimination during project design
- Producing outstanding equipment reliability and zero breakdowns
- Optimizing life-cycle profits and minimizing life-cycle costs
- Using defect elimination throughout the life cycle to make equipment failure-free
- Continually eradicating supply chain, safety, operational, and equipment risk
- Applying new, better individual asset and business-wide solutions to lift reliability
- Making it normal for everyone in the operation to deliver best practice performance

The most powerful asset management results are achieved when you have a holistic, life-cycle mind-set. Figure 6.1 indicates the phases in the life of a physical asset. The life cycle extends from the birth of the business concept to the end of the asset's useful life. Because of the time-dependent and interactive nature of the life cycle, everyone in your organization eventually has a part to play in getting the maximum sustainable availability and utilization of your physical assets.

Although human factors such as leadership, motivation, and culture are intangible assets, they are critical to the successful achievement of physical asset management and require due consideration. They influence how an organization treats its staff, employees, contractors, and suppliers during the asset management life-cycle phases. The purposes and duties of the operations, engineering, finance, and maintenance teams must not be taken in isolation. It is essential that the distributed knowledge they have is shared through interaction to find the best ways to get optimal operating performance. This optimization is only realized when operations, engineering, finance, and maintenance work in partnership across the life cycle with a common responsibility to successfully and quickly achieve the organization's asset management goals.

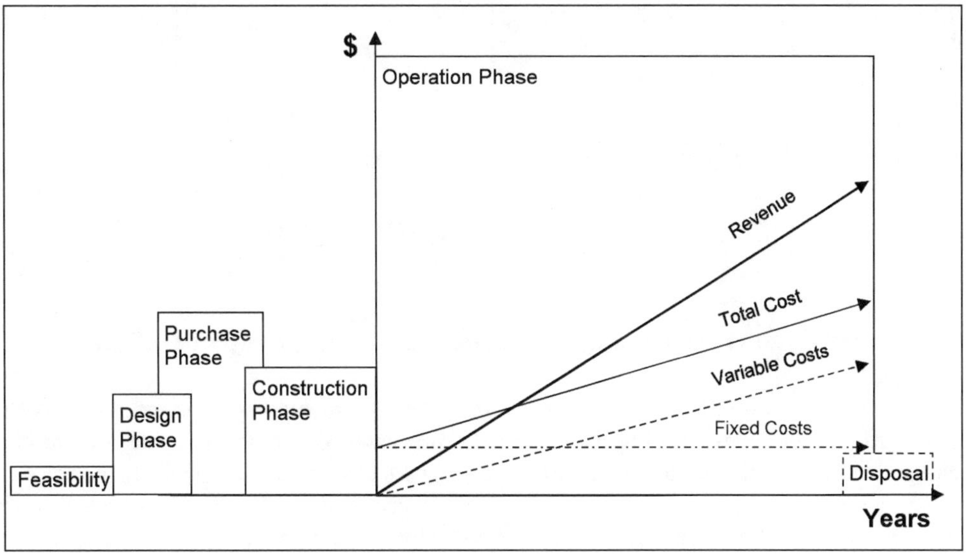

Figure 6.1—Phases of the Asset Life Cycle

The profitability of an operation is determined before a shovel of dirt has been turned. Figure 6.2 shows the typical phases of an industrial operation's lifetime and the points during its life when future operating costs are committed.[2] Some 95% of all operating costs are fixed during the feasibility and capital project phases. The decisions and selections made during the project conception and design phase set the scope and extent of your future operating success. Operational Excellence and return on investment stem from project decisions; future operating profits come from the process and plant design choices that you make. Poor design choices plague the maintenance, production, safety, and financial success of an operation for all of its life. Low operating costs are a business feasibility and project design engineering outcome, not an operational result.

Once you make your financial and engineering choices during the feasibility and detailed design stage, you are stuck with them for the future of the operation. Those first choices are self-fulfilling. If the concept and design are flawed and failure-prone from the start, nothing can fix them when your operation is up and running. Make poor choices early in your operation's life cycle, and you will always have poor operating profits. If you want world-class production success, you have to design it into your business as part of your project financing and engineering choices.

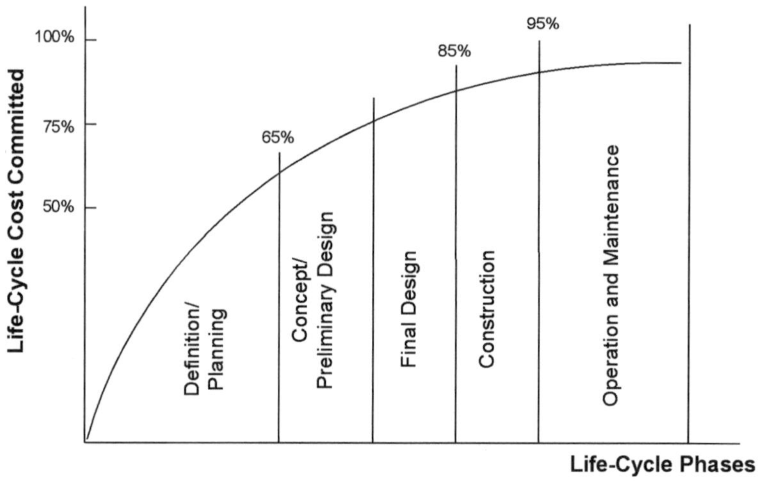

Source: Benjamin S. Blanchard, ***Design and Management to Life Cycle Cost*** (Forest Grove, OR: M/A Press, 1978).

Figure 6.2—Operating Cost Commitments across the Life Cycle

Analyze Operational Consequences during Project Design

The design and selection phase is most critical to the long-term success of an industrial business. The choice of technologies, the choice of production processes, the choice of layout, and the choice of equipment to make the product mix all fix the facility's cost structure. This is the period during which a facility's future profits and options to adapt to changing market forces are set. When you buy an operating asset, you also get its inherent behaviors, performance, and costs of operation. If the equipment chosen requires major upkeep or cannot sustain quality production for long runs, then you get high-cost product and much waste. Operating profit will always be less because part of the profit margin must pay for the extra upkeep of the facility and equipment. There will be less cash available for business and plant improvements to make products more competitively. In time, the products will become uncompetitive and will be replaced as your rivals deliver cheaper, better-quality choices to the market.

Design and Operations Cost Totally Optimized Risk (DOCTOR)

Project groups have the power to build great businesses or just "also-ran" businesses. When project teams design a plant, select its equipment, and build and install it, they are creating the future of a business. You have one chance to get it right—after that, you must live with what you are given. Project groups need a financial tool to visualize the impact of their decisions on the future success of the business they are creating. In the Plant Wellness Way, the tool that is used to successfully improve operating profits by design is called DOCTOR—an acronym for Design and Operations Cost Totally Optimized Risk. Figure 6.3 represents the life-cycle perspective of DOCTOR. It uses TDAF costing to optimize the plant design and selection of equipment based on eliminating future operating costs and failures so that you maximize operating profits as a project design outcome.

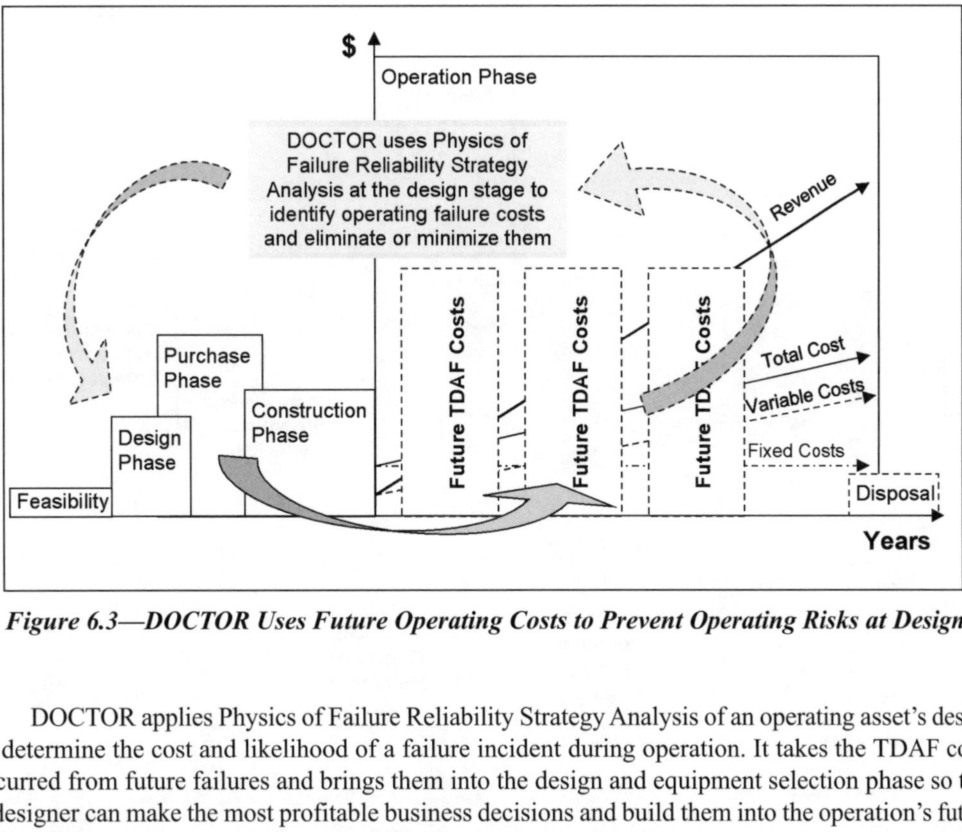

Figure 6.3—DOCTOR Uses Future Operating Costs to Prevent Operating Risks at Design

DOCTOR applies Physics of Failure Reliability Strategy Analysis of an operating asset's design to determine the cost and likelihood of a failure incident during operation. It takes the TDAF costs incurred from future failures and brings them into the design and equipment selection phase so that a designer can make the most profitable business decisions and build them into the operation's future success. Figure 6.4 shows how to use DOCTOR during the project feasibility and detailed design phases.

DOCTOR rates and eliminates operating risk while projects are still on the drawing board. If failures during operation have severe consequences, the possible causes are investigated and removed. When problems cannot be eliminated, effective operating and maintenance strategies are chosen to reduce the likelihood of their occurrence and limit their consequences. Pricing is done using TDAF costing, and the life cycle is modeled with net present value (NPV) methods by the project group. By assuming a failure and building a TDAF cost model, those designs and equipment selections with high failure costs are identified. Investigating the cost of an "imagined" equipment failure lets the

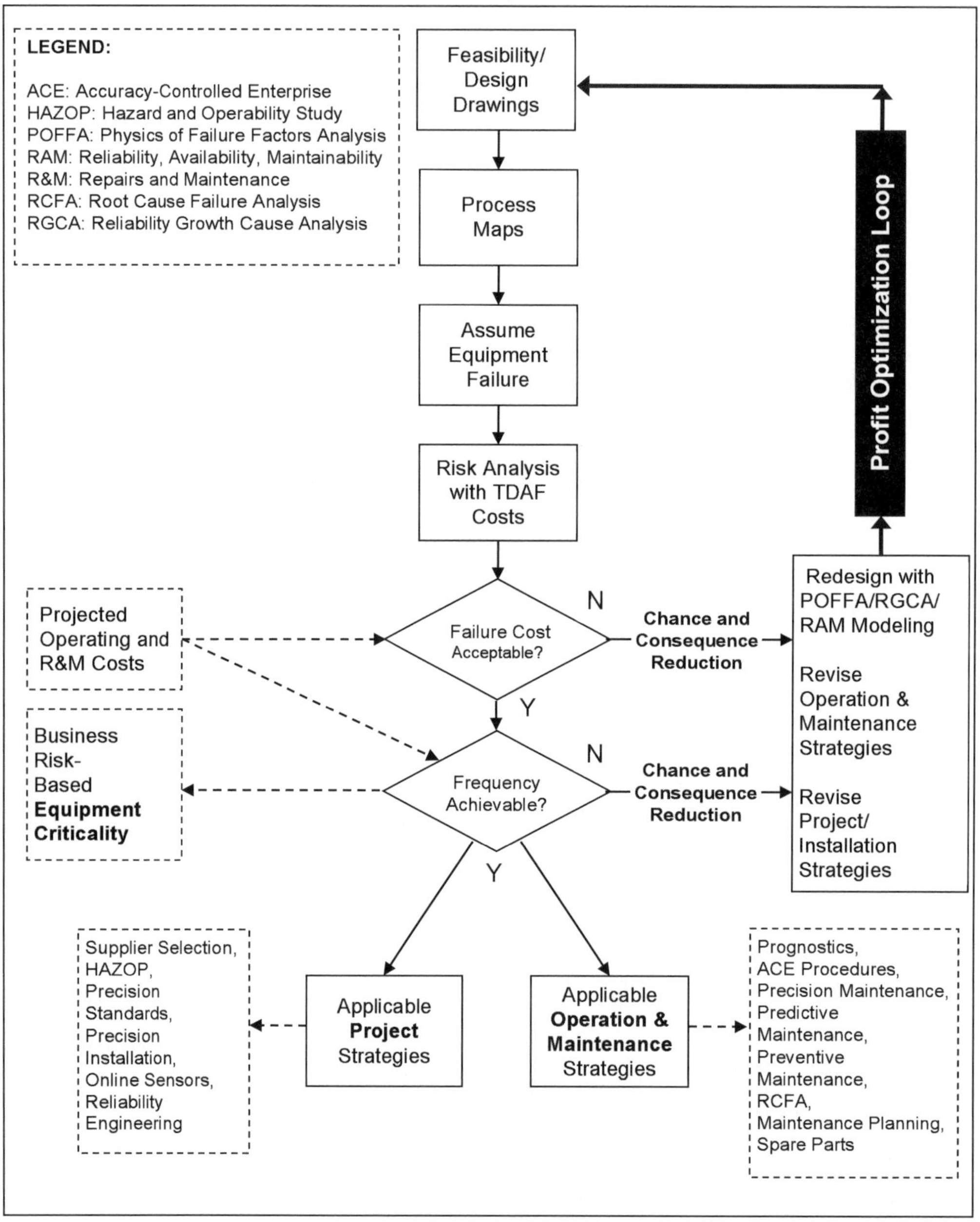

Figure 6.4—Optimized-Profit Operating Risk Management Design Methodology

project designer see whether decisions will harm the business or make it more profitable. The design and equipment choice are then revised to deliver lower operating risk and costs. By modeling the operating and maintenance consequences of capital equipment selection during the project phase, the plant engineering, operations, and maintenance strategies that produce the optimal life-cycle operating profit are part of the business from the start.

Optimizing Project and Operating Costs

Applying DOCTOR allows recognition of the operating cost impact of project choices and the risk they cause to a project's return on investment. The costs of future operating failures are used to rate the robustness of the design decisions. Basing capital expenditure justification on actual operating scenarios makes the estimate of future operating and maintenance costs from project decisions realistic. By getting the project group to look at the impacts of their choices on future operating profit during the capital design and equipment selection phase, future profitability can be optimized. Using TDAF costing in design decisions simulates the financial and operational consequences of equipment use so that plant design, operation, and maintenance can be fine-tuned with choices that get the best plant reliability and operating profit results.

A DOCTOR analysis starts by taking each item of equipment in a project and assuming it will fail; this makes the business-wide impacts of a failure clear. TDAF costing of a failure incident is done by the project group. The expenditure and operating assumptions for modeling are those costs and practices used to run the plant and equipment. It is likely that operating costs are available from the plant owner's financial and maintenance groups. The NPV financial models are developed with the help of the plant owner's operations group to reflect the company's management culture. The designed-in operating costs modeled are put through review and compared against other choices. Consideration of parts stock holding is also developed, and plant maintainability is improved to allow fast maintenance response for low cost. This optimization process is reiterated until future operating costs are minimized.

The DOCTOR process can be applied to every item of plant and equipment, even down to component parts. If the failure costs are unacceptable, then one of three things must happen:

1. A design change is made to eliminate or reduce the chance of failure
2. Risk reduction requirements to reduce opportunity to fail are included in the design
3. Supply chain, engineering, operating, and maintenance practices are changed to control operating risk and cost consequences

Each new decision regarding a design or operating practice is run through the DOCTOR process to compare operating costs with previous results. If a new choice reduces risk, the expectation is that it will lower the operating cost. This iterative process is used to achieve a balance between the lowest life-cycle operating cost and the expense of initial capital cost. Once the operating TDAF costs for equipment are known, a risk analysis can be conducted using a table like Table 6.1 to identify strategies that produce the least operating risks. Use of alternative table layouts for more detailed event risk analysis and costing are at your discretion.

An example of a more detailed risk analysis is available in the Operating Criticality worksheet, which is included in the downloadable materials accompanying this book.

If least capital expenditure is important (as opposed to least operating cost), TDAF cost modeling can optimize for lowest operating costs using least capital expenditure. Alternatively, if some other parameter is important—for example, least environmental costs or least maintenance costs—the TDAF cost model lets you optimize for the least capital cost. TDAF costing combined with DOCTOR is a powerful way to make good business investment decisions. It lets you foresee future operating scenarios during design. It allows the project group to make sound, practical choices using optimal long-term financial judgments about capital equipment selection, plant design, and operations and maintenance practices. DOCTOR reduces the chance of poor capital equipment acquisition and destructive long-term financial decisions from not knowing their operating consequences.

Table 6.1—Risk Identification and Control Table for a DOCTOR Analysis

Equipment ID No.	Equipment Description	Assembly	Subassembly	Parts	Possible Causes of Failure	TDAF Costs of Failure	Risk-Control Plans	Actions to Be Taken	Proof That Actions Have Been Completed

Enterprise Asset Management the Plant Wellness Way

Enterprise asset management is defined as the "coordinated activity of an organization to realize value from assets."[3] It is a corporate-wide solution used to attain the physical asset performance needed to meet business aims. Figure 6.5 is a simplified overview of an enterprise asset management process to deliver an organization's objectives.

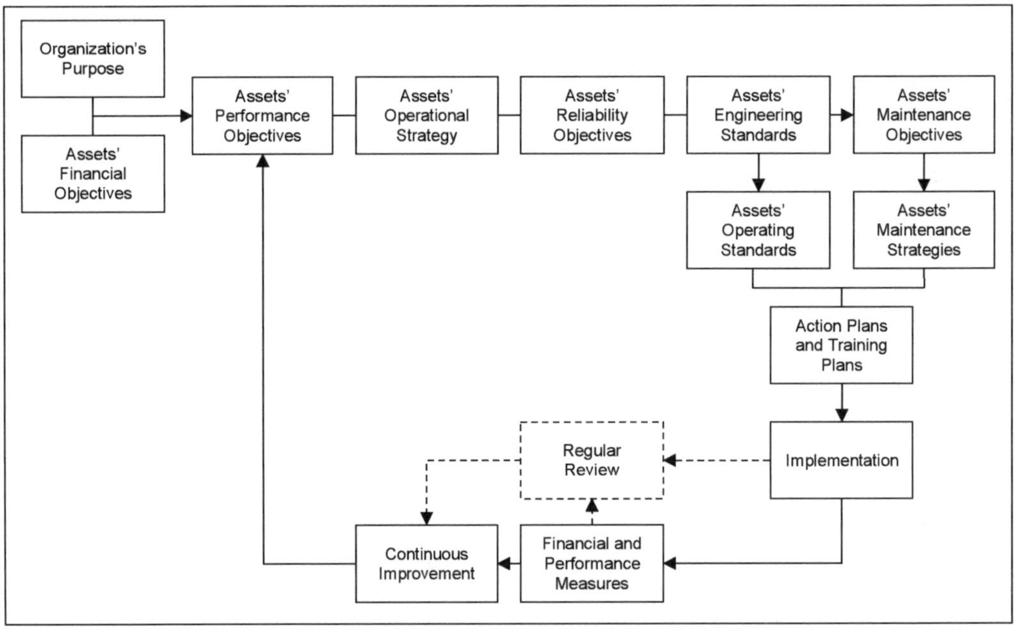

Figure 6.5—Simplified Enterprise Asset Management Model

The appeal of enterprise asset management is its "promise" of maximum life-cycle profit, along with its converse, minimum life-cycle cost. In order to achieve that promise, it is necessary to institute the required practices and systems of physical asset management stewardship throughout the organization and across its lifetime. This is no easy matter in most organizations, especially those that are reactive and those that have failed to stay modern.

Enterprise asset management proposes that businesses follow a path to desired equipment performance by using the foundation elements of systems engineering, reliability engineering, maintenance management, operational management, risk management, and industrial engineering, guided by sound financial management. Numerous internationally recognized industrial and military standards form the documented database of the practices applied by organizations seeking to be world-class engineering asset managers. Practically, the intentions of physical asset management have proven very difficult to attain. The evidence shows that extremely few industrial businesses around the world reach the world-class performance level that enterprise asset management is meant to deliver. There are important factors not yet recognized by current asset management models and methods that every business needs to deal with. This book addresses the "missing links" needed for sure enterprise asset management success by providing the Plant Wellness Way methodology to build a business-wide system of reliability.

The right life-cycle asset management mix requires time for organizations to introduce them in a staged fashion. In large organizations that have successfully introduced asset management, it has taken up to six years to build the necessary culture and skills.[4, 5] For smaller operations, the time is less. In all cases, committed, stable leadership and cultural change management are required to maximize the rate at which enterprise asset management benefits accrue to an organization.

The changes necessitated by enterprise asset management usually require developing new knowledge and skills in the people of the executive, finance, engineering, operations, and maintenance groups. A representation of the organizational practices and business processes used in organizations changing from enterprise asset management to Industrial and Manufacturing Wellness is shown in Figures 6.6 and 6.7. The IMW solution methodology is designed to more than halve the time it takes organizations to reach world-class operating performance.

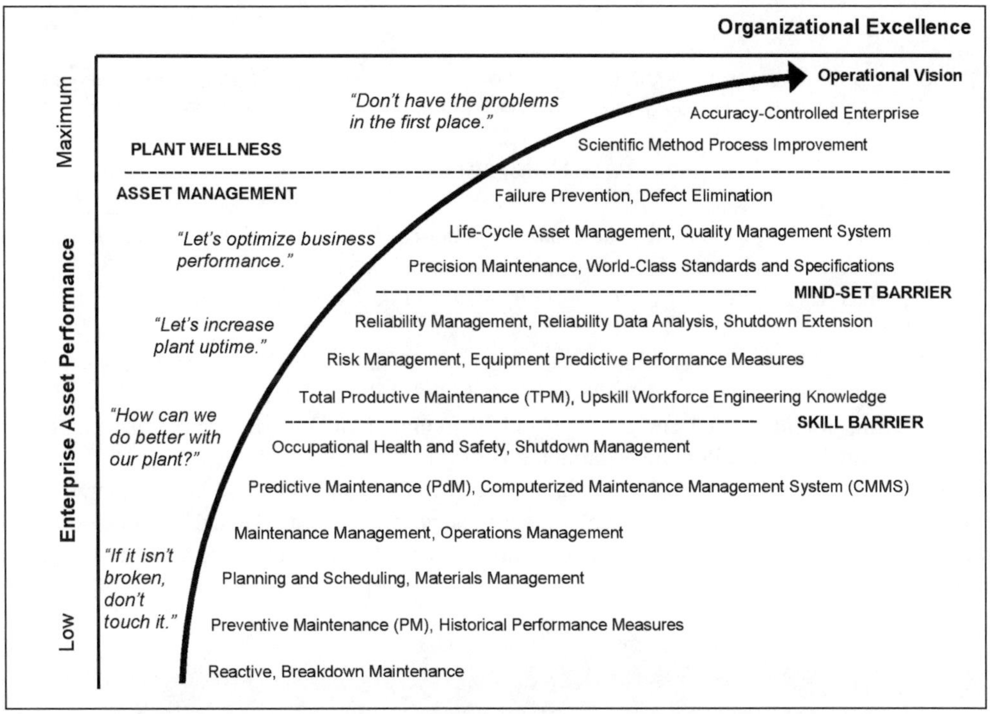

Figure 6.6—Enterprise Asset Management with the Plant Wellness Way

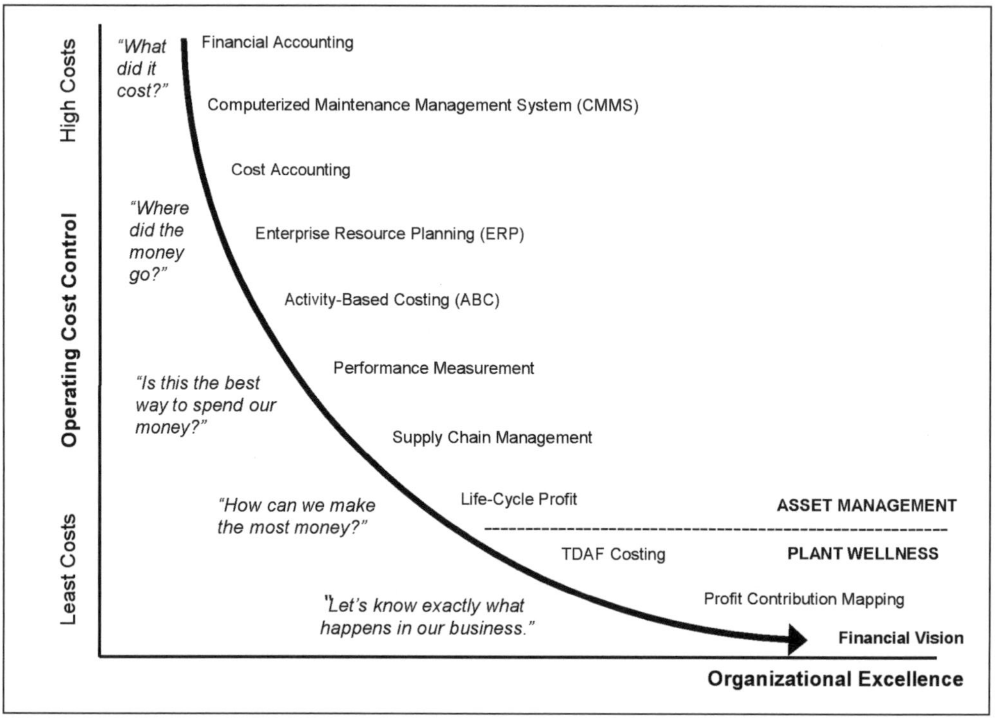

Figure 6.7—Enterprise Asset Management with Plant Wellness Way Cost Control

For more information on this topic, see the document *Introducing Enterprise Asset Management and Plant Wellness into Organizations,* included in the downloadable materials available online.

References

1. Ron Moore, *Making Common Sense Common Practice: Models for Manufacturing Excellence,* revised and updated ed. (Boston: Butterworth-Heinemann, 2002).
2. Benjamin S. Blanchard, *Design and Management to Life Cycle Cost* (Forest Grove, OR: M/A Press, 1978).
3. "ISO 55000: 2014 Asset Management—Overview, Principles and Terminology," accessed at http://www.iso.org/iso/catalogue_detail?csnumber=55088, July 28, 2015.
4. V. J. Flynn, "Maintenance Benchmarking and the Evolution of DuPont's Corporate Maintenance Leadership Team," E. I. Du Pont de Nemours & Co.
5. Nigel Cumerford, "Crow/AMSAA Reliability Growth Plots and Their Use in Interpreting Meridian Energy Ltd's Main Unit Failure Data" (paper presented at the 16th Annual VANS Conference, Rotorua, New Zealand, 2005), accessed at http://www.plant-maintenance.com/articles/Crow-AMSAA.pdf, August 1, 2015.

7

Operating Equipment Risk Assessment

Risk is an amount of loss or gain. A risky situation implies that a range of outcomes exist. The risk of investing money in the stock market brings with it the possibility of great reward as well as the possibility of serious loss. The challenge is to develop methods to increase the likelihood of a good outcome while removing and controlling bad risks. Because risk has such a profound impact on everything to do with business and commerce, it is critical to understand it. Once you have a good perspective on risk, you will be better able to identify the risk management strategies that will provide the greatest financial, production, and safety benefits to your organization.

Risk is virtually impossible to reckon exactly because it is probabilistic—a situation might happen like this, or it might happen like that, or it might not happen at all. Risk management specialists try to model and quantify risk to give it a firm value, but the results are notoriously misleading because real-life situations are unlikely to go the way they were imagined while sitting in the office, unless they follow a well-defined and well-rehearsed script. Identifying the full profile of risk—the range of its total makeup and the depth and the extent of each component risk—is the first step in understanding the scale of risks in a situation. Once you know the size and nature of the risks you face, you can pick effective mitigation strategies.

Estimating Risk

The calculation of risk is straightforward and can be done mathematically in a spreadsheet or determined with the help of a risk matrix. Risk is equal to the likelihood of an event occurring multiplied by its cost consequence should it occur.

Formula 7.1

Risk ($/yr) = Consequence ($/event) x Likelihood of Occurrence (events/yr)

The likelihood of occurrence is measured by the number of times an event actually happens during a period. Usually, a year is used so that risk can be annualized. An event that happens every five years has an annualized likelihood frequency of 0.2 times a year. The consequence of an occurrence is the total financial impact of the event. By calculating the likely frequency of an event per year and valuing the consequence of occurrence, the risk formula calculates the annualized cost of the risk. It is a means to quantify the yearly cost to an organization of every risk event that it might suffer, good or bad. It provides a figure to weigh one risk against another and thus allows you to address your worries in a justifiable order of priority.

The likelihood of occurrence can be further divided, so that the full risk formula is as follows:

Formula 7.2

Risk = Consequence ($) x [No. of Opportunities (/yr) x Chance of Occurrence at Opportunity]

The number of opportunities is how many times a specific situation that could lead to a failure arises per year. The chance of occurrence (or probability) is the odds that a situation results in a failure. Chance equals one (1) if there is a sure failure every time the situation arises and zero (0) if there can never be a failure. Probability takes a value between 1 and 0 because the chance of something going wrong is possible to a larger or a smaller degree. Risk elimination is when you drive the risk to zero by having either no consequences from the event, no opportunities for the event to occur, or no possible outcome even when opportunity arises. Risk mitigation applies risk controls to reduce the consequence, opportunity, or chance but does not completely eradicate the effects of an event.

There are great benefits available to businesses that reduce their risks of failure. When the chance of a failure decreases, money is saved because fewer money-losing events happen during the period. As a simple example, take a failure event that happens twice a year and costs $10,000 each time it occurs. Using the standard risk formula, the risk is calculated as follows:

Cost of Risk = $10,000 per event x 2 events per year = $20,000 per year

By introducing risk reduction strategies that lessen the chance of the event to once every two years, the diminished risk becomes

Cost of Risk = $10,000 per event x 0.5 events per year = $5,000 per year

Mitigation delivers savings of $15,000 per year, year after year. If the cost of reducing the risk to once every two years is less than $15,000 per year, then the company makes money by preventing the risk. The challenge is to select strategies that cost the least but realize the greatest risk reduction.

Equipment Operating Criticality

You need to know how bad an equipment failure can be for your company. Events that harm people or the environment or destroy the business can never be allowed. Your business-wide risk from equipment breakdowns and quality failures is far greater than just the risk to production. *Operating criticality* indicates an asset's severity for your operating future. It measures the business-wide risk that an equipment failure causes the company. It is used to rate equipment in priority order of importance to the continued operation of a facility. Equipment that stops production or causes major production costs when it is failed is considered the most critical. Once criticality is known, resources, engineering effort, operating practices, maintenance, and training are matched to the priority and importance of the asset's continued operation.

The Plant Wellness Way methodology diverges from the standard criticality method by using two meanings in its rating of operating criticality. It first presumes that the worst outcomes will happen, including the death of employees, destruction of the environment, major plant and equipment loss, and plausible "acts of God" such as lightening and serious bad weather damage. The assumption of sure catastrophe makes an asset's initial operating criticality rating equal to the TDAF cost, because the likelihood of failure is taken as certain and so probability becomes 1 in the risk formula. By identifying the worst consequence of a failure, you are aware of the severity of a calamity. It is recorded as "Operating Criticality 1" to differentiate it from the standard criticality that uses historical failure frequency in the risk formula. The standard operating criticality that includes likelihood effects is identified as "Operating Criticality 2" in the criticality analysis spreadsheet. Knowing the full range of an asset's risk profile encourages better life-cycle engineering, operating, and maintenance decisions to be made so that you manage all risks with a higher certainty of success.

When you adopt Industrial and Manufacturing Wellness in your operation, you are required to reduce risk with mitigations that lower the chance of equipment failure to frequencies not expected to happen during three times an asset's working life. The primary strategy is risk elimination, whereby you remove all opportunities for an event to arise by using effective solutions throughout the life cycle. The next strategy is to make the asset so robust and reliable against the event that it does not fail even if the opportunity occurs. The third strategy is to use procedural quality controls in the manufacture, installation, use, and care of the asset.

Selecting actions that only limit the consequential loss are not preferred mitigation options for Plant Wellness Way practitioners. To control risk by reducing consequence and not by eliminating or preventing the likelihood of the failure cause means that a business is hoping that luck will always be in its favor. That is an impossibility. Limiting risk by minimizing the financial consequences of a failure is not risk management based on good judgment and sure risk control. In the Plant Wellness Way, you do not dismiss the use of consequence reductions; they are choices that are a last resort, used in combination with chance mitigations when it is necessary to drive risk as low as possible. For example, if there is a risk that someone could suffer injury, choosing to make an ambulance available is consequence reduction, but providing the ambulance will not keep your people safe. It is not a risk prevention, it is a risk control. The right risk management strategy to use is injury elimination—have no chance of injury in your company. An ambulance can never protect your employees from harm, but it may keep them from dying.

A catastrophic failure event is not acceptable under any circumstances. Examples of intolerable outcomes are immediate or long-term risk to human life, total or substantial operating plant destruction, the loss of a major customer, or a catastrophic environmental incident. It is unnecessary to ponder the frequency of horrific events because they are so bad that everything must be done to stop them. Even if a disastrous event were to happen once in 100 years, it would cause such severe effects that it must never happen. It is impossible to predict when a one-in-10-year, or a one-in-100-year, or a one-in-1,000-year failure will occur. It could be tomorrow. Such failures are controlled by the use of appropriate engineering, fabrication, and construction quality controls, coupled with effective operational and maintenance methods, systems, and practices that provide lasting asset integrity—not by hoping they will not happen.

Beware when standard equipment criticality multiplies consequential cost by a low likelihood of an event occurring. The devastating impact on the business is hidden by the resulting low risk value. If you disregard major events because their frequency is low, you guarantee that catastrophes will happen in your operation from time to time. Because a small risk value makes you discount low-risk events, you can have many perils all around you. In an operation carrying numerous unaddressed low-chance, high-cost opportunities, there will be a steady stream of disasters. The next one is waiting just around the corner. By seeing risk as an annualized cost that looks very low, you are tricked into making wrong decisions—or making no decisions—believing that the risk is sufficiently controlled. You're easily led into not considering that catastrophic incidents eventually transpire if they are not prevented.

If an operation accepts many disastrous risks with low frequency of occurrence, the odds worsen with time that one or more will happen. As the years go by and a failure has not yet occurred, the chance of the event rises as protective systems degrade, uncontrolled modifications are made, management's focus changes, experienced people are replaced by those less experienced, people become complacent, and numerous other reasons that cause unwanted catastrophes. Unless precautions are vigilantly maintained, the worst failure event becomes increasingly possible.

Determining Asset Operating Criticality

Identifying an operating equipment's risk profile is determining its operating criticality—it is an operating risk rating of the asset's effect on the whole business. Operating Criticality 1 is estimated using Formula 7.3.

Formula 7.3

Operating Criticality 1 (\$/yr) = TDAF Cost Consequence (\$) x 1 (/yr)

Formula 7.4 is the standard risk equation used to calculate Operating Criticality 2.

Formula 7.4

Operating Criticality 2 = Business Risk (\$/yr) = TDAF Cost (\$) x Failure Frequency (events/yr)

Operating equipment carries a risk of failure from every critical working part. It also carries risk from external sources such as forklift damage, consequential damage from the failure of nearby equipment, earthquakes, lightning strikes, and so on. Operating criticality is the total business risk incurred by having the equipment, and it is determined by adding together all of the annualized risks that are not mutually exclusive.

Separately identifying Operating Criticality 1 as the TDAF cost of sure failure prevents you from ignoring risks that would be considered minor by traditional rating methods and forces consideration of the necessary precautions to prevent them. Knowing your Operating Criticality 1 value is vital for making fully aware business risk choices. Once Operating Criticality 1 is established, you again review the failure scenarios for an equipment item to calculate Operating Criticality 2 values. This is a standard risk analysis that reviews critical assemblies, subassemblies, and parts using the likely frequency of their failure.

Formula 7.5 indicates how the risks from individual working parts in an asset are summed to determine Operating Criticality 2.

Formula 7.5

$$\begin{aligned}\text{Operating Criticality 2} = \\ &[\text{TDAF Cost Consequence}_{\text{Part 1}}\ (\$) \times \text{Failure Frequency}_{\text{Part 1}}\ (/\text{yr})] + \\ &[\text{TDAF Cost Consequence}_{\text{Part 2}}\ (\$) \times \text{Failure Frequency}_{\text{Part 2}}\ (/\text{yr})] + \\ &[\text{TDAF Cost Consequence}_{\text{Part 3}}\ (\$) \times \text{Failure Frequency}_{\text{Part 3}}\ (/\text{yr})] + \\ &[\text{TDAF Cost Consequence}_{\text{Part 4}}\ (\$) \times \text{Failure Frequency}_{\text{Part 5}}\ (/\text{yr})] + \ldots \\ &[\text{TDAF Cost Consequence}_{\text{Part n}}\ (\$) \times \text{Failure Frequency}_{\text{Part n}}\ (/\text{yr})]\end{aligned}$$

An asset can suffer hundreds of concurrent risks, making the calculation of operating criticality daunting. To reduce the calculation, it is common to include only the bigger risks. Events that might happen once in 100 years or more are assumed to be so rare, and their annualized cost so small, that they are negligible in comparison to large risks. It is a mathematically convenient practice to disregard them in order to allow the quick determination of criticality. But ignoring large-consequence events with little possibility of occurring does not stop them. When you have hundreds of them sitting in your business, you'll be regularly involved in major problems.

Another convenient practice is to consider only failure events that can or may happen in an operation. Such events are failures that have occurred previously in a business or have happened in comparable organizations. Often the problems that can arise in an organization have not yet occurred, and there is no history of failure from which to determine a likelihood of failure specific to the situation. In this case, you research other companies in your industry that have suffered the event, and, after considering your operation's reliability practices, you determine an estimated likelihood to use in the risk calculation.

There is one other class of risk scenarios to be aware of that people usually skip over—those that have never before happened in recorded history. What has not yet happened still can happen. Not everything that can go wrong has gone wrong, and only using historical incidents as justified inclusions in assessing operating criticality neglects other plausible events that have not been experienced but could occur. Risk scenarios that have not happened are known as "black swan," a reference to the discovery of black swans in Western Australia when only white swans had ever been seen. Having a huge Operating Criticality 1 value warns you that every major failure will be a catastrophe for you. When you could lose massive amounts of money from a failure, it is well worth consider all of the things you don't know about a situation and intentionally imagining black swan events so that you can protect against them.

An approach often used by industry is to substitute "cost consequence" with a scale based on the impact of an event's production disruption, for example, no impact = 0, two-hour production loss = 1, half-day production loss = 2, full-day production loss = 3, and so on. The "failure frequency" is derived from the company's maintenance history or industry norms for a similar operational situation. This is not done in the Plant Wellness Way method for operating criticality. A key premise of Industrial and Manufacturing Wellness is that you are building a world-class business. In order to make the best business decisions, it is necessary to know the real financial worth of the business-wide losses, waste, and costs of a failure, not just its impact on production uptime. Unless the true and total business-wide costs are included in determining operating criticality, the real risk of an equipment failure to the business is not appreciated. A failure that lost $1,000,000 in profit means more to people than the same failure described as a two-day outage. Using TDAF costing gives an accurate value of consequential loss and leads to a truer, more relevant operating risk assessment.

One of the great business benefits from using operating criticality is showing everyone how much company profit is destroyed by a failure. Once you have a complete total of all costs, waste, and losses for equipment downtime, you ought to publish the hourly operating income lost for each critical equipment that breaks down. The value lost with every hour of breakdown focuses people on doing what is important for both the business's well-being and protecting their own job and income.

Talking to a company's senior management about how to make money will catch their attention. When you have the full cost of a failure for an asset and its expected frequency of loss, you can do economic modeling of risk reduction choices and pick the most profit-making option. Knowing the true worth of business losses from equipment breakdown helps you make a powerful business case to management to remove the failure causes because you know the real benefits and new profits that the proposed risk reductions will bring to the operation.

Conducting a Risk Analysis

When doing an operating criticality analysis, a competent team of people is brought together to identify risks for a facility. Normally, a database of TDAF costs for the operating assets is developed.

That database is used to populate calculation spreadsheets and makes the analysis quicker and easier. Typically, the review group consists of the operators, maintainers, and engineers of the plant who contribute their knowledge and experience. The group reviews process maps, the production process, the engineering design drawings of the item, and manufacturer's information of its equipment. For each operating equipment, the group analyzes the business-wide consequences of failure and develops a table showing each equipment's Operating Criticality 1 and 2 ratings. It is convention that the final arbiter of a choice is the operations or production group, as it must live with the consequences and costs of a failure.

The people involved with the equipment risk assessment need to do the following:

- Understand the equipment operation and design—the operator manuals, maintenance manuals, and manufacturer's design drawings contain this information
- Comprehend the impact severity on production of losing use of the equipment; this information is contained in plant drawings, process flowcharts, and process and instrumentation diagrams
- Know the business-wide financial loss from each forced outage to put a true value to the TDAF costs for each downtime event
- Know the effects on business reputation and the impact on clients of forced outages
- Review and adopt the risk assessment methodology in reputed risk management standards
- Calibrate the risk matrix using the information developed from the foregoing steps and the advice of experienced and senior people in the operation

Using the Risk Matrix to Gauge Risk

Gauging operating risk requires an estimate of the chance that a particular piece of equipment will fail. Quantifying chance involves calculating the probability of occurrence. This is a difficult requirement unless you are trained in probability mathematics and methods and you have kept a complete and thorough failure history for every failed component and replacement part installed during the life of an asset. If you have, then you can quantify the likelihood of equipment failure and readily determine the operating risk of the equipment. If you have not trained in probability and statistics, or the historical failure data do not exist, or your data is incomplete and unreliable, you can instead use the likelihood scale of a risk matrix for allocating equipment failure frequency.

In the Plant Wellness Way, the risk matrix is used to see the financial impact of risk before and after its mitigation. The matrix is a visual financial decision-making device. The financial worth of consequences can be calculated and their value shown on the risk matrix. The likelihood of events is the frequency of failure incidents. The two charted together—consequence and likelihood—cross at an identifiable cell in the matrix. Once a cell is identified, you have a risk level—acceptable, low, medium, high, or extreme—and a risk value from the cell for the event. Plotting current and mitigated risk on the matrix allows you to see risk move from the current point to a future point as you make choices about how to control your risks.

8

PROCESS 2–ORDER RISKS BY IMPORTANCE

Summary Description of Process 2: Rating Operating Risk

Not all of your operating assets are equally important to your company's success. The grading of operating asset risk is known as Operating Criticality Analysis. With it, you can match operating risk controls to the importance an asset has in your operation.

In Process 2, you follow the flowchart in Figure 8.1 to determine each asset's Operating Criticality 1 and Operating Criticality 2 values. Process 2 also requires you to develop an appropriate enterprise asset management policy and a supporting maintenance management policy. These policies articulate why physical asset management and maintenance are important to your business and give legitimate reasons for the use of business resources to implement them.

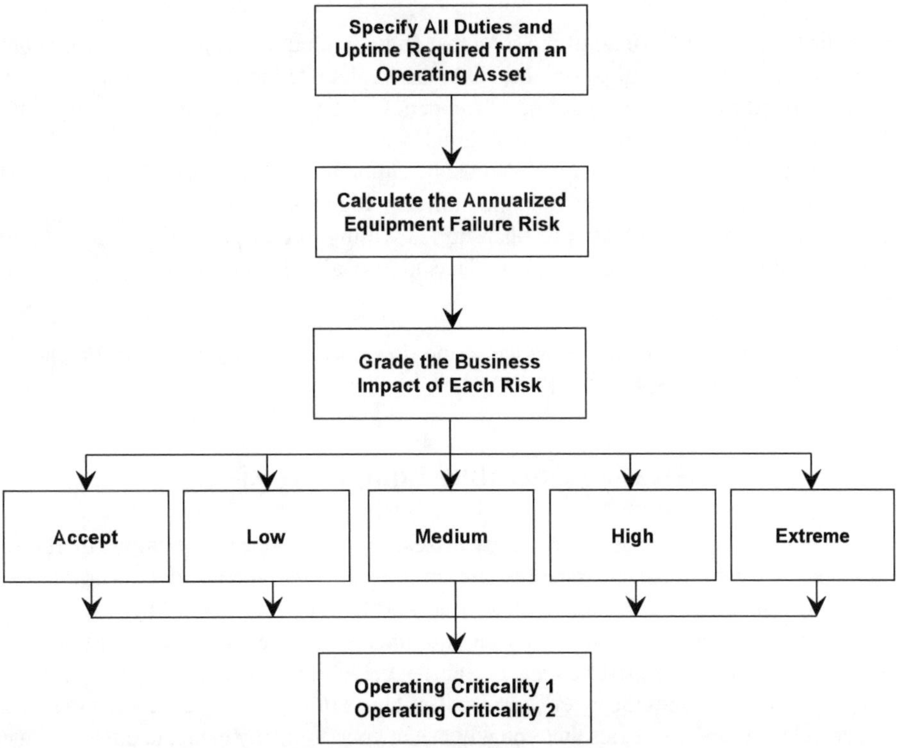

Figure 8.1—Process 2 Steps

Specify the Asset Performance That Delivers the Operational Requirements

Why is an operating asset in your organization? What is its purpose in the business? For each physical asset, number it uniquely and give it a three- or four-word descriptive name that explains what it is and what it does for your operation—for example, "Finished Product Supply Pump" or "Workshop Overhead Gantry Crane." Specify the complete range of duties that each asset must perform when it is in service. Select a measurable indicator that represents each duty. Typically, this is a performance indicator that you will use to determine the adequacy of the asset's operation. Record on the process map the required performance measures for each duty the asset must deliver.

Also note on the flowchart the service time that each asset must be available. For example, a stand-alone pump used to move product from a vessel to a storage tank must run when it is needed and deliver a desired flow at a minimum pressure using a motor of sufficient power. On the process map, the required functions and their minimum performance (e.g., pump flow, pressure) and the needed availability are noted. For example, for a pump with service duty of at least 1,000 liters per minute flow rate, at 500 kilo Pascal (kPa) pressure, used three times a day for six hours duration each time, you would note "1,000 l/min, 500 kPa, 3 x 6/24 hours" on the process map. This information helps determine the significance of an equipment item. If the pump cannot do its minimum duty as required during the working day, there will be consequential effects on production and throughout the business. Such service duty and availability information is needed for every item of equipment in your operation to clarify the purpose and use of each item.

Determine the Business Risk from Operating Failures

Operating criticality is the sum of all of the risks that an item of equipment poses for your company. In the operating criticality determination, the TDAF cost calculated in Process 1 is the consequence value. It is multiplied by the yearly likelihood frequency of a failure to estimate the annualized financial loss you would suffer if the equipment failed.

The risk matrix you use (see Chapter 4) must be calibrated to the impact that risk has on your business. You need to know what a low risk, medium risk, high risk, or extreme risk costs your business. Identify the risk boundary that your operation is willing to carry (i.e., the boundary between acceptable risk and low risk) before putting in place additional risk control strategies and actions.

The sample 16 x 13 risk matrix provided in the downloadable spreadsheet accompanying this book is calibrated using a $10,000 per year risk boundary.

Assessing Operating Equipment Risk

In the Plant Wellness Way, operating criticality arises from component criticality; therefore, risk analysis is done at the equipment subassembly and component levels and not at the whole asset level. Because an asset fails when a critical part fails, you need to know the size of the risk that your organization carries from each component in your equipment. You then tailor the risk mitigations to reduce unacceptable component risk to an acceptable level while also aiming for no more than one failure in a period of three times the asset's service life. Those mitigations are the asset management, maintenance, and operating strategies that you will use in your company to create outstanding equipment reliability.

In order to understand the consequences of failed assemblies in your equipment, each asset is subdivided into its major assemblies. If a major assembly contains substantial numbers of individual equipment, these are further divided into subassemblies and key components. The annualized sum of failures for subassembly critical components is used for the likelihood frequency.

A business makes money if a risk is prevented for less than the risk's total consequential cost. The greatest opportunity for a business to reduce risk for the least cost is to identify the methods, systems, and practices that prevent a risk, or minimize the chance of a risk arising, and then implement them with energy and vigor across the organization. Maintenance is one way to reduce the risk of equipment failure, but it is typically used as a consequence reduction strategy in response to failure. There are also numerous engineering and operational choices, which are usually more cost-effective over the equipment life cycle than maintenance because they are chance reduction strategies that stop failures from starting in the first place.

Plant and equipment risk analysis applies the risk formula using historical financial and failure information for the asset under review. Table 8.1 shows typical column headings for a risk assessment spreadsheet used to gauge current operating equipment risk.

Table 8.1—Typical Risk Calculation Spreadsheet Layout

Ref No.	Equip ment Tag No.	Equip Descript ion	Failure Events or Causes	Cost Consequence of Failure ($)	Years Equipment in Service or Expected	No. of Item Historical Failures at This Site or Expected	No. of Annualized Failure Events Due to Cause (/yr)	Likelihood of Failure Event (1–0)	Estimated Current Risk ($/yr)
1	2	3	4	5	6	7	8	9	10

The "Equipment Tag No." (Column 2) is the identifying number given to each item of equipment at a site. Every tag number is included—machinery, electrical equipment, instrumentation, piping, even the buildings and each functional area in a building—in order to determine a site's total risk. Additional columns are added when subassemblies or components need to be individually identified.

The "Equipment Description" (Column 3) is the official descriptive name for the equipment or its subassemblies. List the assemblies, subassemblies, and parts that have failed in the past or have a fair possibility of failing in the future.

"Failure Events or Causes" (Column 4) are the many separate ways in which an item of equipment has failed or could fail.

The "Cost Consequence of Failure" (Column 5) is the TDAF cost impact of an item's failure.

"Years Equipment in Service or Expected" (Column 6) is the number of years the item has been in use and is expected to be in use. For new equipment, the expected years in service is used. For existing equipment, it is the sum of life to date plus expected years remaining in service. Work in years to an accuracy of a calendar quarter.

The "No. of Item Historical Failures at the Site or Expected" (Column 7) is determined for each identified failure event or cause by looking at the equipment history in the operating records and maintenance management system. If actual site failures are not available, then the industry average adjusted for the on-site reliability culture is adopted. If there is a good reliability culture and standard industry maintenance and care practices are applied well, use the industry average as the event frequency; in a poor reliability culture, assume a substantially worse outcome.

The "No. of Annualized Failure Events Due to Cause" (Column 8) is calculated by dividing Column 7 by Column 6.

Table 8.2—Determining the Likelihood of Equipment Failure at a Site

Descriptor	Description	Indicative Frequency (expected to occur)	Actual Failures per Year (historic evidence basis)	Likelihood of Failure per Year (opportunity for failure basis)	
				Opportunities	Probability of Failure
Certain	Failure event will occur at this site annually or more often	Once a year or more often	1/year or more	Count every time the situation arises when a failure event could occur	1 if failure results every time the situation arises
Likely	Failure event regularly occurs at this site	Once every 3 years	1 in 3 years = 0.33/year		0.1 if failure results 1 in 10 times the situation arises
Possible	Failure event is expected to occur on this site	Once every 10 years	1 in 10 years = 0.1/year		0.01 if failure results 1 in 100 times the situation arises
Unlikely	Failure event occurs from time to time on this site or in the industry	Once every 30 years	1 in 30 years = 0.033/year		0.001 if failure results 1 in 1,000 times the situation arises
Rare	Failure event could occur on this site or in the industry but doubtful	Once every 100 years	1 in 100 years = 0.01/year		0.0001 if failure results 1 in 10,000 times the situation arises
Very Rare	Failure event hardly heard of in the industry; may occur but under exceptional circumstances	Once every 1,000 years	1 in 1,000 years = 0.001/year		0.00001 if failure results 1 in 100,000 times the situation arises

The "Likelihood of Failure Event" (Column 9) is a probability assessment using the frequency of historical events. It is also described using the terms listed in Table 8.2 developed from international risk management standards and industry guides.[1, 2]

In the last column is the item's "Estimated Current Risk," which is calculated by multiplying the values from Columns 5 and 9 using the standard risk formula (see Chapter 7, Formula 7.1).

Determining the likelihood of failure is fraught with uncertainty. The opportunity for failure may rise often but never proceed to conclusion. Counting historical failures is easy because there are records. But counting an opportunity for failure that does not progress to a failure is open to speculation. For example, one opportunity for failure is overload upon equipment start-up. The likelihood of failure for a part known to fail as a result of high-stress overload during start-up can be calculated using Formula 8.1. The opportunity for this failure is a count of the average number of starts between failures. The likelihood of failure formula is as follows:

Formula 8.1

$$\text{Likelihood of Failure} = \frac{\text{No. of Failures}}{\text{Average Number of Starts between Failures}}$$

For an operation running continuously with 10 starts a day and failures averaging every six months, or twice a year, the likelihood of failure is calculated as follows:

$$\text{Likelihood of Failure} = \frac{\text{1 Failure}}{\text{1,825 Starts}} = 0.000556$$

With a TDAF cost of failure of $25,000, the risk calculated using Formula 7.2 (see Chapter 7) is as follows:

Risk = Consequence ($) x [No. of Opportunities (/yr) x Probability of Failure]
= $25,000 x [3,650 x 0.000556] = $50,000/yr

The $50,000 annual risk estimated by first finding the probability is the same as that estimated using the standard risk formula ($25,000 x 2/yr). When failures have happened, it is easier to count the average failures per year from historical evidence and use that number in the standard risk equation. Historical failures are used because they already reflect the risk present in an operation. Future failure rates will remain the same as in the past until better risk management strategies are put in place. You use the opportunity for failure approach in Formula 8.1 if you know how often a failure situation actually arises. But if the number of opportunities is uncertain, use the historical average of failures per year for the site.

The "Estimated Current Risk" shown in Table 8.1 is the yearly cost for the existing risk in the business. The value is used to rate and gauge the size of a risk and compare it with others. For those risks that a business does not want to carry, mitigations are put in place to eradicate or reduce them to acceptable levels if they cannot be eliminated.

Risk Assessment and Risk Mitigation Templates

A Risk Identification and Assessment Template completed manually can be used to find and list the operating risks to each equipment, assembly, and subassembly. Identifying failure events and grading their risks can be done using Table 8.3. Alternatively, a spreadsheet can be developed to replace the template. For equipment and assemblies being assessed, a calibrated risk matrix is used to categorize the consequence, likelihood, and risk level for each risk event.

Table 8.3—Risk Identification and Assessment Template

EQUIPMENT OR ASSEMBLY	EVENT OR FAILURE *What can happen?*	SOURCE *How can this happen?*	IMPACT *of event happening*	CURRENT CONTROL STRATEGIES *and their effectiveness* (A) – Adequate (M) – Moderate (I) – Inadequate	CURRENT RISK LEVEL			ACCEPTABILITY (A/U)
					LIKELIHOOD	CONSEQUENCE	CURRENT RISK LEVEL	
1	2	3	4	5	6	7	8	9

A Risk Treatment Schedule and Action Plan Template completed manually can be used to identify actionable activities to reduce risk. The template in Table 8.4 is used to list actions to mitigate the risk and to judge their effectiveness.

Table 8.4—Risk Treatment Schedule and Action Plan Template

EQUIPMENT OR ASSEMBLY RISK	POTENTIAL TREATMENT OPTIONS	COSTS AND BENEFITS	TREATMENTS TO BE IMPLEMENTED (Y/N) *and their effectiveness* (A) – Adequate (M) – Moderate (I) – Inadequate	RISK LEVEL AFTER IMPLEMENTATION			RESPONSIBLE PERSON	TIMETABLE *to implement*	MONITORING *strategies to measure effectiveness of risk treatments*
				LIKELIHOOD	CONSEQUENCE	TARGET LEVEL			
1	2	3	4	5	6	7	8	9	10
		FINAL Cumulative Risk Level after Treatment							

At the end of the risk assessment and review, all of the risk mitigation actions for an asset become part of that asset's risk management plan.

Performing an Operating Criticality Analysis

Table 8.5 shows the start of an example of a Plant Wellness Way Operating Criticality Analysis.

In the Plant Wellness Way, an operating asset's risk assessment is done using the Operating Criticality worksheet, which is included in the downloadable materials accompanying this book.

The analyst who does the risk assessment needs to be a person who knows the engineering of the asset being investigated and the ways in which its components can fail. When necessary, gather a team of subject matter experts to ensure the completeness and thoroughness of the analysis. Each column heading in the spreadsheet is addressed as accurately as possible during the analysis. When accurate information is not available, adopt the considered recommendations of the subject matter expert team.

Table 8.5—Operating Criticality 1 and 2 Analysis

Ref No.	Equipment Tag No.	Equipment Description		Parent Assembly		Subassembly	
1	TRN-01	250 KVA Power Supply		Transformer			
					1	Tank with Fins	
					2	Tap Changer	
					3	Oil	
					4	HV Termination Bushings	
					5	Insulation	
					6	Core	
					7	Windings	
					8	Oil Conservator	
2	SWB-01	Infeed Switchboard		Switchboard			
					1	Panel	
					2	Bus Bar Connections	
					3	Panel Connections	
					4	Drive Rack	
					5	Motor Starter	
3	PMP-01	Circulation Pump-Set					
			1	Power Cable			
			2	Electric Motor			
					1	Terminal Block	
					2	Motor	1
							2
							3
							4
							5
					3	Stator	
							1
							2
					4	Rotor	
							1

References

1. "ISO 31000:2009—Risk Management," accessed at http://www.iso.org/iso/home/standards/iso31000.htm, August 3, 2015.

2. Richard M. Robinson et al., *Risk and Reliability: An Introductory Text,* 7th ed. (Melbourne, Australia: R2A, 2007).

9

REMOVING RISKS AND RAISING RELIABILITY

Without getting into the math, equipment reliability is a measure of the odds that an item of equipment will last long enough to do its duty. It is the chance of a thing remaining usable until a particular point in time. When equipment operates at duty capacity for as long as expected, it is considered reliable. When the period between out-of-service episodes is too short, the equipment is considered unreliable. You measure the reliability of equipment by its trouble-free time. If the equipment is meant to last for 10,000 hours (about 14 months of continuous operation), and it does last that long, it is 100% reliable to 10,000 hours. But if after 10,000 hours there is an occasional failure, its reliability beyond 10,000 hours is less than 100%. When we talk about reliability, we must also say what time period is involved.

Table 9.1 defines the words and terms commonly used in reliability improvement.

Table 9.1—Reliability Improvement Terms and Definitions

Term	Definition
Critical Item	A part or assembly for which a failure mode(s) remains and has not been designed out. These items require operating and maintenance strategies to ensure a long, trouble-free life.
Criticality	A measure that combines severity (the cost and inconvenience of the failure) and frequency (how often a failure occurs) to indicate the overall risk caused by the item if it fails.
Failure	Any unwanted or disappointing behavior of an item.
Failure Effect	The impact on performance of the item's failure.
Failure Mechanism or Failure Cause	The processes by which the failure mode(s) arise, including physical, mechanical, electrical, and chemical causes or other processes and their combinations. Knowledge of a failure mechanism provides insight into the conditions that cause failures.
Failure Mode	How a part or combination of parts fails. Failure modes can be electrical (open or short-circuit, stuck at high), physical (loss of speed, excessive noise), or functional (loss of power gain, communication loss, high error level).
Failure Site	The physical location where the failure mechanism is observed to occur; it is often the location of the highest stress and lowest strength.
Redundancy	The use of duplicate or more items so that upon the failure of one, another takes over the role. It presumes that the replicated item is ready and available to immediately take on the full service.
Reliability Improvement	Documented explanation of why known failure modes occur and how to address them. This is the basis for engineering, operating, and maintenance strategies for a part or assembly.

To a great extent, you can choose how long you want equipment to be reliable. You get high equipment reliability by ensuring that the chance of parts failing is extremely low. The secret to remarkably long-lived, trouble-free machines is keeping their components at low stress so that there is no risk of microstructure deformation and providing a healthy contacting environment so that the material does not degrade. If there is nothing to cause a failure, a failure will not happen, and your equipment will continue in service at full capacity and full availability.

Equipment reliability needs to be seen as more than just a time span. Reliability is a business imperative for companies that use physical assets. You need highly reliable plant and equipment if you are to build a business that is a world-class performer. High-reliability organizations expect equipment to last a long time, and they are unhappy when it does not. Not only are they unhappy, but also they take effective measures to learn from and improve as a result of failures.

Identifying Equipment Reliability Growth Opportunities

You must deeply want the production and profit benefits that equipment reliability brings before you will do what is necessary to get them. You get plant and equipment to operate trouble-free for a long time only when you do those activities that cause high reliability and do them well enough to produce high reliability. To get outstanding equipment reliability, you must control failures in your equipment across the life cycle and operate and maintain equipment with masterly precision. If you want high production uptime and low operating costs, you will need to implement engineering, purchasing, supply chain, storage, operating, and maintenance regimes and practices that deliver the reliability and life-cycle costs you want.

A great reliability strategy describes in detail how to ensure high reliability for your operating assets. Plant reliability can only be improved in one of two ways: (1) by using redundancy so that another unit takes over when the duty unit fails (see parallel arrangements in Chapter 1) or (2) by making each equipment in the operation supremely reliable so that it is failure-free during service. Reliability solutions involve applying the three Series Reliability Properties and building in parallel arrangements. The Series Reliability Properties and parallel arrangements are foundational concepts for reducing the risks to your operating equipment and work tasks. With them, you can create the reliability you want by embedding in your engineering, procurement, supply chain, operating, and maintenance processes the best practices and right methods that are sure to deliver it.

Apply Series Reliability Property 1

The reliability of a series system is no more reliable than its least reliable component.

Unreliable equipment are known as "bad actors." They cause a lot of production troubles. Series Reliability Property 1 advises you to fix your bad actors or else they'll stop your business from performing at its best and generating high operating profits. To get high equipment reliability, you need to ensure that every critical part of the equipment is highly reliable. Figure 9.1 is a top-level process map for a centrifugal pump-set when it is in operation. This item of equipment will be used as an example for developing a plan to make effective reliability improvements.

Series Reliability Property 1 requires you to identify whether the reliability of the least reliable items needs improvement. For the pump-set example, a minimum series reliability of 0.9999 is expected from the equipment—that is, there is a chance of one failure in 10,000 opportunities. In a continuously operating process plant that runs the pump 10 times a day, 10,000 opportunities represent 1,000 days without a failure—about three years of service. To get that level of reliability from the pump-set system, each item depicted in the process map needs greater individual reliability.

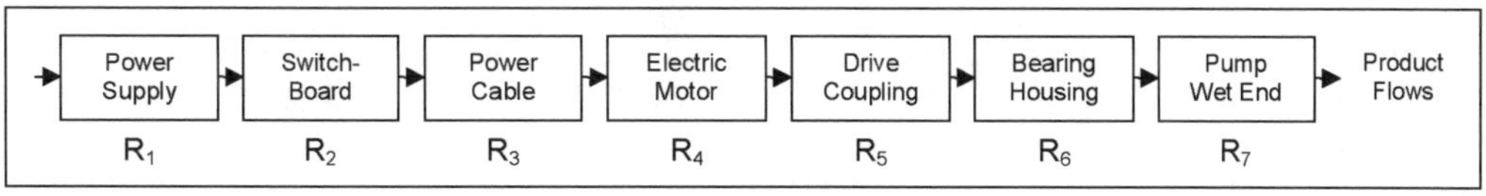

Figure 9.1—A Centrifugal Pump-Set Process Map

We can estimate the scale of the individual item reliabilities required by using Formula 1.2 from Chapter 1 and assuming that the first four electrical items in the process have equal reliability and that the three mechanical items have 10 times less reliability. Industrial mechanical equipment is less reliable than electrical equipment, with mechanical equipment typically incurring 10 times more failures in the same period than electrical equipment.[1]

$$R_{\text{Pump-set}} = R_1 \text{ x } R_2 \text{ x } R_3 \text{ x } R_4 \text{ x } R_5 \text{ x } R_6 \text{ x } R_7 = 0.9999$$

$$R_{\text{Pump-set}} = 0.9999971^4 \text{ x } 0.999971^3 = 0.9999$$

A pump-set reliability of 0.9999 requires an individual electrical item reliability of 0.9999971—about three failures in every 1,000,000 opportunities for failure—and a mechanical item reliability 0.999971—three failures every 100,000 opportunities. Each electrical item must have only one failure every 100 years and each mechanical item one failure every 10 years in order for the pump-set to have only one failure in three years.

One hundred years of failure-free electrical system operation is a daunting requirement. Such reliable electrical equipment cannot yet be made with guaranteed certainty of success using the technologies now available. With good maintenance, it is not impossible for electrical transformers, switchboards, and cables to remain in operation for 50 to 100 years, although failure rates for their components will rise as they age in service. Already it is clear that predictive and preventive maintenance will be needed for parts that degrade during the service life of the system. Some equipment, such as the electric motor and transformer tap changer, will require total overhaul several times during the decades ahead. Replacement with new parts will be necessary for many of the electrical items at least once, and in some cases twice, during the next 100 years.

Mechanical items in the equipment will need to be replaced a number of times to get a century of failure-free service from the pump-set. One failure in 10 years is a challenging but not impossible goal for the majority of mechanical parts in nonabrasive or corrosive services. The most unreliable machinery component is the mechanical seal in the pump. These seal types can experience variable and unsteady operating conditions that cause them to fail sooner than once in 10 years—including cavitation, pump vibration, pump frame distortion, water hammer, shaft misalignment, torque overload, poor assembly on installation, corrosion, wear and impact, and chemical decomposition of elastomeric items.

To get higher reliability for the pump-set system, the reliability of the mechanical items must be raised. Series Reliability Property 1 advises you to work on new ways to make the least reliable mechanical components much more reliable. The mechanical items in the pump-set require better solutions to prevent degradation and stress in the components' materials of construction in order to achieve higher reliability. This is where the process map helps identify more reliable options than those currently in place.

Localized reliability improvement of individual process steps or equipment parts is vital when the improved reliability delivers a strong return on investment. These improvements can include use of better-engineered designs, use of parts that are more robust, better installation practices, better operating practices, better maintenance strategies, and even complete equipment replacement with a more reliable item. The economics of the situation will drive which choices you make—higher costs and losses justify spending more money to prevent the problem. It is sensible to do a life-cycle costing model to understand the effects of TDAF costs, replacements, and capital expenditures on the future of the operation.

Figure 9.2 shows quality assurance tasks and precision activities added in parallel with the mechanical items. When done correctly, they greatly increase component lifetimes. Adding the life-cycle requirements of high engineering quality specifications to remove modes of failure, equipment installed with precision procedures to prevent deformation, shafts correctly aligned to precision standards, and the pump-set precision-operated to minimize component stress and degradation will produce long-lived assemblies and a failure-free pump-set during its working life. These practices are learned skills and techniques that prevent stresses and errors from being introduced, and their use lifts equipment reliability. Paralleling precision skills with high work accuracy and defect removal raises the reliability of each mechanical item. The process map makes it easy to see where to add parallel life-cycle tasks to improve operating equipment reliability.

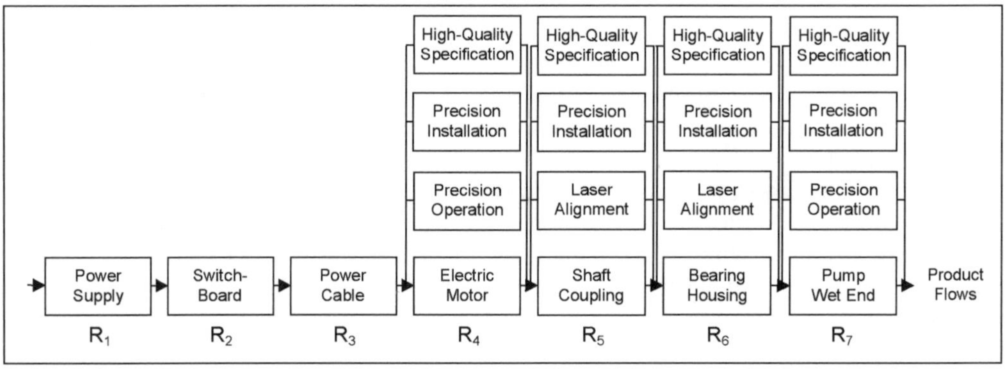

Figure 9.2—Centrifugal Pump-Set Reliability Improved by Parallel Tasks

Very few operations will use the same equipment for 100 years. Practically speaking, most businesses will build another plant in 30 to 50 years or modernize equipment when productivity degrades too severely and continue operation with brand-new assets. The most progressive companies will continually upgrade their equipment with the latest model when the current equipment has repaid its investment or when new equipment becomes available with a sufficiently high return on investment to justify a change-out. This strategy ensures that they will never have "old" plant in use, thereby guaranteeing high reliability. It is an effective way to constantly gain the benefits and higher profits of reliability improvement from the equipment manufacturer's technological and engineering innovations. Nonetheless, while equipment is in a location, you want failure-free service for its entire working life, regardless of how long that life is.

Apply Series Reliability Property 2

Add "k" items into a series system of items, and the probability of failure of all items in the series must fall by an equal proportion to maintain original system reliability.

The requirement here is to simplify your processes by combining steps and removing steps so that there are fewer opportunities for things to go wrong. Series Reliability Property 2 gets you to remove unnecessary components from the system or redesign the system so that a smaller number of items are used. By reducing components or steps, the system becomes more reliable because there are fewer possibilities for failure. Figure 9.3 asks what would happen if the drive coupling was removed from the centrifugal pump-set. Is there technology available to eliminate the drive coupling?

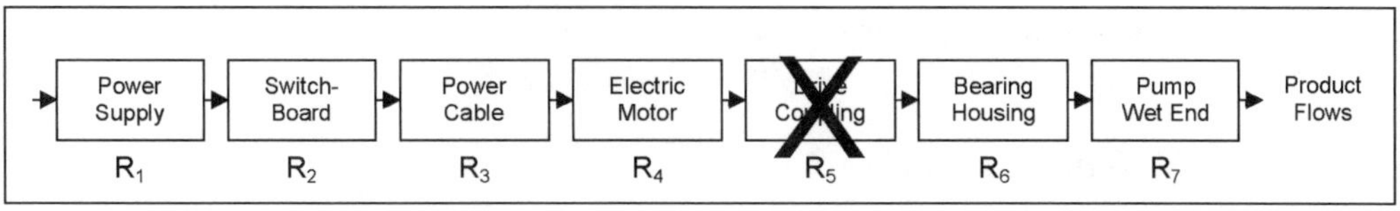

Figure 9.3—Centrifugal Pump-Set Reliability Improved by Removing Drive Coupling

Figure 9.4 shows two such technologies: canned motor pumps and magnetic drive pumps. Both pump types do not have a drive coupling.

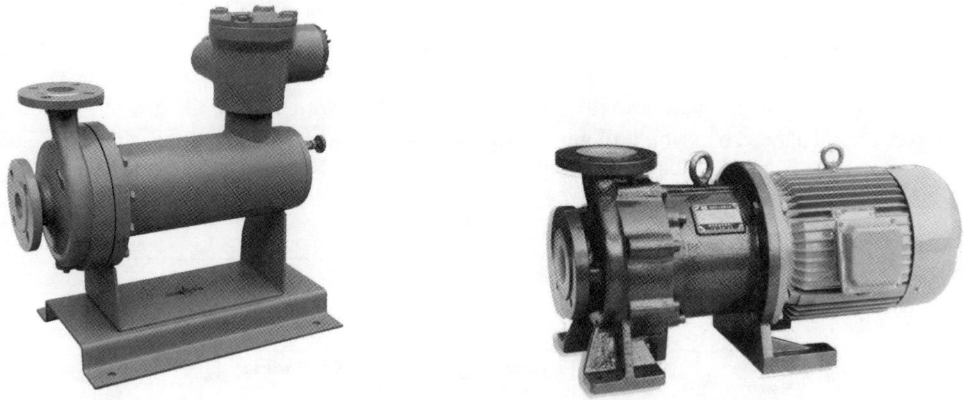

Figure 9.4—Canned Motor Pump (left) and Magnetic Drive Pump (right)

With the drive coupling removed, the system reliability (assuming that the other items maintain their individual reliability) is calculated as follows:

$$R_{Pump\text{-}set} = R_1 \times R_2 \times R_3 \times R_4 \times R_6 \times R_7 = 0.9999$$

$$R_{Pump\text{-}set} = 0.9999971^4 \times 0.999956^2 = 0.9999$$

Removing the drive coupling allows less reliable items to be used while still achieving system reliability—the mechanical assemblies can have more than four failures per 100,000 opportunities. To confirm the effect of simplifying the system, the bearing housing is also removed. This means the pump wet end is directly mounted on the electric motor shaft. The system reliability then becomes,

$$R_{Pump\text{-}set} = R_1 \times R_2 \times R_3 \times R_4 \times R_7 = 0.9999$$

$$R_{Pump\text{-}set} = 0.9999971^4 \times 0.999912 = 0.9999$$

The wet end can fail nine times per 100,000 opportunities, and the system reliability is unchanged, at one failure in three years. In simplifying your system, you can reduce your capital costs with poorer-quality equipment and still get the required system reliability. But every breakdown will cause large TDAF costs. Even a planned corrective repair causes lost production and adds maintenance costs. By using equipment that suffers nine failures in the time that high-quality equipment has three failures the operating and maintenance costs rise substantially and continually recur. If instead you get more reliable equipment, thereby eliminating six failures, you will make significantly more operating profit year after year. This is how having a corporate policy to use high-quality equipment and making economic replacement of assets before they "age" makes fortunes for industrial operations.

Apply Series Reliability Property 3

An equal rise in the reliability of all items in a series causes a much larger proportionate rise in system reliability.

System-wide reliability improvements return far more profit than step improvements. Using Series Reliability Property 3 is an improvement choice that delivers astounding operating profits. Figure 9.5 shows the introduction of a company-wide policy to limit operating loads to 90% of an item's design duty. Another option is to "oversize equipment" and get the next-higher duty model so that its components operate at below design duty. Any additional capital cost to get the heavier-duty model will be more than repaid through a higher return on investment from the greater productivity and lower maintenance costs you will get throughout its life.

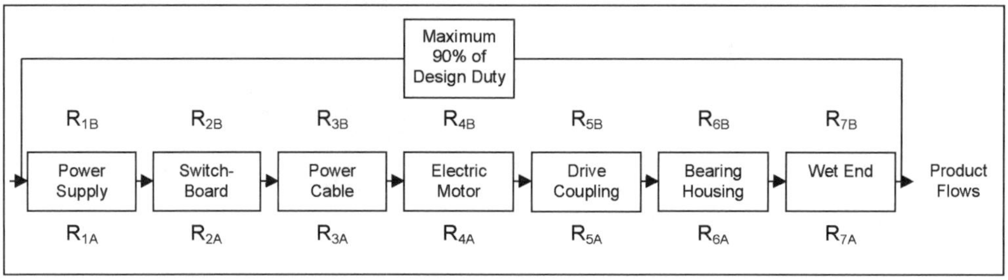

Figure 9.5—Pump-Set System Reliability Improved by a System-Wide Policy Decision

A business policy to run equipment at 90% of design duty is a system-wide decision that will cause stress reductions in all production equipment. It is an especially valuable strategy if your operating plant is of low-quality design, made of weak materials, or poorly installed or maintained. In high-fatigue situations, reducing fluctuating loads by 10% can increase parts' fatigue life up to 10 times.[2] Running at no more than 90% of design duty load can mean nine fewer failures during the service life. The choice to run equipment at 10% below design duty ensures that all parts throughout the entire operation get higher reliability. With one asset management policy decision, you stop equipment failures everywhere in a company and convert fortunes in TDAF cost losses into new operating profits.

Wherever possible, apply Series Reliability Property 3. It will bring the greatest benefits and amazing return on investment. But do not neglect the other two Series Reliability Properties. Apply Series Reliability Property 2 during design by using the DOCTOR (Design and Operations Cost Totally Optimized Risk; see Chapter 6) to get project managers to create robust, highly profitable business processes. Use Series Reliability Property 1 with your maintenance engineers, plant operators, and maintenance technicians to constantly make local reliability improvements to your "bad actors."

Current Maintenance Strategy Selection Methods Do Not Prevent Failure

Maintenance strategies and activities can be chosen by using top-down or a bottom-up methodologies. Top-down strategy selection is based on equipment failure history. Maintenance records are reviewed to identify which plant assets fail and what range of causes produce the failures. Strategies are then selected to reduce equipment failure rates. The bottom-up approach identifies causes of equipment failures and then selects maintenance strategies and practices to maximize equipment reliability. Depending on the risk from the event and the certainty of failure detection, you might choose predictive maintenance, preventive maintenance, breakdown maintenance, or failure-finding tests. The better methodologies also include cost–benefit analysis to make sound economic decisions about the best mix of maintenance strategies to use in order to maximize operating profit.

The standard maintenance strategy selection methods currently used in industry do not prevent failures. They are designed to find and stop disastrous events. They promote failure by letting defects become serious problems before you act to prevent a breakdown. If you only look at your machines for problems to fix, then problems are what you will always get. Although top-down and bottom-up methods let you arrive at some sort of maintenance strategy, they cannot maximize reliability because they need failures to fix. They cannot minimize maintenance costs because they generate maintenance. Their limitations allow preventable failure causes to remain in your equipment until they become failures.

Developing a Highly Successful Equipment Risk Prevention Plan

The Plant Wellness Way recognizes that there is a dependency hierarchy of risk that extends from the operating stresses and local environment affecting a component's microstructure, such as healthy lubricant, correct lubricant film thickness, proper interference fit between mating parts, trueness of component shape, fineness of surface finish, subsurface stress levels, adequate load distribution, and so on, through to a catastrophic failure of the plant.[3] Every equipment failure can be traced back to what was done to its components' materials of construction during their lifetimes.

Figure 9.6 shows these dependency links between an initial Physics of Failure mechanism through the component, subassembly, assembly, parent asset, operating process, and, finally, business system levels. The state of health at each level of the hierarchy immediately causes risks at that level, which can become problems at the levels above. Ultimately, the reliability of every physical asset, and the subsequent risks they cause to a business, depends on how well you eliminate and prevent the causes of Physics of Failure mechanisms throughout its parts' life cycles.

Current maintenance strategies look at the asset and assembly levels of the dependency hierarchy for problems. This makes maintenance ineffective at preventing failures because it allows the effect of microstructure failure mechanisms to become so severe as to be detectable in the behavior of the parent asset. So long as failure mechanisms are present in equipment, they will trigger the risk of failure. Only when there is no risk at the Physics of Failure mechanism level of the dependency hierarchy can you ensure that an asset has maximum chance of exceptionally long lifetime reliability resulting in the least risk of operating failure.

Getting high equipment reliability is within the power of every business. Reliability is malleable by choice of policy and quality of practice used at each level of the dependency hierarchy. At the

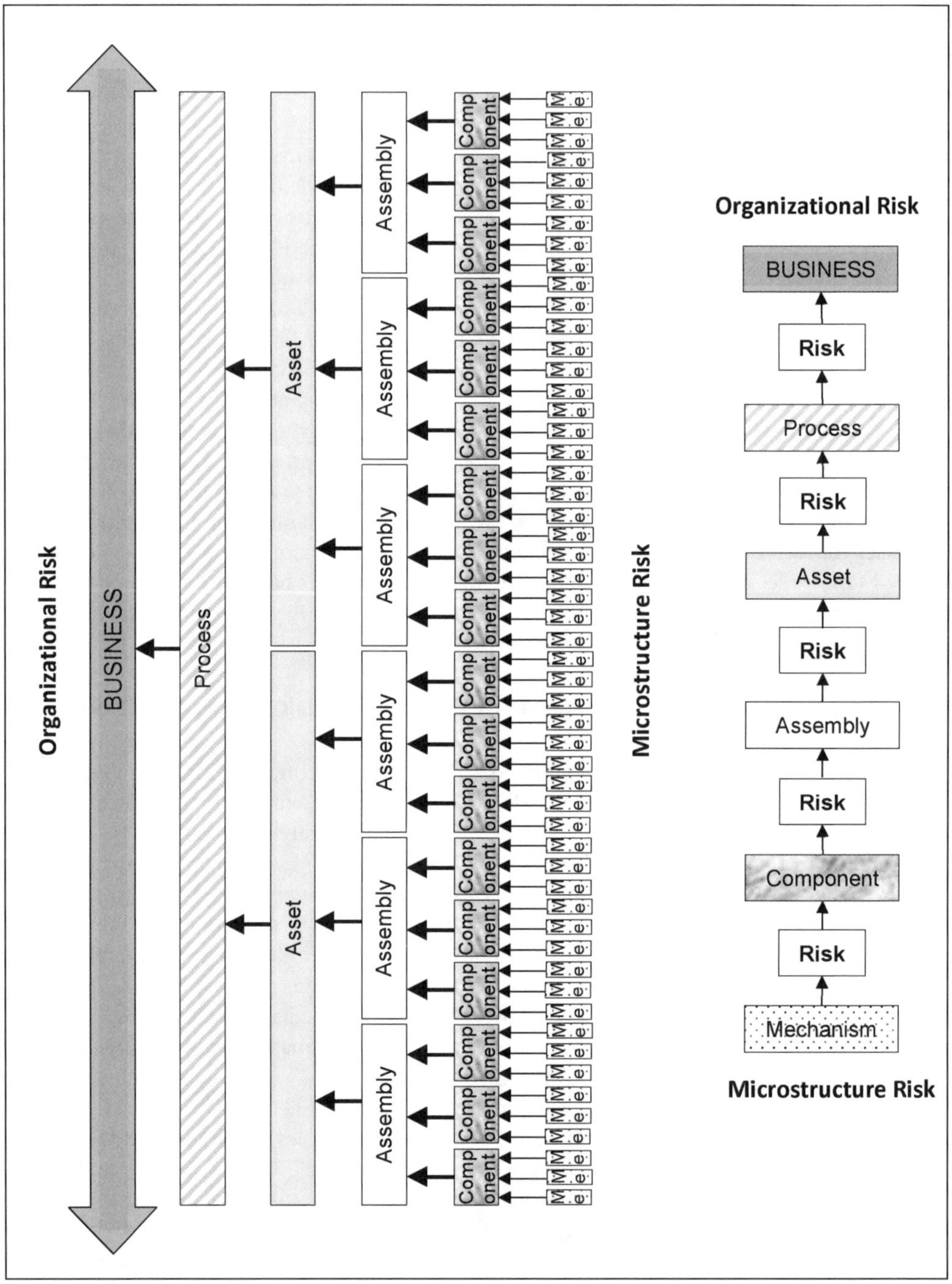

Figure 9.6—Hierarchy of Risk from Component to Organization

mechanism level, set demanding quality standards for equipment parts so that you get outstanding reliability during operation. At the component level, use precision that delivers long service lives to parts and apply only the highest-quality operating and maintenance practices. At the assembly level, use more robust, durable materials for components so that they take greater stresses and don't degrade. At the asset level, replace assets early with the newest models rather than wait too long and get numerous failures and lower productivity. Be willing to pay for higher-quality equipment so your company can make fortunes from fewer failures. At the process level, have right knowledge everywhere and use only the right practices, ensuring that they are done correctly. These requirements need to be incorporated into corporate thinking when you make engineering, asset management, operational, or maintenance decisions. The people who undertake new business initiatives and select new capital assets or replacement equipment only have one chance to take the best business option—after that, everyone has to live with the choice for years to come.

The most beneficial life-cycle strategies are those that eliminate Physics of Failure mechanisms to reduce the total risk that a part carries during operation. The lower down the dependency hierarchy that good decisions are made and precision quality controls installed, the more comprehensive is the risk control strategy. In the Plant Wellness Way, the technique used to select the right engineering

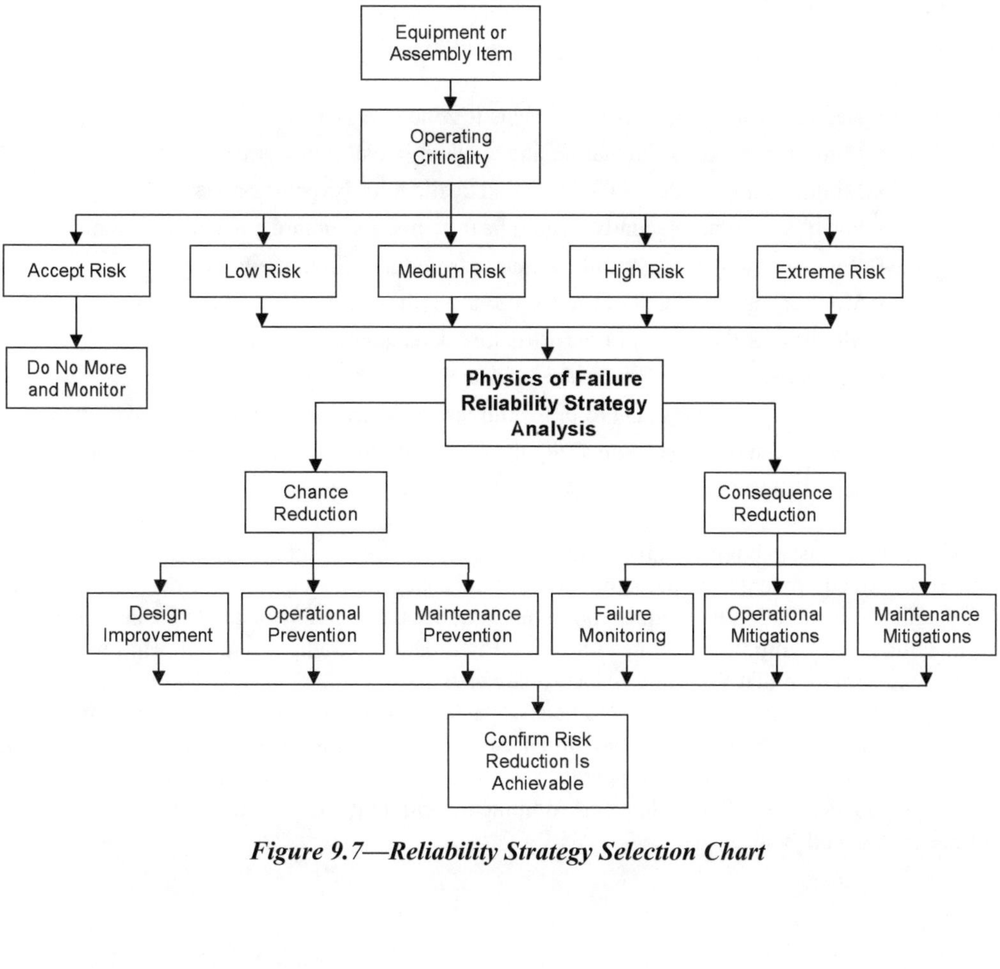

Figure 9.7—Reliability Strategy Selection Chart

design, manufacture, supply chain, maintenance, and operating requirements throughout the dependency hierarchy is called Physics of Failure Reliability Strategy Analysis. The technique follows the Stress-to-Process Model (see Chapter 4), starting with the risks to the component microstructure and working up the hierarchy to specify the correct strategy, knowledge, skills, and practices required in the business and life-cycle processes used to provide component reliability.

Each mechanism cause of a critical part's failure is identified and addressed one by one until the part's lifetime risk control plan is complete. The risk control plan covers everything that will be done during the life cycle to remove or significantly reduce operating risk. It lists the mix of design, manufacture, supply chain, operations, and maintenance activities to lower the risk of microstructure destruction and deliver high equipment reliability.

Figure 9.7 shows how Physics of Failure Reliability Strategy Analysis is used in the process of choosing operating risk reduction actions. Mitigation and prevention actions will fall to the maintenance and operations groups, and design improvements will go to the engineering group. Design changes are carried out by a professional discipline engineer or a competent technical person who understands the equipment's purpose and construction.

Plant and equipment reliability are improved only if parts do not fail. By doing Physics of Failure Reliability Strategy Analysis at the component and mechanism levels, you identify the engineering, manufacturing, operational, and maintenance issues that must be addressed at each level of the dependency hierarchy for maximum asset reliability. The choices available to prevent component failure and subsequent equipment breakdown are as follows:

- Keep total stress well inside the elastic zone everywhere in the microstructure
- Minimize intended fluctuating and cyclic microstructure stresses
- Eliminate unintended fluctuating and cyclic microstructure stresses
- Establish and permanently sustain healthy microstructure contact environments
- Remove microstructure failure causes from throughout the life cycle
- Monitor for the causes of microstructure failure and remove the causes
- Monitor for the onset of microstructure damage and correct the issues
- Replace microstructures before failure
- Change the part's design to prevent microstructure overstress or degradation
- Change the processes impacting the microstructure to prevent overstress or degradation

If you want outstandingly reliable plant and machinery, you cannot be maintaining and repairing them. You need equipment with parts that do not fail. That requires engineering designs, manufacturing methods, installation techniques, operating practices, and maintenance tasks in which component failure is not initiated. You need a reliability creation paradigm and not a failure-focused maintenance paradigm that finds and fixes problems.

Figure 9.8 highlights the difference between a failure-focused perspective and a machinery health–focused view. In failure-focused maintenance, you fix machinery problems as you find them. In wellness-focused maintenance, you create and sustain machine health so that there are no problems. To get wonderfully reliable plant and equipment, continually deliver the conditions that cause their health and well-being.

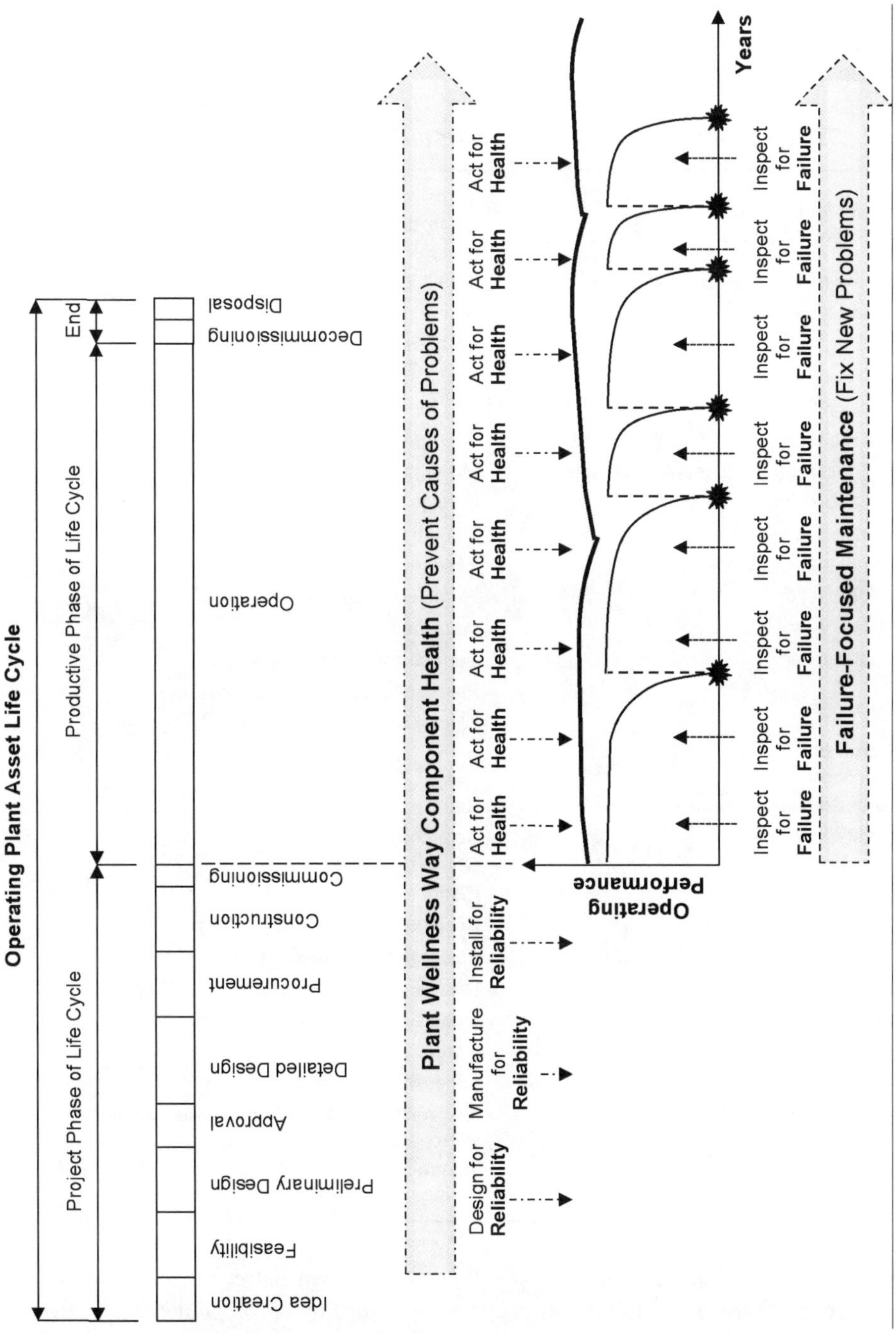

Figure 9.8—Failure-Focused versus Wellness-Focused Reliability Strategies

Physics of Failure Mechanisms and Their Failure Causes

There are just two ways that physical matter can be parted: (1) materials of construction fail when microstructures are destroyed by physical means, or (2) the atomic bonds disintegrate. The principal ways to destroy atomic bonds and microstructures are listed in the left-hand column of Table 9.2. These are called the Physics of Failure factors.

Table 9.2—Physics of Failure Factors and Failure Causes Guidewords

Principal Mechanisms of Solids Microstructure Failure	**Component Manufacturing and Rebuild Events**
Compressive force overload	Metallurgy error
Tensile force overload	Formulation error
Shear force overload	Process conditions error
Cyclic stress fatigue	Chemical composition error
Melt molecular structure	Interference fit tight
Crack in molecular structure (dislocation)	Interference fit loose
Material missing from molecular structure	Misalignment
Material ripped from molecular structure	Foreign inclusion
Wrong atoms in molecular structure	Thin cross-section
Electromagnetic radiation	Weld penetration
Chemical reaction	Flame cut stresses
Crystal lattice attack	Machined surface stresses
Electrical discharge	Surface finish stresses
	Pressed/formed bend stresses
	Weldment stresses
Additional Mechanisms of Plastics Microstructure Failure	Surface porosity
Depolymerization decomposition	Burrs
Ultraviolet radiation	Scratches
	Gouges
	Residual metal chips
	Residual nonmetallic dirt particles
	Absence of specified feature
	Incorrectly located feature
	Incorrect machining of a required feature
	Incorrect machining tool used
	Excessive surface waviness
	Incorrectly assembled componets due to interchanging of parts
	Incorrectly assembled due to misorientation or mislocated of parts

The event and condition columns in Table 9.2 record known causes of atomic bond breakage and microstructure destruction. Collectively, they are called the Physics of Failure cause mechanisms. You can add new triggers of material-of-construction failure to the list as they are identified in your operation. The table is a repository for corporate knowledge of what causes equipment failures.

A comprehensive Physics of Failure Factors and Failure Cause Guidewords list can be found in the downloadable materials accompanying this book.

To prevent material-of-construction failure, we need to ensure that the situations and events that can cause atomic structures to fail do not arise. The elimination of "the cause of the cause" of failure is a powerful concept for creating reliability because it lets you identify, deliver, and sustain the conditions needed for long-lasting microstructure health. Whereas other asset management methodologies focus on identifying failures, the Plant Wellness Way focuses on producing the conditions needed for outstandingly long material-of-construction lifetimes.

Physics of Failure Factor Analysis for Reliability Strategy Selection

Identifying health-focused reliability strategies requires strategy selection methods that get you away from equipment failure spotting and toward lifetime health creation. Physics of Failure Factors Analysis (POFFA) is such a technique. It is part of a suite of investigative techniques making up the Physics of Failure Reliability Strategy Analysis methodology. It identifies all of the ways in which a component can physically fail and then lets you select the most effective life-cycle strategies for maximizing its reliability. It is a scientifically based, holistic, and cost-effective methodology for life-cycle profit maximization by proactively ensuring that components have lifelong health and wellness.

The POFFA approach analyzes failures at the microstructure and atomic levels. The technique identifies how many ways the microstructure of a machine component can be physically destroyed. Once all microstructure failure mechanisms are listed, you then identify all of the causes of each mechanism during the part's lifetime. With the causes known, you place defect elimination and failure prevention activities throughout the life cycle to proactively prevent each cause.

In POFFA, you don't use failure modes. Using failure modes installs unreliability into equipment by design. Instead, the analysis puts you in the time period *before* failures arise so that by design you can prevent the situations that cause a part's microstructure to fail. You give components' materials of construction operating longevity. POFFA identifies all of the causes of atomic structure and microstructure destruction that a part can suffer and guides you to select and use those practices that create and sustain health and minimize stress. You do not address failure—you eliminate it. You determine the conditions needed for exceptional component reliability and establish reliability creation causes throughout the life cycle to produce long equipment operating lifetimes.

POFFA provides a list of issues to address in order to get maximum parts reliability. It is the investigative phase of a Physics of Failure Reliability Strategy Analysis that leads you to choose the strategies and practices to use across the life cycle to prevent microstructure failure and thus deliver the most reliable parts and equipment for your operation.

Doing a Physics of Failure Factors Analysis

Physics of Failure Factors Analysis is a particularly effective means to scrutinize any equipment part in full detail. The technique can analyze every part down to its microstructure. You use it like an "intellectual microscope" to examine a component at any location. You "point" the microscope at the region of structure you want to interrogate. For that location, you identify all of the principal mechanisms that could destroy the microstructure. In Figure 9.9, the intellectual microscope is focused on the rolling element of an electric motor bearing. It could have been pointed at other parts of the bearing—the inner ring, outer ring, retaining cage, or even the lubricant between the surfaces.

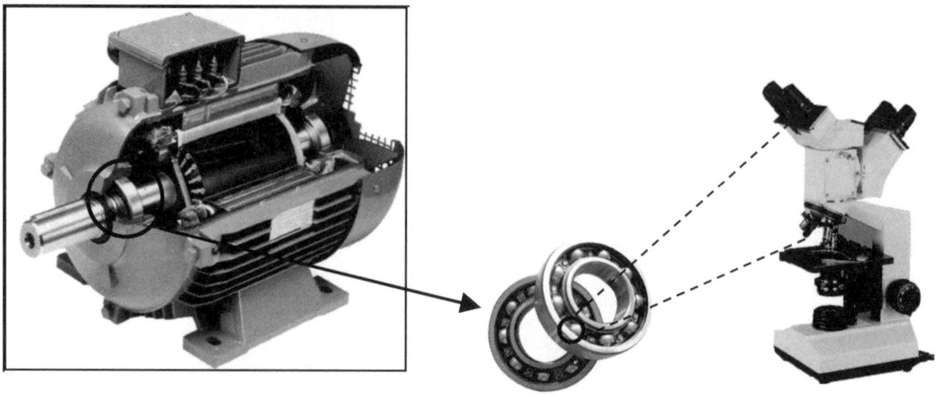

Figure 9.9—Analyzing an AC Motor Deep Groove Ball Bearing

POFFA is performed on all critical components by a person knowledgeable in the design and use of the component and its parent equipment. An experienced discipline maintenance, electrical, electronic, mechanical or structural engineer, or industrial engineer; a discipline design engineer; or a career-long serving electrical or mechanical maintenance discipline supervisor will have the level of engineering and equipment knowledge needed to do the analysis. If there is no one with such expertise and experience available, a team containing the necessary range of knowledge and know-how is assembled. Once the analysis is completed by the appointed person or team, a second competent person or team in the same discipline must be "paralleled" to do a review of the findings for accuracy and completeness.

The analysis starts with a spreadsheet in which you list the complete set of "Principal Mechanisms of Solids Atomic or Microstructure Failure" factors for the part being analyzed. In the example POFFA in Table 9.3 for a rolling element in a ball bearing, each principal material-of-construction failure has a series of questions about its risk to the part's microstructure. The question choices are restricted to yes or no answers to identify whether a risk truly exists. When the answer to a microstructure destruction effect question is no, the factor is no longer considered. When the answer to a question is yes, the factor remains valid and the next question is asked. If the failure factor turns out to be a real issue, it is necessary to identify each of its failure mechanisms and all their causes so that effective strategies and controls can be established during the component's life cycle.

Eight Life-Cycle Questions to Answer

Your business is built from people, processes, and plant interacting together in some coordinated way to produce a product or service for paying customers. Plant and equipment risks prevent companies from being world-class operations by continually causing breakdowns, waste, and losses. You use the eight life-cycle questions to identify the risks to a critical component and where the causes and defects come from during its lifetime. It is a simple and quick way to find your future troubles.

Table 9.3—Physics of Failure Factors Analysis for Rolling Element in Deep Groove Ball Bearing

Equipment: Process Acid Pump Drive Motor

Critical Component: Rolling Element of DE Motor Bearing

Material of Construction: Case hardened steel chrome alloy, ductile inner core

	Microstructure Destruction Effects				Eight Life-Cycle Questions (including effects of component aging)								Physics of Failure Cause Mechanisms and Conditions			
Mechanisms of Atomic or Microstructure Failure (horizontal, vertical, axial)	**Could Mechanism Destroy Molecular Structure?**	**Can Mechanism Occur to the Component?**	**Can Mechanism Result in Component Failure?**	**Is Failure of Component Critical to Equipment?**	**1. Business-Wide TDAF Costs Are Unacceptable?**	**2. Frequency of Failure Is Unacceptable?**	**3. Can Microstructure Be Overstressed?**	**4. Can Microstructure Be Fatigued?**	**5. Can Microstructure Be Degraded?**	**6. Can Human Error Cause a Failure?**	**7. Can Business Process Design Allow a Failure?**	**8. Can an Engineering Design Decision Allow a Failure?**	**What POF Causes Produce the Principal Failure Mechanism?**	**What Must Be Done to Prevent Each Cause of Microstructure Failure?**	**How Are the Proper Component Health Conditions Created?**	**What Actions Must Be Performed to Ensure Component Health?**
Compressive force overload	Yes	Yes	Yes	Yes	Yes	Yes	Yes	Yes	Yes	Yes	Yes	Yes	1-Housing deformation (bend, twist, squash) 2-Shaft misalignment 3-Outer ring shrinkage 4-Inner ring expansion 5-Inner or outer ring tight fit 6-Temperature differential	The bearing rolling element requires precision location and operation	1-Correct fits and tolerance 2-Correct thermal growth 3-No bearing housing deformation 4-Precision shaft alignment	1-Micrometer measurement of size, shape, and form of housing 2-Micrometer measurement of size, shape, and form of shaft journal 3-Correct lubrication selection 4-Correct lubrication performance 5-Duty load within bearing design boundary 6-Shafts aligned accurately
Tensile force overload	Yes	Yes	Yes	Yes	Yes	Yes	Yes	Yes	Yes	Yes	Yes	Yes	Shaft thermal growth	Run equipment at design temperature		
Shear force overload	Yes	Yes	Yes	Yes	Yes	Yes	Yes	Yes	Yes	Yes	Yes	Yes	Shaft thermal growth	Run equipment at design temperature		
Cyclic stress fatigue	Yes	Yes	Yes	Yes	Yes	Yes	Yes	Yes	Yes	Yes	Yes	Yes	1-Shaft misalignment 2-Housing deformation 3-Out-of-balance mass	Duty loads are below life fatigue stress level	1-Shaft precision aligned 2- Machine frame shape keeps components within design tolerance 3-Bearing housing of correct shape and form 4-Rotating components finely balanced	1-Micrometer measurement of size, shape, and form of housing 2-Micrometer measurement of size, shape, and form of shaft journal 3-Duty load within bearing design boundary 4-Shafts aligned accurately 5-Rotating parts balanced so stresses are below life fatigue values
Shock force overload	Yes	Yes	Yes	Yes	Yes	Yes	Yes	Yes	Yes	Yes	Yes	Yes	1-Mechanical impact 2-Shaft misalignment 3-Hammer impact			
Punch hole in molecular structure, e.g. wear particle	Yes	Yes	Yes	Yes	Yes	Yes	Yes	Yes	Yes	Yes	Yes	No	1-Foreign particle 2-Hammer impact			
Melt molecular structure	Yes	Yes	Yes	Yes	Yes	Yes	Yes	Yes	Yes	Yes	Yes	No	1-Electric current 2-Electric discharge 3-Metal to metal contact			
Crack in molecular structure	Yes	Yes	Yes	Yes	Yes	Yes	Yes	Yes	Yes	Yes	Yes	No	1-Housing deformation (bend, twist, squash) 2-Shaft misalignment 3-Outer ring shrinkage 4-Inner ring expansion 5-Inner or outer ring tight fit 6-Temperature differential 7-Metallurgy error			
Material missing from molecular structure	Yes	Yes	Yes	Yes	Yes	Yes	Yes	Yes				No	Metallurgy error			
Material ripped from molecular structure	Yes	Yes	Yes	Yes	Yes	Yes	Yes	Yes	Yes	Yes	Yes	No	1-Metal-to-metal contact 2-Lubrication failure			
Wrong atoms in molecular structure	Yes	No										No				
Electromagnetic radiation	No											No				
Chemical reaction	Yes	Yes	Yes	Yes	Yes	Yes			Yes	Yes	Yes	No	1-Corrosion 2-Product ingress			
Crystal lattice attack	Yes	Yes	Yes	Yes	Yes	Yes	Yes	Yes	Yes	Yes	Yes	Yes	1-Corrosion 2-Product ingress			
Electrical discharge	Yes	Yes	Yes	Yes	Yes	Yes	No	No	Yes	Yes	Yes	No	1-Lightning 2-Electric current earths through item			

Economic Justification: The answers to the economic questions determine whether a full analysis is required for a subassembly or part. When equipment TDAF costs, including opportunity costs, are low and the event frequency is low, so that the risk falls into the acceptable level on the risk matrix, the default choice is to adopt a run-to-failure strategy and rectify the damage in a timely manner. The two exceptions are if maiming or death of people, or destruction of the environment can occur. Even when the consequential costs of such events are affordable, they cannot be allowed to occur. If the economics or the safety or the environmental consequences of the situation are unacceptable, then the analysis continues.

1. Are the business-wide TDAF cost consequences of a failure acceptable?
2. If failure is acceptable, how frequently can it occur before it becomes unacceptable?

Physics of Failure Causes of Parts Failure: These questions highlight the many causes and combinations of causes of an equipment part's failure. There are hundreds, maybe thousands, of ways that can combine to create risks inside your machines. The best protection against equipment failure is to eliminate all possible causes and paths of parts breaking.

3. Can the microstructure be overstressed?
4. Can the microstructure be fatigued?
5. Can the microstructure be degraded?

You factor component age into the consideration through the growing risks that ongoing use causes your operation. Equipment parts accumulate stresses in their microstructure and "age," resulting in decreased strength. Parts that are in service for a long time suffer more environmental degradation of their surfaces. Consequently, the frequency of repairs and associated maintenance costs rise, and the equipment becomes progressively unsafe to use. When reviewing old assets, compensate for the effects of degradation and stress by increasing the frequency of failure of old components by a reasoned proportion based on the asset's time in service and how carefully the asset has been used and maintained during its operating life. The maintenance work history is evidence of past repair frequency. For wear-out parts, you need to increase the failure rate to compensate for an increasingly fatigued asset; for randomly failing parts, reflect the effect on reliability of the asset management culture present in the organization—a culture that is not yet world class will only get worse, and older assets will suffer higher rates of stress induced random failures. If you are fortunate to have kept good component failure data, then reliability engineering analysis can be done to quantify the effects of age and abuse to give you a more certain consideration of the rising risk with the time that assets are in use.

Organizational Factors in Parts Failure: Human error and weak business process design are the major categories in equipment failure. For whole-life protection, the mistakes attributable to people and process design must be prevented. It is necessary to learn where in the life cycle it is best to include useful error-proofing strategies.

6. Can human error cause a critical part to fail?
7. Can business processes allow a critical part to fail?
8. Can a design decision allow a critical part to fail?

Developing a Physics of Failure–Based Reliability Strategy

For each principal Physics of Failure factor identified, list all of the cause mechanisms of atomic or microstructure destruction noted in the event and condition columns that can produce the factor during a part's lifetime. You seek all possible lifetime causes of a component's material-of-construction failure so that they can be eliminated in all phases of the life cycle. Some cause mechanisms will arise many times in the life cycle. For example, whenever a part is installed, whether it be during original equipment manufacturing, scheduled overhaul, or breakdown maintenance repair, the same bad installation practices producing the same mechanisms can exist at all three times. Whether you do anything about a mechanism and what you do depends on the size of the risk to the business.

The Physics of Failure Factors Analysis makes you think through what you must do to keep component microstructure healthy. You are not after the root cause of failure. A failure can result from dozens of root causes, many of which may never be identified. You do not focus on rectifying a problem—instead, the focus is on not having the problem in the first place. The answers to the eight life-cycle questions will generate engineering, manufacturing, supply chain, maintenance, and operational actions to stop a microstructure failure mechanism by designing ways to prevent the causes of those situations in every phase of the life cycle. For each way that a part loses integrity, you introduce suitable means to prevent it. This approach lets you design asset management, operational, and maintenance processes and activities that drive out risks and increase equipment reliability to get failure-free service.

Once you have identified the ways your parts can be destroyed, you need to prevent those causes from arising. This is when you set your asset reliability strategy and select your plans to make your machines and equipment highly reliable. The POFFA outcomes feed the Physics of Failure Reliability Strategy Analysis, in which you choose the ideal reliability tactics and actions for each part.[4]

In a POFFA, you use the Stress-to-Process Model to recognize how your equipment parts' microstructures can be failed. You work up from the bottom of the risk dependency hierarchy to eliminate failure at all levels in an asset. Consequently, you design the right processes and build an organization that can produce the world-class reliability needed for endless Operational Excellence.

References

1. David Sherwin, "Introduction to the Uses and Methods of Reliability Engineering with Particular Reference to Enterprise Asset Management and Maintenance" (presentation, Perth, Western Australia, 2007).

2. Robert C. Juvinall, *Engineering Considerations of Stress, Strain, and Strength* (New York: McGraw-Hill, 1967).

3. Paul B. Price, reliability consultant and condition monitoring specialist, United Kingdom, 2014, e-mail correspondence.

4. An alternative way to use the Physics of Failure Factors to identify how component microstructures fail and arrive at viable solutions is the Deformation–Degradation Analysis introduced in the Appendix, which is included in the downloadable materials accompanying this book.

10

Chance Reduction Risk Management

You might think that it does not matter how you choose to reduce risk, but you would be wrong. The standard risk formula can take the following form:

Formula 10.1

$$\text{Risk} = \text{Consequence} \times \text{Chance}$$

Reduce the chance of an event occurring or lower the consequences of an incident, and risk is reduced. It seems as if either strategy—consequence reduction or chance reduction—will produce the same effect. Halving the consequences is as equally good as halving the chance. The math says it is. In reality, however, the two paths to reducing risk have totally different impacts on the prosperity of an organization. The application of basic accountancy is sufficient to explain why the best risk management path is to reduce the chance of failure and not its consequence.

Impact of the Choice of Risk Reduction Strategies

Figure 10.1 shows the "death by a thousand cuts" production breakdown model. This model is explained more fully in the document *The Instantaneous Cost of Failure*, included in the online materials accompanying this book.

Each breakdown causes a loss of production time and a business-wide cost surge. Companies using consequence reduction strategies minimize their losses by learning to fix breakdowns quickly. They hold lots of spare parts in store, set up a parts cache by machine, use "gun" repairers to fix things speedily, and change access to equipment to do fast repairs. Figure 10.2 shows the reduced production time loss when you follow a consequence reduction strategy. A comparison of Figures 10.1 and 10.2 confirms that reducing downtime increases profit. Losses are minimized when the plant gets back to production quickly. Consequence reduction strategies do reduce risk.

What's interesting about the consequence reduction model is that although costs are lower, there will be much frantic activity and "firefighting" as people go from one failure to the next. In these operations, you wait for failures because they may not happen. But there is a serious misunderstanding in this logic. One risk may not occur, but if you have 10,000 potential risks in your business, some will always arise. Today you might fight fires and put them out, but tomorrow different ones will blaze. It will be the same next week, next year, and for your whole career. If there are 100,000 risks in your operation, you have a never-ending nightmare of a business. In organizations that primarily use consequence failure management, things are always going wrong.

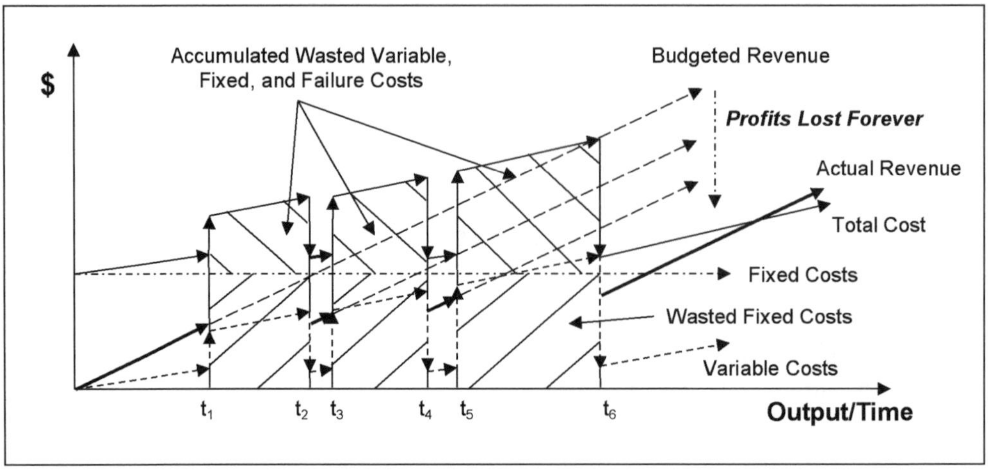

Figure 10.1—Effects on Profitability of Repeated Failure Incidents: Death by a Thousand Cuts

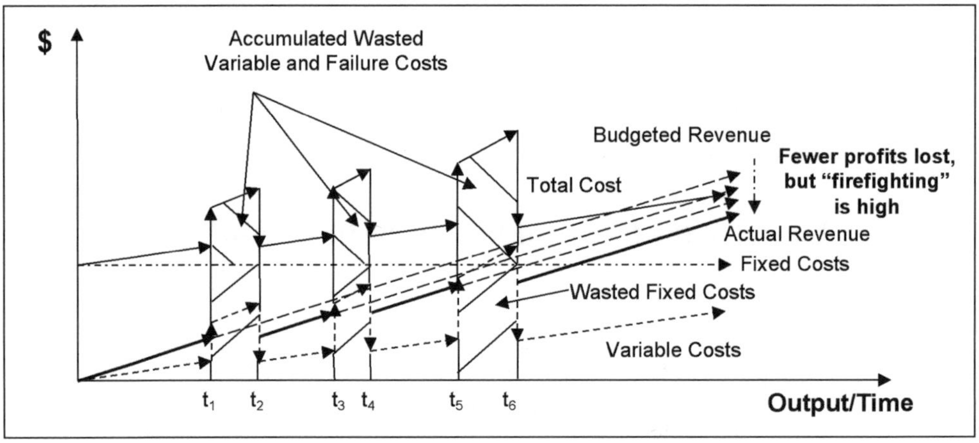

Figure 10.2—Effects on Profit by Reducing Consequence Only

Minimizing risk by reducing its consequence means accepting failure as a normal way of doing business. Sites that use such a strategy instill a reactive culture in their operation. If you walk about in these companies, you see that everyone is busy, but little of their time or effort adds value to the operation—the time spent fixing problems only adds more cost. The breakdowns repeat over and over. Reducing only the consequence of risk still makes work for everyone because reliability is no better. Instead of improving the business so that it can be more profitable, people waste time, money, and effort fixing failures and repeatedly correcting the same problems.

An alternative risk management strategy is to apply chance reduction techniques to proactively eliminate the possibility of failure. With this approach, you identify failure scenarios and prevent their causes. As more and more failure causes are eliminated, fewer production stoppages result. Figure 10.3 indicates the effect of using chance reduction strategy to reduce breakdowns from three in

Figure 10.1 to only one incident during the same period. Less profit is lost when there are fewer failures because TDAF costs are not incurred. Chance reduction strategy is far more lucrative than consequence reduction strategy because failures don't waste your resources and rob you of profit. Improving reliability by stopping opportunities to start failures and by making processes resistant and resilient to failure are profit-making strategies.

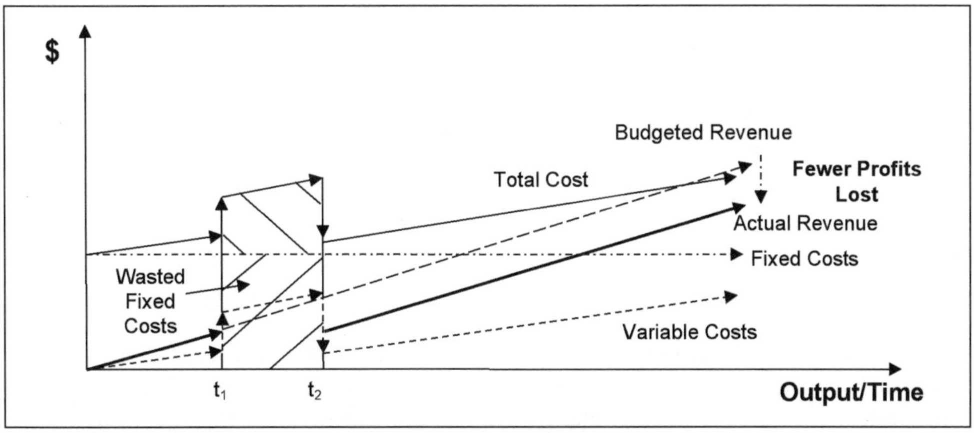

Figure 10.3—Effects on Profit by Reducing Chance Only

Your choice of risk management strategy matters because only chance reductions stop problems, whereas consequence reductions can never do that. They can lower your losses from an incident, but they can't create reliability so that there is no incident. Controlling consequences is important. The fire brigade, ambulance, police, insurance companies, personal protective equipment, and condition monitoring of machinery are all worthwhile consequence reduction strategies. They do not reduce the chance of failure, but they can save you fortunes in operating losses and stop a bad event from becoming a major disaster. Even though consequence management costs money, it makes a difference to the total cost of a risk, and there is a place for it in workplace safety, asset management, and maintenance management strategy.

A complete business risk management strategy is to use both chance reduction and consequence reduction to maximize profit. It is far better not to have a failure, but if one does happen, you need to quickly minimize its impacts. Your business processes need to be good at doing both strategies well. The benefit of using the combined strategy is evident in Figure 10.4, in which both lost time and failure frequency are reduced. When risks exist, pursuing a combined risk management strategy delivers the least profit loss. When no chance of risks exist, you don't have to do anything because nothing goes wrong. That makes chance reduction strategies the best business choices by far because they eliminate consequences and retain operating profit.

Table 10.1 lists some of the current methods available to address risk. The methods are classified as chance reduction or consequence reduction strategies. Several observations arise when viewing the two risk management philosophies. Consequence reduction strategies require a failure event to respond to. They use lagging indicators as triggers for action. They come into play at the end of the cause-and-effect path when no risk control options other than loss minimization and corrective actions are left. In contrast, chance reduction strategies focus on the early elimination of failure causes and

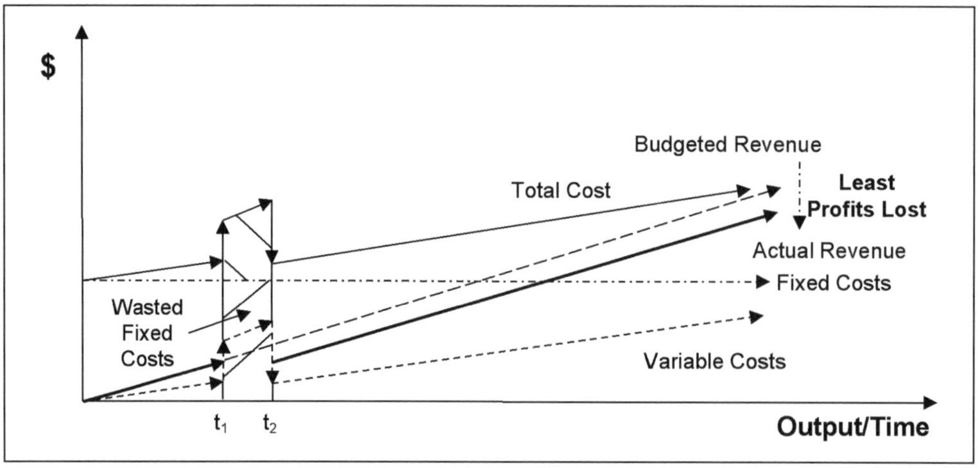

Figure 10.4—Effects on Profit by Reducing Both Chance and Consequence

on making business system changes to prevent or remove the opportunity for failure. They seek to eliminate and prevent the roots of risk. These methodologies work to improve business process success rather than improving failure detection methods. They expend time, money, and effort early in the cause-and-effect path to identify and stop problems so that the chance of failure is eliminated or at least drastically minimized.

Table 10.1—Risk Management Processes and Methods

Operating Risk = Consequence x Chance

Consequence Reduction Strategies	**Chance Reduction Strategies**
• Preventive maintenance • Corrective maintenance • Breakdown maintenance • Predictive maintenance o Nondestructive testing o Vibration analysis o Oil analysis o Thermography o Motor current analysis • Prognostic analysis • Emergency management • Computerized maintenance management system • Root Cause Failure Analysis • Key Performance Indicators • Risk-based inspection • Operator watch keeping • Financial accounting • Stores and warehouses • Maintenance engineering	• Engineering and maintenance standards • Design-out maintenance • Precision maintenance • Physics of Failure Reliability Strategy Analysis • Standardized operating procedures • Failure mode effect criticality analysis • Reliability Growth Cause Analysis • Hazard and operability study • Hazard identification • Training and upskilling • Quality management systems • Planning and scheduling • Continuous improvement • Supply chain management • Accuracy-Controlled Enterprise • DOCTOR • Total productive maintenance • More robust, durable materials • De-rate/oversize equipment
Reactive response to failure	*Proactive prevention of failure*

Both risk reduction philosophies are necessary for optimal protection. But a business with a chance reduction paradigm proactively prevents defects, unlike one with a consequence reduction focus, which will only fix defects. Those organizations that primarily apply chance reduction strategies set up their business to ensure a decreasing number of failures. As a result, they will get outstanding plant and equipment reliability and reap all of the business benefits that brings.

Power Law Implications

Equations of the risk and loss type are known as *power laws* and take the general form $x = zy^n$, where x is the outcome, z is an influential factor, y is a second influential factor, and n is the exponent. For the standard risk formula, n is assumed to equal 1. Power laws have particular properties. For example, they are "scale-free." In the case of risk, consequences are not linear; although one incident may cost only a few dollars, when it happens another time, it could cost an immense sum.

Power law outcomes are "typically a signature of some process governed by strong interaction between the 'decision-making' agents in the system."[1] This implies that risk does not arise entirely randomly; rather, it is affected by the decision makers present in interacting processes. These "influencers" change the chance of events toward the final outcome. Situations that follow power laws have a higher number of large events occurring than in a normal distribution. For risk, this means that catastrophic events will occur more often than they would by pure chance. In power-law-mirrored events, a few factors have huge impacts, while all the rest have little effect. When it comes to risk situations, a small number of key factors influence the likelihood of catastrophe. Control these, and you will increase your chance of success.

The left side of Figure 10.5 is a graph of the risk formula on a normal linear-linear graph.[2] The risk plots as curves. You develop the risk curves by keeping the value of risk constant and then varying the frequency and the consequence. Anywhere on a curve is the same risk ($1/event x 100 events/yr = $100/yr, or $100/event x 1 event/yr = $100/yr). The right side of the figure shows the log of the risk equation plotted on a $\log_{10}$-$\log_{10}$ graph. The fact that the base-10 logarithm of the risk equation plots as straight lines has special significance. It is an example of how power laws have an uncanny ability to reflect the real world. The insurance industry uses such curves to set insurance premiums because they closely represent what actually happens in human endeavors.

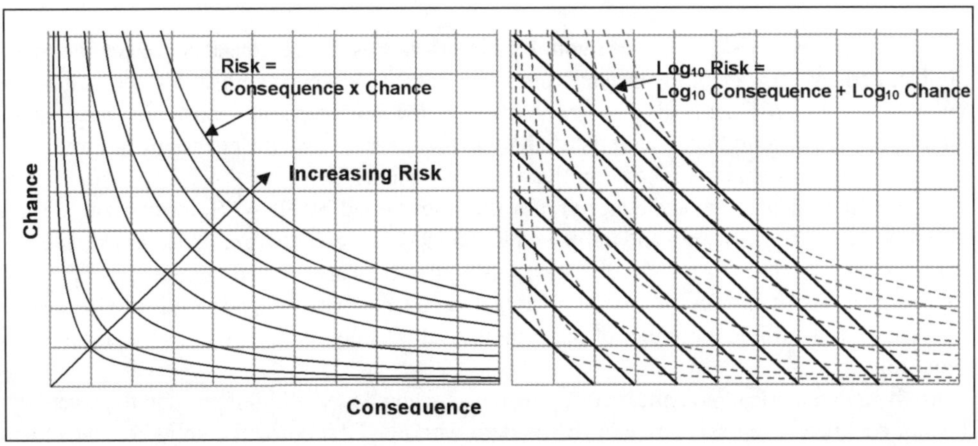

Figure 10.5—Risk Curves on a Linear Graph Become Risk Lines on a Log-Log Graph

Power laws that reflect the human world also tell us much about the situations from which they arise. Perhaps the most important understanding is the presence of "decision-making agents" within the system to which the power law applies. Philip Ball, in his book *Critical Mass,* points out that "physicists' long experience with power laws . . . leads them to believe that such laws are the universal signature of interdependence. A power law generally emerges from collective behavior between entities through which local interactions can develop into long-range influences of one entity on another."[3] Our simple risk and loss equations now take on far greater and menacing implications.

Risk reflects the presence of "agents" working in an uncoordinated fashion within a system. The effects of these independent agents move through the system in unknown ways, and the results of their uncoordinated—and most likely perfectly justifiable—efforts is to increase risk. We now have another reason chance reduction strategies are more successful than consequence reduction strategies in reducing long-term organizational risk: chance reduction strategies work on controlling the systems in a business. They coordinate people and information, thereby removing the random influence of independent agents acting unwittingly to increase the causes of failure and loss. Gradually and continuously, chance reduction strategies act to regulate and organize the efforts of these mysterious independent agents playing unscripted parts so that random actions and their effects are reduced and, eventually, removed.

Chance reduction strategies are the opposite of consequence reduction strategies, which consider risk and failure normal. Instead, chance reduction strategies reduce risk forever. Because they strike at the random behavior of the independent agents in a system, they align people, decisions, actions, and behaviors into an overarching system for achieving organizational outcomes using a specific, agreed-upon approach. Chance reduction strategies remove randomness and unplanned interactions from business systems.

It is in your organization's best interest, and will consistently generate the most profit for the least amount of work, to focus on chance reduction strategies. Consequence reduction strategies are still important and necessary—once a failure sequence has initiated, you must find it quickly to minimize its effects so that you lose the least amount of money. But consequence reduction will not help your organization achieve world-class success because it expends resources. Only chance reduction strategies reduce the need for resources because they proactively eliminate failure incidents through defect elimination and failure prevention.

Similarity between Safety Incidents and Equipment Failures

Nothing is certain with risk; it changes with the circumstances. Some consequences of risk will be negligible, perhaps only an annoyance at worst, while others will be severe, and some will be catastrophic. Controlling risk demands that an organization develop the culture and habits that guarantee continuous, rigorous compliance with risk reduction practices, or else the chance of failure will rise over time as systems degrade to the point that the worst can happen.

Figure 10.6 updates the accident pyramid first developed by H. W. Heinrich from his early 1900s workplace safety research and published in his 1931 book, *Industrial Accident Prevention: A Scientific Approach.* It shows that for every serious injury, there are many minor incidents preceding it. The incidents are themselves preceded by numerous events that are opportunities to become disasters. If there is a sufficient number of opportunities, it is likely that one will cause serious injury at some stage.

Analysis of historical industrial safety data not available in 1931 highlights that the safety pyramid is not completely representative of the modern workplace.[4] It correctly embodies the situation for minor injuries, where reducing the number of safety incidents leads to fewer minor injuries. But new data indicate that reducing the number of incidents does not correspondingly reduce the number of serious injuries. This is in line with the realization that risk is a power law and influenced by the

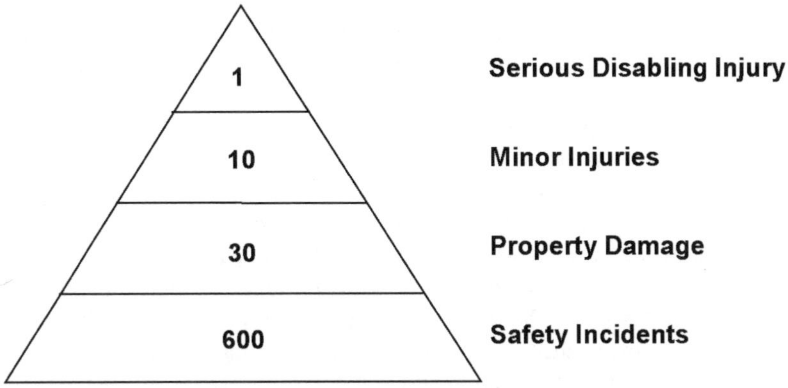

Figure 10.6—The Updated Heinrich Accident Pyramid

decision-making elements within a system. Serious injuries are not accidental but rather the result of systematic failure caused by unintentional outcomes of uncoordinated decision makers within the system. Current best practice in workplace safety is to identify serious injury-causing situations before they happen and immediately act to stop them from ever leading to a real injury.

There are equivalent industrial data for the number of equipment failure opportunities needed before a serious production breakdown occurs. The concept of an equipment failure pyramid also applies, with many small errors at the bottom leading to greater consequences higher up. Figure 10.7 depicts a failure pyramid for equipment failures.[5]

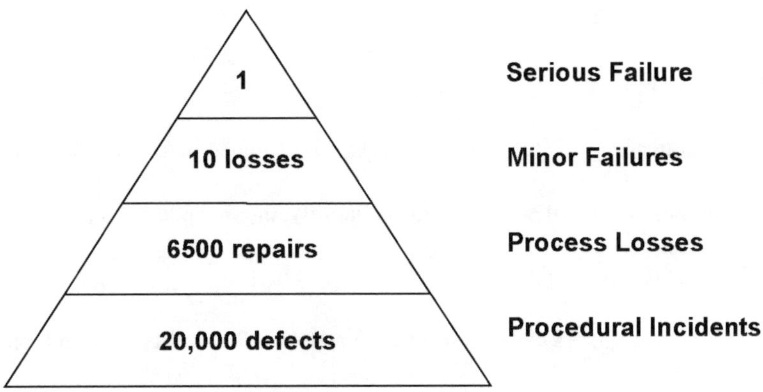

Figure 10.7—Equipment Failure Pyramid

The nature of risk, with independent actors all playing unscripted parts, changes the frequency with which situations arise. This implies that basing risky decisions on things not changing for long periods of time is fraught with danger. It is highly unlikely that frequency will remain constant because factors that are unknown and unknowable caused by decision-making agents are always altering the future. Risk's scale-less quality allows small concerns to cascade into major problems from the same events. What worked for us one day to prevent a failure may not work the next day because failure has found a different route. Our best protection against risk is to be vigilant in looking for its presence—look for its warnings, proactively eliminate the chance of defects, imagine the hazards and remove them, and be prepared to respond with the right knowledge and skills when risk finds new ways into the organization.

As with the accident pyramid, the failure pyramid reflects a power law, and stopping minor failures does not prevent catastrophic failures. Catastrophic loss is not controllable until the random decision-making elements in a system are regulated. Like minor safety injuries, minor equipment failures can be reduced by preventing the numerous and frequently occurring defects and errors that precede them. But in order to address catastrophic failures, you must intentionally imagine the worst outcomes and put into place measures to prevent them from ever happening. Physics of Failure Reliability Strategy Analysis uses that logic. Proactive measures are put in place throughout the life cycle to stop or lower the chance of risks arising later. You intentionally prepare your business by embedding standardized risk elimination and risk control practices in life-cycle processes so that randomness is restricted and situations that can precipitate disaster are eradicated.

There is one more concept regarding risk that is worth understanding and further justifies managing risk by chance reduction rather than consequence reduction. Serious risk events require many occurrences to coincide. Catastrophic events—those in which lives are lost or great costs result—do not often happen. A catastrophic loss requires the failure of a number of overlapping protective systems. A bad incident happens when situations align in such a way that the incident becomes possible: an opportunity for disaster now exists where it should not. Your best protection is to use reliable systems designed and built to scream that the right practices are not being done well before the catastrophe happens.

Three-Factor Risk Analysis

The standard risk formula has two components: consequence and chance. The complete risk equation consists of three components: consequence, opportunity, and chance. The relationship is shown in the following formula:

Formula 10.2

Risk ($/yr) = Consequence ($) x [Opportunity (/yr) x {1 – Chance of Success at Each Opportunity}

By replacing the "chance of success" with reliability, the formula becomes,

Formula 10.3

Risk ($/yr) = Consequence ($) x [Opportunity (/yr) x {1 – Reliability at Each Opportunity}

Reduced to a simpler form, we can write risk and its three components as follows:

Formula 10.4

Risk ($/yr) = Consequence ($) x [Opportunity (/yr) x {1 – Reliability}]

The risk you carry is directly proportional to the size of the event consequence (bigger losses bring greater risk) and the number of opportunities for the event to occur (more opportunities to fail means more risk) and negatively proportional to reliability (higher reliability brings lower risk). Risk

Operating Risk = Consequence of Failure x [Opportunity to Fail x (1 – Reliability)]

Consequence Reduction Strategies	Opportunity Reduction Strategies	Reliability Improvement Strategies
Strategies presume failure event occurs and act to minimize consequent losses	**Strategies prevent opportunities for the causes of a failure event to arise**	**Strategies reduce probability of failure initiation if opportunity is present**
• Preventive maintenance • Shutdown maintenance • Predictive maintenance o Nondestructive testing o Vibration analysis o Oil analysis o Thermography o Motor current analysis • Total productive maintenance • Prognostic analysis • Emergency management • Computerized maintenance management system • Key Performance Indicators • Risk-based inspection • Operator watch keeping • Financial accounting • Logistics, stores, and warehouses • TDAF costing • Maintenance engineering	• Physics of Failure Reliability Strategy Analysis • Accuracy-Controlled 3T practices • Design and Operations Cost Totally Optimized Risk (DOCTOR) • Reliability Growth Cause Analysis • Engineering/maintenance standards • Statistical process control • Degradation management • Lubrication management • Risk analysis • Hazard and operability study • Hazard identification • Failure design-out maintenance • Failure mode effects analysis • Root Cause Failure Analysis • Precision maintenance • Precision operation • Training and upskilling • Quality management systems • Planning and scheduling • Continuous improvement • Supply chain management • Reliability Engineering	• Precision maintenance • Training and Upskilling • Oversize/de-rate equipment • More robust, durable materials • Segregation/separation • Controlled atmosphere environment (e.g., +ve /-ve pressures, explosion-proof atmosphere)
Done to reduce the cost of failure	**Done to reduce the frequency of failure**	

Table 10.2—Risk Management Options with the Three Factors

can be driven to zero if there are no consequences from a bad event, if there are no opportunities for a bad event to occur, or if there is perfect reliability, meaning that no failure can happen. You can lower risk by using three different techniques—consequence reduction, opportunity prevention, and reliability improvement.

In Table 10.2, the two lists in Table 10.1 are broken down into the three risk categories: consequence, opportunity, and reliability. Having three components provides a wider perspective on your risk reduction options and lets you pick more effective solutions for a situation. With three-factor risk analysis, you open up more possibilities to address a risk innovatively and simply, as your thoughts are not constrained by considering only two factors.

Opportunity elimination and reliability improvement are the prime risk control principles used in the Plant Wellness Way methodology. The use of consequence reduction techniques is not preferred because they do not remove the causes of risk, while you mistakenly believe that you have controlled it.

References

1. Philip Ball, *Critical Mass: How One Thing Leads to Another* (New York: Farrar, Straus and Giroux, 2005).

2. Peter Buckland, extract from "Boss, We Need a New Switchboard" presentation to the Australian Asset Management Council, 2005.

3. Ball, Critical Mass, 324.

4. Miguel Angel Mariscal Saldaña, Susana García Herrero, Miguel Angel Manzanedo del Campo, and Dale O. Ritzel, "Assessing Definitions and Concepts within the Safety Profession," *International Electronic Journal of Health Education*, vol. 6 (2003): 1–9.

5. Winston Ledet, *The Manufacturing Game* (Humble, TX: Ledet Enterprises, 2002).

11

SELECTING RELIABILITY STRATEGY

Equipment is not reliable because you maintain it. Maintenance has only a little to do with getting high equipment reliability. All industrial operations in the world maintain their equipment, yet the vast majority suffer from poor plant reliability. Even doing condition monitoring or preventive maintenance according to a planned schedule will not deliver highly reliable machines. High reliability is mainly attributable to factors other than the maintenance you perform on your equipment.

The maintenance normally undertaken to care for operating plant arises from the following:

1. Core maintenance activities required by the equipment design
2. Core maintenance activities required by the business process in which the equipment is used
3. Additional maintenance activities resulting from equipment failures

Nowhere in the list is the business imperative to deliver high reliability. Reliability is not a result of maintaining and repairing your equipment. Every maintenance intrusion produces downtime—the opposite of reliability. The right maintenance focus is producing uptime. You want your maintenance crew to focus on creating and sustaining the ideal conditions for the health and well-being of all vital equipment parts so that they don't need maintenance.

There's a story that came out of the power generation industry on the east coast of Australia about a power station's decision not to overhaul a steam turbine for more than five times longer than recommended by its manufacturer.

It passed that three years after a new steam turbine electrical power generation set was installed, its manufacturer knocked on the power plant manager's door to let him know it was time to do the recommended overhaul. The manager asked his plant maintenance engineer's opinion. They looked at the operating performance data, the trends from the inbuilt condition monitoring sensors, and the equipment's maintenance history and decided not to do the overhaul because there was nothing wrong. The manufacturer told the plant manager that the warranty was voided, and he would wait for a call to repair the breakdown when it happened.

Three years later, without a call during that time, the manufacturer returned to the power generator to advise that it was now time to do the overhaul. The plant manager and maintenance engineer looked at the last six years of operating performance data, the trends from the inbuilt condition monitoring sensors, and the equipment's maintenance history and decided not to do the overhaul again because there was nothing wrong. Three years later, the manufacturer returned and was sent away. Then again three years after that, and again after that. It was 17 years before the turbine was overhauled for the first time, and that was only after there was evidence that an overhaul was necessary.

When you have a successful plant reliability strategy, your maintenance program prevents the causes of equipment failures. Once the focus is on creating healthy parts inside of assets instead of caring for sick assets, a different list of maintenance work results.

1. Core reliability activities required for failure-free equipment
2. Core reliability activities to optimize equipment performance in the process in which it is used
3. Additional reliability creation activities to eliminate all equipment failures

Maintenance Is a Risk Control Activity

Removing perceived operating risk is seen by some as sufficient justification to do maintenance.[1] Maintenance tasks are initiated in the belief that doing maintenance will prevent a risk. Condition-based and preventive activities are wrongly founded on the question, how can my machine fail? Maintenance will end up carrying out inspections and intrusive actions on perfectly healthy equipment. This approach to maintenance selection wastes resources, effort, and money. If instead we ask what failures the equipment will actually experience, the answer will be, fewer than what you think might happen! Not all the systems and components of the equipment shown in Figure 11.1 will fail, although they could.

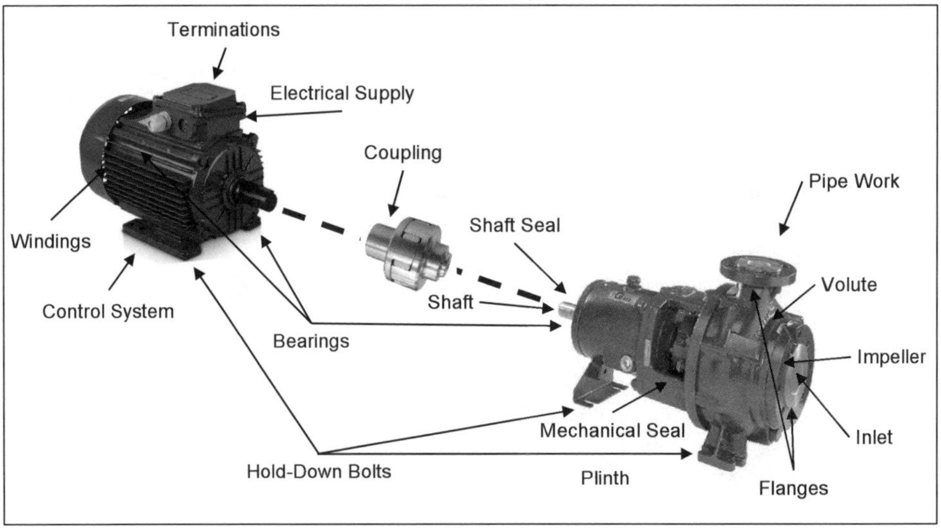

Figure 11.1—Equipment Can Fail in Many Ways

Because a maintenance strategy developed from an "all that may fail" approach primarily focuses on imagined failures, the maintenance performed does not match what is required to prevent the real causes of risk. Figure 11.2 shows what happens in the maintenance management system when real risk is not considered during the maintenance strategy selection process.

It is the actual threat that is the essential element in choosing the management strategy and actions to address operating risk and workplace safety. The requirement to assess where operational risks actually come from cannot be left undone because it is those failure events that occur that will

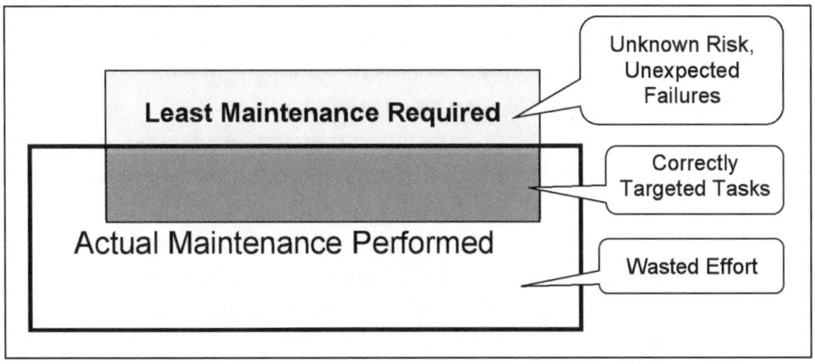

Figure 11.2—Doing the Wrong Maintenance Is Wasteful

put life and the business in danger. The expenditure of maintenance dollars on operating risk management (e.g., condition monitoring, process control, etc.) needs to be directly related to the probability of failure events and their consequent effects. This is how to be sure your maintenance expenditure gets the best return for the investment.

Standard risk assessment methodology applies to all risk situations. The risk management phases as normally applied to plant and equipment maintenance are shown in Figure 11.3.

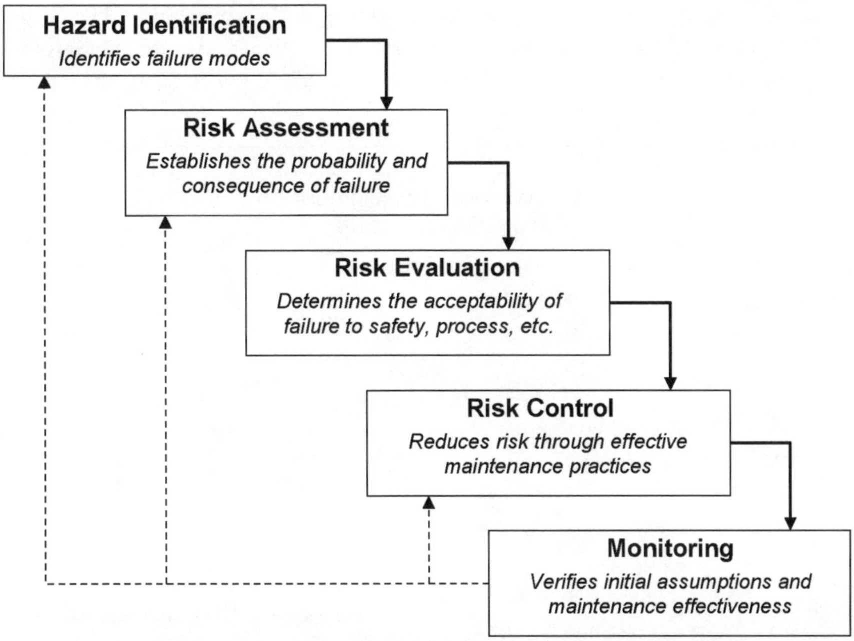

Figure 11.3—The Application of Risk-Based Principles to Maintenance

This risk review approach to maintenance activity selection aims to identify the ways in which an item of equipment may fail. Then, depending on the criticality of the asset, consider whether it is possible to detect and measure how the failure process develops (from which predictive maintenance arises), or consider its previous failure history and use hindsight to identify whether age fatigue or normal use will lead to failure (from which preventive maintenance arises). Those failures having no operating, safety, or environmental effects are allowed to run to failure (from which breakdown maintenance arises) and then repaired.

Frequently, reasonable judgments based on experience can be made without the rigor and expense of exhaustive risk analysis. Sometimes a formal math-based quantitative risk assessment using event probabilities is necessary, and decisions are made based on those outcomes. This is often the case in statutory-related risk management.

The decision tree shown in Figure 11.4 is typical of the experience-based maintenance activity selection processes generally used in industry. If the answer is that no evidence of failure is detectable, then, depending on the criticality or risk, either planned preventive maintenance, functional tests, or breakdown maintenance is applied. If the answer is yes and the criticality justifies the cost, then condition-based maintenance will be applied. If the answer is yes but criticality does not justify it, then planned preventive or breakdown maintenance is chosen. The approach thus far requires that every item of plant (system, machine, component) be reviewed, criticality considered, and a decision made about the maintenance it will get—repair on failure, scheduled replacement, proof test, or condition-based failure detection. In most operations, the maintenance actually performed is aimed at failure detection (condition monitoring) and failure correction (repair before failure). As a result, maintenance activities are focused on the equipment's current failure state—even though such activities are not effective at preventing failures from arising in the first place.

Targeting maintenance to critical operating equipment failures focuses on failure detection strategies that identify when a risk of failure exists. Because maintenance is now better aligned with risk, the maintenance management system delivers the outcome shown in Figure 11.5. A good proportion of the activities undertaken will reduce risk, but not all plant and equipment problems will be prevented and a significant proportion of the maintenance effort will continue to be pointless.

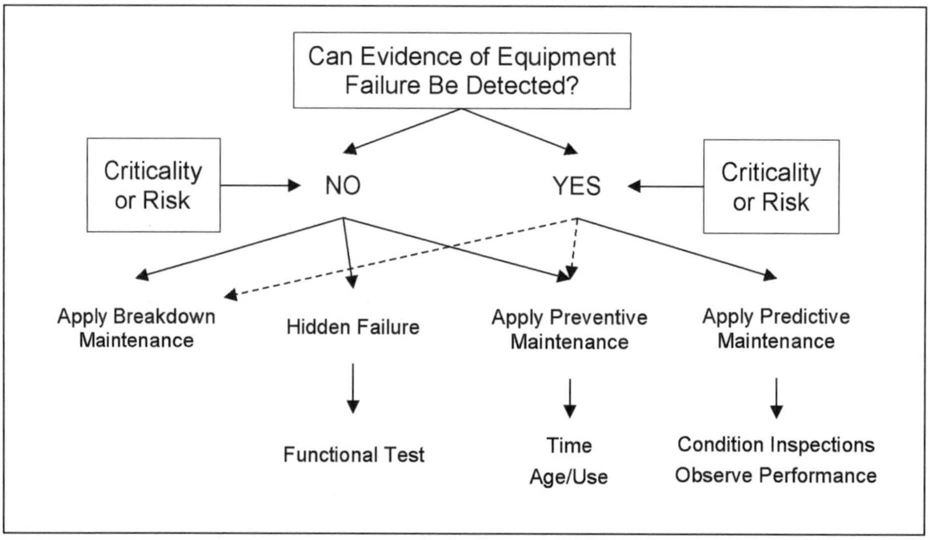

Figure 11.4—Common Risk Decision Tree for Maintenance Tasks

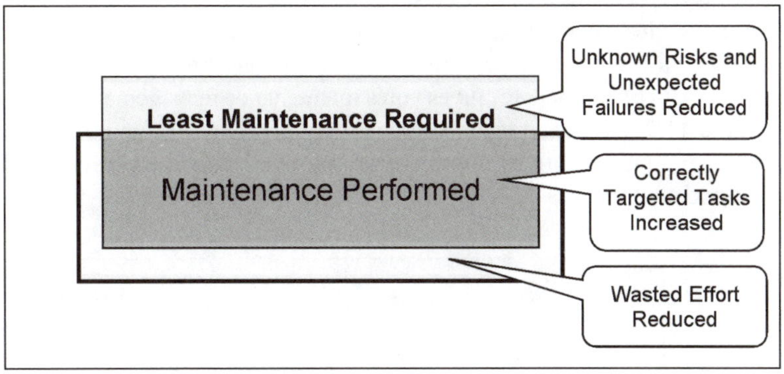

Figure 11.5—Failure Detection Focus Minimizes Maintenance

The Failure Prevention Focus of Plant Wellness

The Plant Wellness Way takes a different approach to protecting plant and equipment—it asks, what needs to be done to keep component microstructure healthy and in full strength condition? The risk assessment is done early in the life cycle, as indicated in Figure 11.6, and identifies the necessary

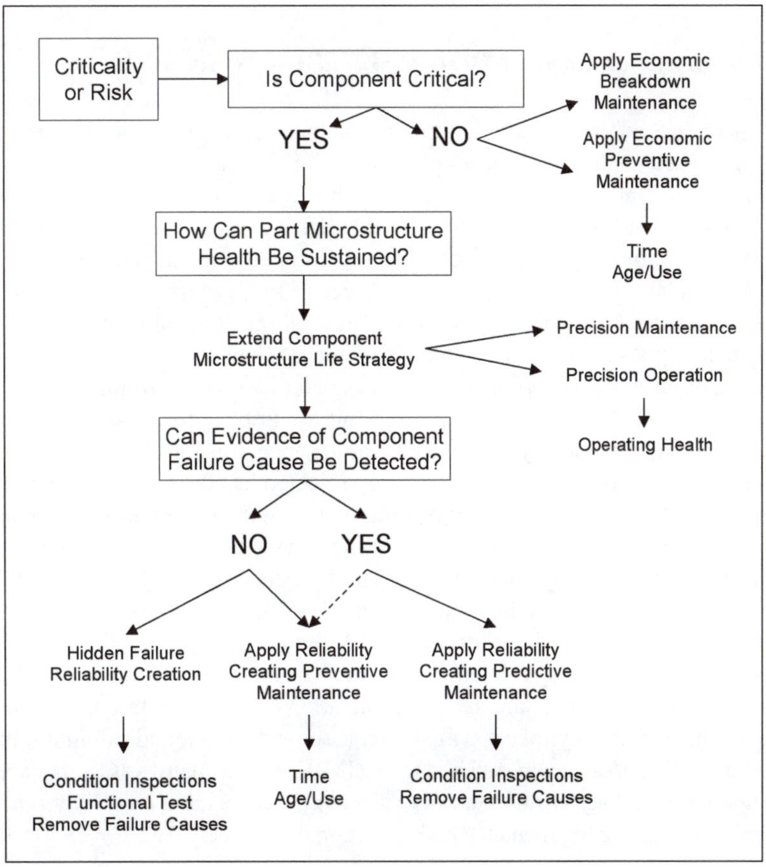

Figure 11.6—Risk Decision Tree for Plant Wellness Tasks

risk management activities that must be introduced throughout an equipment component's lifetime to ensure the least operational risk when the equipment is in use.

Using a life-cycle defect elimination and failure prevention strategy focused on ensuring the endless health of components gives you a maintenance management system with the outcome shown in Figure 11.7. Now maintenance proactively prevents risk and eliminates its causes. Every task brings operational benefit by improving uptime and maximizing throughput from highly reliable machines and equipment.

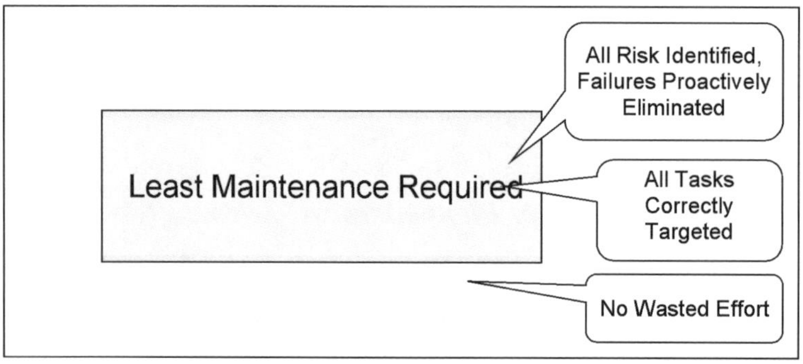

Figure 11.7—Maintenance Minimized to Sustain Equipment Health

Maintenance Cannot Deliver World-Class Reliability

Figure 11.8 is representative of the maintenance strategies available and the plant uptime they can achieve in a continuous process plant operation. It hints at the answer to getting highly reliable plant and equipment. Point A, near 90% availability, can never be reached in reactive operations. These unfortunate businesses are doomed to always have costly maintenance and poor production performance. A combined strategy of preventive, predictive, and planned maintenance might achieve around 90% availability. These businesses will get better production performance and lower maintenance costs because their maintenance efforts are well coordinated with production needs. Of the standard methodologies currently available, combining reliability growth, precision maintenance, and defect elimination strategies brings the greatest operating profit margin.

Point B shows that availability—and, by implication, equipment reliability—has little to do with how much money you spend on maintenance. It is the choice of strategy that matters most. For identical maintenance costs, the strategy selected has the greatest effect on production and maintenance performance. Because Industrial and Manufacturing Wellness requires you to starts with a fully designed, life-cycle-long, and business-wide "system of reliability," it is unlike any existing enterprise asset management strategies and produces exceptional productivity almost immediately.

Figure 11.9 highlights that the Plant Wellness Way gets you to world-class performance far faster than any other viable strategy. IMW produces highly reliable plant and machinery because it is a holistic business system containing the best methods and the most effective solutions by design. It proactively identifies and removes operational risks from throughout the life cycle to get utmost equipment reliability, plant availability, asset utilization, and minimum production unit cost from the start. The board room decision to use IMW as a company's asset management and reliability creation methodology ensures that its business processes and reliability practices will quickly create and bring lasting maintenance and operational success.

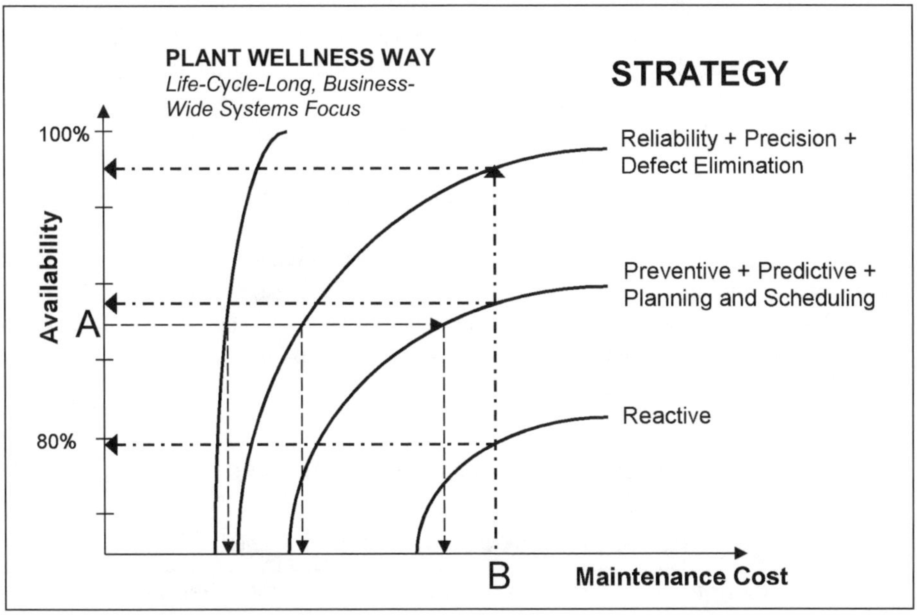

Figure 11.8—Availability and Maintenance Cost Is a Result of Asset Management Strategy

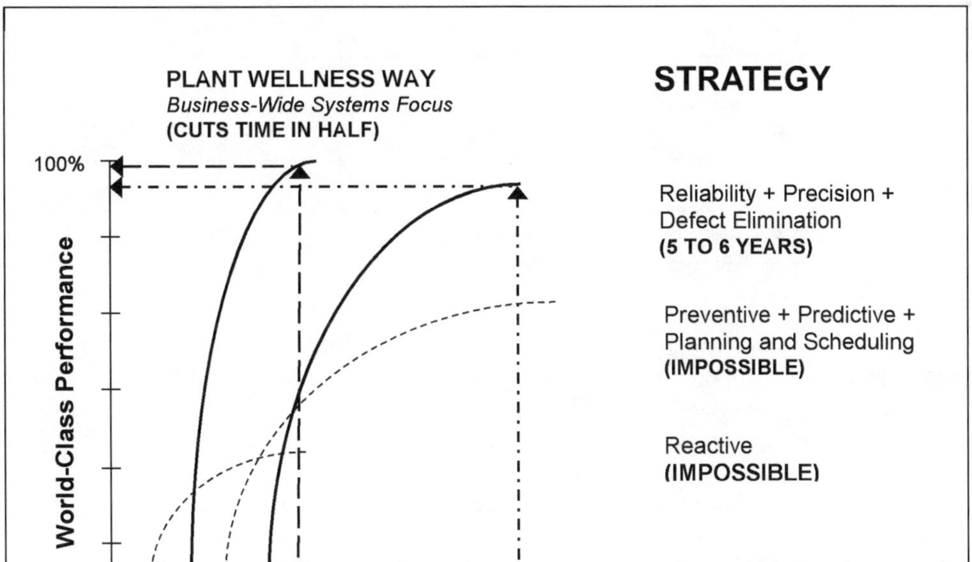

Figure 11.9—Speed to World-Class Performance Is a Result of Asset Management Strategy

Table 11.1 lists the number of days of lost production in a continuous operation for different values of availability. To go from a reactive operation with 80% availability to a site with 90% plant availability means halving the number of days lost to downtime. To go from 90% to 95% availability requires downtime to be halved again. To get to 99% availability from 95% means regaining another 14 days of downtime. It is impossible to halve current downtime losses using a company's current processes. The downtime that an industrial operation suffers is the result of the processes it uses. Halving your downtime requires using life-cycle and business processes that inherently produce the necessary uptime. To go from 80% availability to 90%, then to 95%, and then to 99% means a total replacement and rebuild of the vision, systems, paradigms, knowledge base, and skill sets used in a company. For each jump in availability, the organization must first develop, and then properly use, the right reliability creation and asset management processes with the capability to make and sustain the next jump. Only when an organization recreates itself and builds effective processes to achieve each level of success can the subsequent results be gained.

Table 11.1—Downtime versus Availability

For Continuous Operation Plant		
Availability (%)	**Downtime** (days/yr)	**Uptime** (days/yr)
80	73	292
85	55	310
90	37	328
95	18	347
98	7	356
99	3.7	361.3
99.5	1.8	365.2
99.9	8.8 hrs	364.7

Reactive maintenance is an easy but disastrous business decision for an organization to make. It leads to poor equipment utilization, much production downtime, and high operating and maintenance costs. Unless your products sell at high premiums and you can easily afford the excessive operational costs of using reactive maintenance practices, your business will continually struggle with high production losses and unsatisfactory operating profit.

Combining preventive, predictive, and planned maintenance is expected to lead to higher plant availability and improved operational performance. This maintenance strategy involves being prepared and organized to do the necessary rectifications before there is an equipment breakdown. Replacing parts with new ones before they fail can bring higher equipment availability from the restored equipment service life. This approach minimizes plant downtime and reduces equipment maintenance costs by preventing unplanned production loss. But planned outages stop the production of saleable product. If there were no outage, you would have made product to sell. Maintenance shutdowns and turnarounds may be planned and scheduled to minimize lost production time, but the total cost includes the lost opportunity to make more operating profit.

Another drawback of planned and scheduled maintenance is that the work performed may not actually be effective in lifting plant availability. You can have a good and efficient maintenance planning and scheduling process, yet the work done on an asset is shoddy, wrong, or damaging to component health. Excellent maintenance planning and scheduling that makes efficient use of resources and production time is a requirement for getting high plant availability, but it cannot create reliability. The maintenance maybe planned and scheduled, but it is the precision quality of the work performed that causes the equipment to be reliable.

The reliability growth, precision, and defect elimination strategy is not really a maintenance strategy; it's a reliability creation strategy. You embed reliability growth, precision quality, and defect prevention practices into a plant reliability management system that extends across the asset operating lifetime. That choice leads to the use of practices and methods that produce outstanding plant and equipment availability because assets are never allowed to fail. You use effective, economic preventive, predictive, and planned maintenance activities focused on sustaining asset health and not on finding faults to fix. IMW delivers effective operating plant reliability management strategies and requires good maintenance planning and scheduling to be done, but its focus on component microstructure health creates a life-cycle-long system of reliability in which operating risk is eliminated at every phase of the life cycle, not only in the operating phase.

Physics of Failure Reliability Strategy Analysis

Figure 11.10 shows how the Plant Wellness Way is used to select an effective strategy mix for proactive risk reduction of equipment parts. Operating criticality determines which equipment component failure brings too much risk. For these parts, you identify the causes of their material-of-construction destruction using Physics of Failure Factors Analysis.

A Physics of Failure Reliability Strategy Analysis takes the POFFA failure causes for each critical component and identifies how to get exceptional lifelong equipment reliability by preventing all of the causes of failure. With Physics of Failure Reliability Strategy Analysis, you choose chance reduction strategies and defect elimination solutions to prevent failure mechanisms and embed them into your business and plant life-cycle processes. By doing so, you develop a coherent, complete, and standardized strategy to get utmost reliability from the parts throughout your facilities, equipment, and infrastructure.

In doing a Physics of Failure Reliability Strategy Analysis, all of the causes of microstructure failure are addressed with the right plans and the right actions at the most appropriate places in the component's life cycle. The plans are implemented by introducing necessary changes and practices into a company's processes, workplace procedures, and supply chain vendor management. Planned maintenance work orders detail the breakdown, preventive, predictive, and precision maintenance activities needed to control the level of equipment risk. Engineering and maintenance activities incorporate the failure prevention, defect elimination, and precision tasks that stop failure causes. The operations group's procedures contain activities to reduce and control variation so that stable operation below microstructure threshold stress levels is standard practice. Equipment makers, suppliers, and service providers deliver quality machinery and workmanship. In this way, everyone reduces operating equipment risk by staying within the constraints of an asset's Physics of Failure limits.

Once developed, your reliability creation strategy and plans are transformed into effective actions that are approved by all key stakeholders and embedded in documents across the operation. The stakeholders are the board, senior executives, finance, engineering, operations, procurement, and maintenance department managers and their work team supervisors. They review the equipment

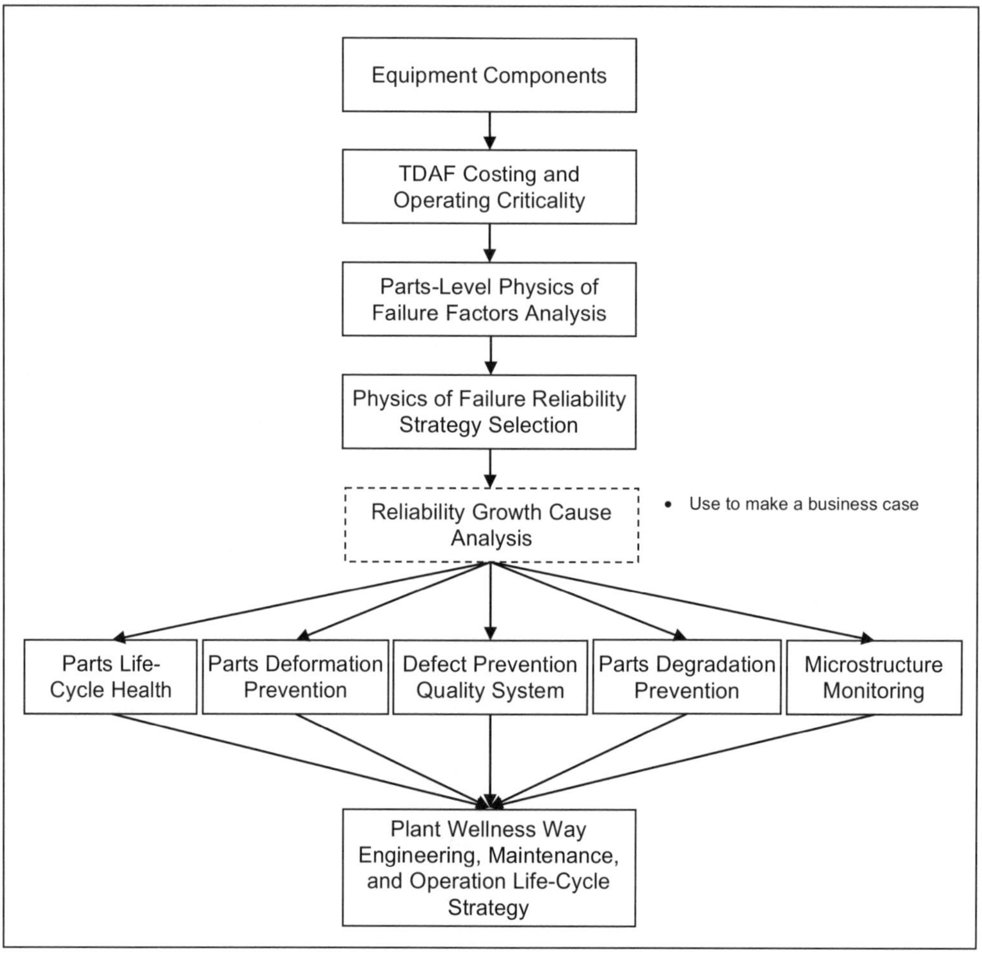

Figure 11.10—Developing Plant Wellness Strategy for Risk Management

reliability plans and include anything else they feel is necessary. Meetings are held with relevant workplace groups to explain and discuss the plans and the roles the groups will play in their achievement. Providing avenues of communication and opportunity for discussion helps gather support from the people who need to endorse or implement the new strategies.

Life-Cycle Tasks, Work Procedures, and Critical Spares

The Physics of Failure Reliability Strategy Analysis develops mitigations to eliminate the causes of component risk. Risk reduction strategy selections throughout the equipment life cycle are permitted. These include choosing the specific work quality controls to be applied during a critical part's construction; selecting condition monitoring inspections; specifying particular component health maintenance; picking the most economical preventive maintenance option; choosing equipment replacement policies and timing (i.e., replace with new on failure or near end of life); justifying critical spares holding; and stipulating breakdown recovery strategies for the part.

Preferred mitigations are those that significantly remove the frequency of equipment parts overload stress and stress accumulation. These are considered to be most effective at reducing the risk of failure.[2] Other acceptable abatements include those that substantially lower the opportunity for a defect to occur and those that make a component far more robust and durable in service. Less favored solutions are those that only reduce consequences, although they, too, are used when they clearly save substantial amounts of money and effort.

During manufacture, precision work and component quality are crucial to eliminate defect creation. Upon installation, precision and work quality are vital to prevent distortion. During operation, low-stress operating practices are used to minimize degradation. When parts are stored, apply top-class stores management practices to retain reliability. Procedures for each asset need to be written down, specifying the reliability creation activities to be done by operators and maintainers and stipulating the precision quality methods to be used. All work orders are to be fully and accurately planned. The work orders contain the risk reduction and reliability creation activities and quality standards needed to keep risk levels low and reliability high. They are scheduled to be done before the chance of failure rises above a low risk level and put into the computerized maintenance management system so that details are never forgotten. All work orders are done when due so that you get the maximum risk control benefits and production successes from your reliability creation strategy.

Because you are creating highly reliable equipment, you can expect a marked decrease in the range and number of critical spare parts you need to keep on hand once the plans are properly implemented. Critical spares are kept because their availability lowers your risk. During a breakdown, they ensure that you can recover and return to operation quickly. They are also used for proactive replacement of a severely aged working part. The correct critical spare stocking practice is to let the economics of carrying a spare drive the holding decision. You should only buy a critical spare and store and maintain it if that is the least expensive TDAF cost choice.

A useful precaution against error and misunderstanding when choosing mitigations for high and extreme risk levels is to gather a team of competent people who are knowledgeable about the situation. Team agreement is best when revising event frequency or likelihood for catastrophic scenarios. A group decision that is well debated and discussed benefits from the "wisdom of crowds" for arriving at a good and reliable consensus.[3]

Gradually, you build a proactive engineering, maintenance, and operational strategy with the practices that deliver highly reliable equipment. No longer is there mystery as to what maintenance to do and why it must be done, or why plant and machinery are only operated in specified ways that minimize stress, or why limits are set for particular production process variables to reduce microstructure degradation. Now the type and content of engineering, operating, and maintenance done in your operation matches those needed to maximize reliability, productivity, and business success.

Verifying Business Benefits

In order to accept that a suggested improvement is effective, it must be unquestionable in its ability to reduce risk levels by a substantial percentage over what would have been the case without it. Proposed risk controls need to be proven that they will be effective. The prevention strategies identified to limit the chance of failure and any actions chosen to minimize the consequence of failure need to *actually* reduce risk to the required level. Estimating the extent of risk reduction can be done using a table formatted with the headings shown in Table 11.2. If necessary, tests and trials can be conducted on equipment to confirm the stress reduction gained by a suggested mitigation. Measurements such as reduced electrical power use, lowered equipment vibration levels, decreased operating temperatures, and other appropriate indicators of lower stress can be used.

Table 11.2—Equipment Risk Reduction Layout

Ref No.	Equipment Tag No.	Equipment Description	Failure Event or Causes	Estimated Original Risk ($/yr)	Engineering, Maintenance, and Operational Activities to Reduce Risk	Years Equipment Remaining in Service or Expected to Be in Service	Current No. of Historic Failure Events Due to Cause (/yr)	No. of Failure Events or Expected Due to Cause after Risk Reduction	Annualized Likelihood of Failure Event after Risk Reduced (/yr)	TDAF Cost of Failure Event ($)	Estimated Remaining Risk ($/yr)
1	2	3	4	5	6	7	8	9	10	11	12

Many maintenance activities only add cost to a business and do not actually profit the business by making equipment reliable. If doing a maintenance activity does not reduce operating risk, then the work is a waste of time and resources with no business benefit whatsoever. A maintenance activity must reduce the cost of failure or the frequency of failure—preferably both. To show the benefit of a risk reduction decision, a calibrated risk matrix like that in Figure 11.11 is used to test a mitigation's value to the organization. On the matrix, you show the current TDAF cost risk if the part fails, and then you show the new risk, presuming that the selected Physics of Failure Reliability Strategy Analysis mitigations are in place and properly done. Provided that an improvement substantially reduces an event's frequency or substantially reduces its consequent cost, it will show up as a real business benefit on the risk matrix. Once a mitigation proposal is proven financially sound, it gives you strong reason and justification to make the necessary changes in your business processes and practices.

Documenting Your Plant Reliability, Maintenance, and Operating Strategy

The development of a strategy starts by stating the outcomes required. The outcomes may not be easily achieved, but by continuously improving your process designs, they will be realized.

Set the Business Objectives

Set measurable objectives based on the asset management and maintenance policies. For example,

- Halve current plant maintenance costs as a share of replacement asset value by adopting the Plant Wellness Way methodology.
- Reduce breakdown maintenance costs to less than 5% of total maintenance cost for the plant by making defect elimination a part of maintenance.
- Double the time between planned shutdowns and turnarounds by operating the process and its equipment for least degradation.

Plant Wellness Maintenance Management Strategy Process

Figure 11.12 is a flowchart that summarizes the maintenance management issues addressed during Physics of Failure Reliability Strategy Analysis. An operational risk control strategy is developed for each critical equipment identified in the operational criticality. Taking one asset at a time, you

Likelihood of Equipment Failure Event per Year				TDAF Cost per Event		$30	$100	$300	$1,000	$3,000	$10,000	$30,000	$100,000	$300,000	$1,000,000	$3,000,000	$10,000,000	$30,000,000	$100,000,000	$300,000,000	$1,000,000,000
Event Count per Year	Time Scale	Descriptor Scale	Historical Description			C1	C2	C3	C4	C5	C6	C7	C8	C9	C10	C11	C12	C13	C14	C15	C16
						1.5	2	2.5	3	3.5	4	4.5	5	5.5	6	6.5	7	7.5	8	8.5	9
100	Twice per week	Certain		L13	2	3.5	4	4.5	5	5.5	6	6.5	7	7.5	8	8.5	9	9.5	10	10.5	11
30	Once per fortnight	Certain		L12	1.5	3	3.5	4	4.5	5	5.5	6	6.5	7	7.5	8	8.5	9	9.5	10	10.5
10	Once per month	Certain		L11	1	2.5	3	3.5	4	4.5	5	[illegible]	[illegible]	[illegible]	[illegible]	[illegible]	[illegible]	8.5	9	9.5	10
3	Once per quarter	Certain		L10	0.5				3.5	4	4.	[illegible]	[illegible]	[illegible]	[illegible]	[illegible]	7.5	8	8.5	9	9.5
1	Once per year	Almost Certain	Event will occur on an annual basis	L9	0					3.5	4	[illegible]	[illegible]	[illegible]	[illegible]	6.5	7	7.5	8	8.5	9
0.3	Once every 3 years	Likely	Event has occurred several times or more in a lifetime career	L8	-0.5						3.	[illegible]	[illegible]	[illegible]	5.5	6	6.5	7	7.5	8	8.5
0.1	Once per 10 years	Possible	Event might occur once in a lifetime career	L7	-1							[illegible]	[illegible]	4.5	5	5.5	6	6.5	7	7.5	8
0.03	Once per 30 years	Unlikely	Event does occur somewhere from time to time	L6	-1.5								3.5	4	4.5	5	5.5	6	6.5	7	7.5
0.01	Once per 100 years	Rare	Heard of something like it occurring elsewhere	L5	-2									3.5	4	4.5	5	5.5	6	6.5	7
0.003	Once every 300 years			L4	-2.5										3.5	4	4.5	5	5.5	6	6.5
0.001	Once every 1,000 years	Very Rare	Never heard of this happening	L3	-3											3.5	4	4.5	5	5.5	6
0.0003	Once every 3,000 years			L2	-3.5												3.5	4	4.5	5	5.5
0.0001	Once every 10,000 years	Almost Incredible	Theoretically possible but not expected to occur	L1	-4													3.5	4	4.5	5

Reduce Consequence
Reduce Chance
New Operating Profit

Figure 11.11—Prove Maintenance Strategy Brings Value to Your Business

identify all of its threats to your operation's success and eliminate them. Failure cause after failure cause is removed from critical assemblies and components. It takes effort and time to do such detailed risk assessment and risk control selection, but the thoroughness protects your business by greatly reducing the chance of catastrophe over the operating lifetime of an asset. Once a reliability strategy to get the utmost operating profits is set, it is vital to fully execute the risk reduction activities.

Example Equipment Risk Reduction Strategy for a Pump-Set Bearings

Developing a maintenance strategy to prevent the failure of a centrifugal pump-set starts by drawing a process map for the equipment. Each of its assemblies is then analyzed by using Physics of Failure Reliability Strategy Analysis. During the analysis, a maintenance strategy is developed that delivers high reliability for all components. An example of the final operational risk reduction strategy for a pump's bearings is shown in Table 11.3. If the proposed operational and maintenance activities are carried out properly, they will ensure that the pump bearings have a long, failure-free life. The pre-

Table 11.3—Example Equipment Operation and Maintenance Strategy Development

Operational Risk Reduction Strategy for Centrifugal Pump Bearings

Equipment Tag No.	Current Failure Events	Failure Events Frequency	TDAF Costs of Failure	Risk Reduction Activity	Improvement Expected	Frequency of Activity	Cost (/yr)	Failure Event Reduction
Pump 1	Bearings fail	2 years	$35,000	Laser shaft alignment to precision practices every time the pump is installed	A precision alignment is expected to deliver 5 years between bearing failures	Every strip-down	$200	Failure interval likely to be greater than 5 years
				Oil and wear particle analysis every 1,000 hours of operation	Oil and wear particle analysis can indicate the start of failure several hundred hours prior the event	1,000 hours or six months	$600	Failure will be prevented by a predictive planned condition monitoring task
				Visual inspection by the operator each shift of the oil level in the sight glass	Visual inspection of the oil level ensures the bearings are always lubricated	Every day shift	No cost	Failure will be prevented by operator condition monitoring
				Operator physically touches pump bearing housing each week to feel for changed temperature and vibration	Touching the bearing housing will identify problems before they cause failure	Wednesday day shift	No cost	Failure will be prevented by operator condition monitoring
				Motor load monitoring using process control system to count overloads	Monitoring the electrical load will identify how badly and how often the equipment is stressed by overload	Continuous with monthly report to operations manager	$100	Poor operating practices will be identified and personnel trained in correct methods
				Pump performance monitoring of discharge flow and pressure using process control system	Monitoring the pump performance will indicate gradual changes of pump internal clearances affecting service duty	Continuous with monthly report to operations manager	$100	No direct impact on reducing risk of pump failure, but identifies performance drop and allows planned maintenance to rectify internal wear

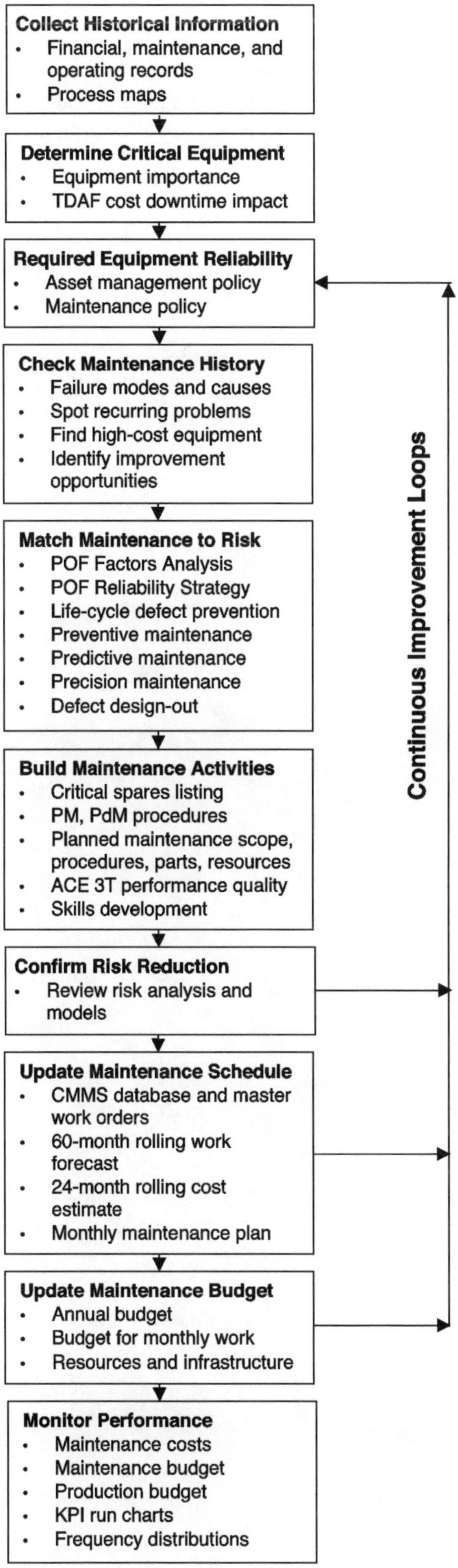

Figure 11.12—Turn Plant Wellness into Planned Maintenance

cision maintenance laser alignment removes the chance of overstressing parts, and the inspections remove the risk of unknown environmental and operational degradation. The likelihood of a bearing failure on the risk matrix goes from "likely" to "very rare" and the criticality goes from high to low.

References

1. Thanks to Peter Brown, Industrial Training Associates, Perth, Western Australia, for the use of his concepts.

2. David J. Smith, *Reliability, Maintainability, and Risk, 7th ed.* (Boston: Elsevier, 2005).

3. James Surowiecki, *The Wisdom of Crowds: Why the Many Are Smarter than the Few and How Collective Wisdom Shapes Business, Economies, Societies, and Nations* (New York: Doubleday, 2004).

12

Process 3–Numerate Risk Elimination Options

Summary Description of Process 3: Select Risk Control Strategy

The operating risk control strategies you use or do not use directly impact your plant reliability and operating costs. Adding maintenance routines to control risks causes maintenance costs to rise. The added maintenance is beneficial only if it brings more operating profit. Burdening your company with extra maintenance because of poor capital project engineering selections adds still more operating costs. Running and maintaining plant and equipment in ways that destroy their reliability causes breakdowns and huge TDAF losses. To maximize operating profit, it is essential to apply a life-cycle asset management methodology that produces effective engineering, operations, and maintenance plans that guarantee operating success. The Plant Wellness Way uses Physics of Failure Reliability Strategy selection and Reliability Growth Cause Analysis to design a reliability-creation strategy and prove that it will get world-class reliability and utmost operating profits.

Select Operating Risk Control Options in Projects

Design and Operations Cost Totally Optimized Risk (DOCTOR) analysis (see Chapter 6) is used for all capital projects or plant upgrades to maximize future operating profits. It slashes operating costs by letting project designers, managers, and engineers eliminate and reduce operating risks when making equipment and design choices.

Set Equipment Lifetime Reliability Targets

Zero breakdowns during the service lifetime of your equipment is what you want. Determine and set the breakdown-free equipment service life you need. This does not mean there will be no maintenance—rather, your maintenance and operating strategies will change to ensure that parts' microstructures operate at least stress conditions and in healthy environments in which failures cannot happen. If an unwanted failure has initiated, it is identified and corrected so that equipment does not break down.

For more information on this topic, see the document *Quality Standards for Failure Prevention*, which is included in the supplementary materials available online.

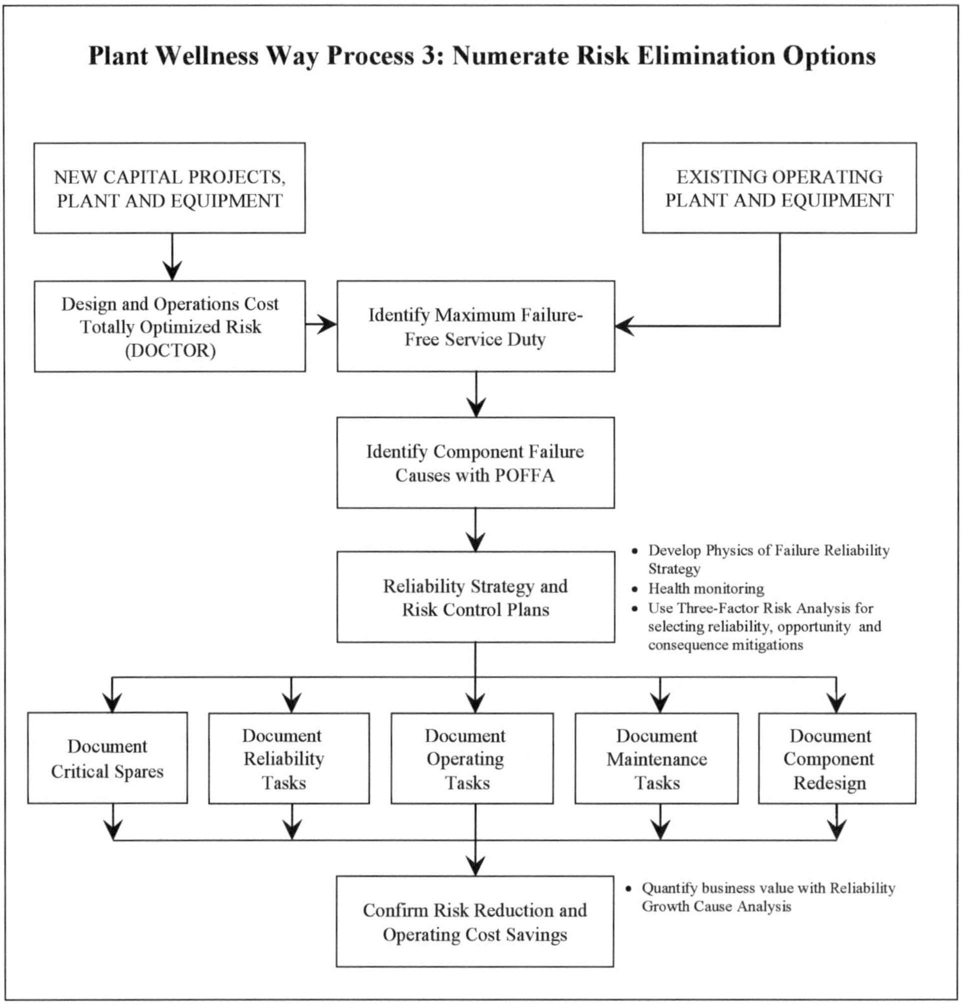

Figure 12.1—Process 3 Steps

Identify Causes of Failure Using Physics of Failure Factors

You go deep into the detail of what causes equipment component failures in your operation to find and understand the failure cause mechanisms at play. Once you understand the events that start a failure, you're better able to select solutions to eliminate its root causes. You identify all possible microstructure destruction mechanisms and their causes using Physics of Failure Factors Analysis (POFFA) so that you know which situations that lead to a part's failure to remove or prevent.

Select Reliability Strategy for Operating Plant

Operating sites that want long reliability and low costs need to eliminate the reasons for poor reliability and high costs. Useful maintenance tasks are those that stop risks from becoming failures. The very best maintenance activities are those that remove the opportunity for failure—those that prevent the cause of failure so that events cannot arise and consequences cannot occur. Once failure causes are identified using POFFA, you complete a Physics of Failure Reliability Strategy Analysis to select component design, manufacturing, supply chain, operating, maintenance, reengineering, and defect elimination strategies to use across the life cycle. To get a fuller appreciation of the reliability creation choices, you can use a Three-Factor Risk Analysis to investigate and select the most effective opportunity elimination, reliability improvement, and consequence reduction strategies.

Document Your Plant and Equipment Risk Management Strategy and Plans

The entire Physics of Failure Reliability Strategy strategic asset management plan is written into the operation's processes, all applicable procedural documents company-wide, and work orders established in its computerized maintenance management system (CMMS).

Confirm Extent of Risk Reduction Using Reliability Growth Cause Analysis

It can be 12 to 18 months before the effects of new reliability improvement strategies reduce the failure frequency so that savings show up in monthly reports. You need a way to check that your proposed strategies will remove—or at least substantially reduce—the risk of each failure. Reliability Growth Cause Analysis proves that your equipment reliability strategies will create reliability and deliver least operating costs before you implement them by using a risk matrix or financial calculation to show the expected reduction in risk.

Plant Wellness Way Risk Elimination and Reliability Methodology

Use the following approach to develop effective asset life-cycle and operating risk reduction activities to create reliability.

A Physics of Failure Reliability Strategy Analysis provides you with a structured methodology to select and place successful risk elimination and controls in the equipment life cycle. With it, you identify which parts fail, how they are failed, and where the chance of failure arises during their lifetimes.

For each part, a full range of engineering quality control standards are set for all Physics of Failure mechanisms that destroy its reliability—flatness, roundness, straightness, hardness, surface finish, porosity, cleanliness, fastener tension, fits and tolerances, lubricant condition, operating practices, installation requirements, and all other causes of Physics of Failure risk to the component—so that they are within the range that brings a long, healthy service life. The suite of quality standards defines the component precision operating zone for its entire lifetime. All Physics of Failure risks throughout the component life cycle are either eliminated or addressed and controlled to a level of risk that brings the parent equipment's total risk to within the acceptable range on the risk matrix or represents no more than one failure of the parent equipment in a period equal to three times its service life.

Those parts with risks from wear-out and use are put on preventive maintenance routines in which the item is changed, or returned to its precision zone standards, when quality requirements exceed the allowed tolerance. Those parts that suffer risk from random failure events, typically as a result of high imposed stresses, go on a predictive maintenance condition monitoring plan to check for changed conditions that deviate from the required precision standards. When an item's condition is beyond tolerance, it is returned to its precision standards before failure initiation. For those parts that can suffer infant mortality and be destroyed by human error, error-proof engineering, operating, and maintenance procedures are written to achieve top-class results in all life-cycle processes impacting the part's life.

In the procedures, the applicable precision quality standards are specified and the work quality controls and proof tests needed to get components into their precision zones are set. Equipment overloads and process variables causing microstructure degradation are monitored when the plant operates, and every out-of-tolerance incident, along with its duration and the operating conditions at the time, is recorded for analysis to identify how to keep the equipment operating stably within its design envelope.

Identifying Critical Components

An Operating Criticality Analysis exposes the risks of equipment breakdown from component failure. Those parts for which the risk is too high are the equipment's critical parts—the components that must never fail during the service life. Later, you will develop strategies to prevent their failure. The analysis process to determine whether a part is critical follows. Figure 12.2 shows a roller bearing on a shaft. The item's components are identified using a design logic process map (also called a reliability block diagram) of the assembly. In Figure 12.3, a portion of the bearing assembly mounted on the shaft is drawn as a reliability block diagram mimicking its logical sequence of operation.

***Figure 12.2—
Ball Bearing and Shaft Assembly***

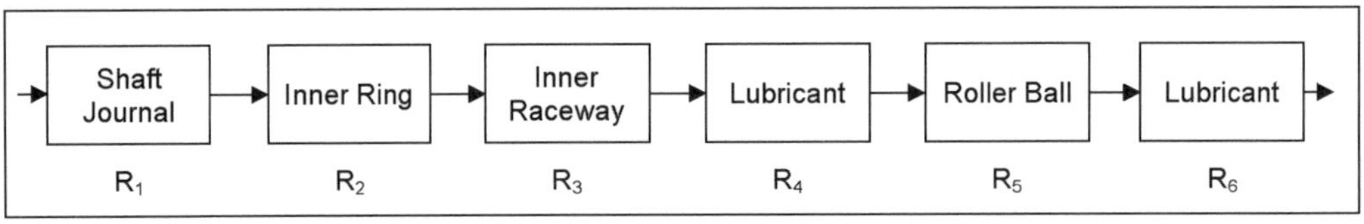

Figure 12.3—Process Map for Roller Bearing on Shaft

The flowchart shows that each part is critical because each item is part of a series arrangement. The loss of any part, such as the failed raceway in Figure 12.4, will cause the roller bearing to fail and the equipment to stop.

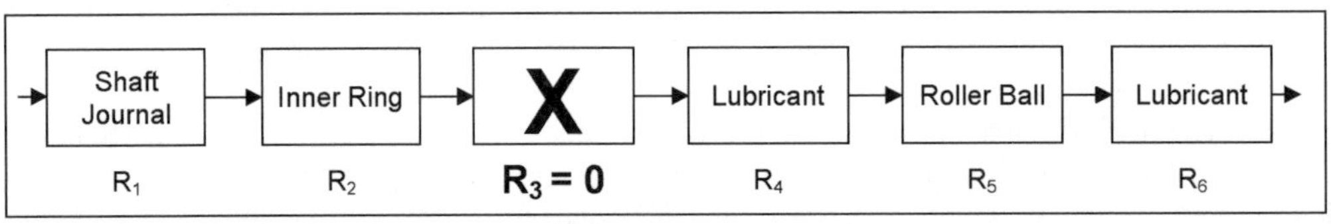

Figure 12.4—Process Map with Roller Bearing Raceway Failure

If you know the engineering of an equipment item, it may not be necessary to draw process maps for assemblies containing only a few parts, as the manufacturer's assembly drawings can be viewed directly and the critical parts identified by sight—the bearings, rubbing shaft seals, gears, and drive components taking start-up and operating loads are typical examples of critical parts. For more complex assemblies, the manufacturer's drawings will not explain how the equipment works as an operating construction, and you will need to develop process maps. For example, it is impossible to be certain from the drawing how the parts in the gearbox shown on the exploded general assembly in Figure 12.5 interact when the gearbox is put together. When there is uncertainty about a part's na-

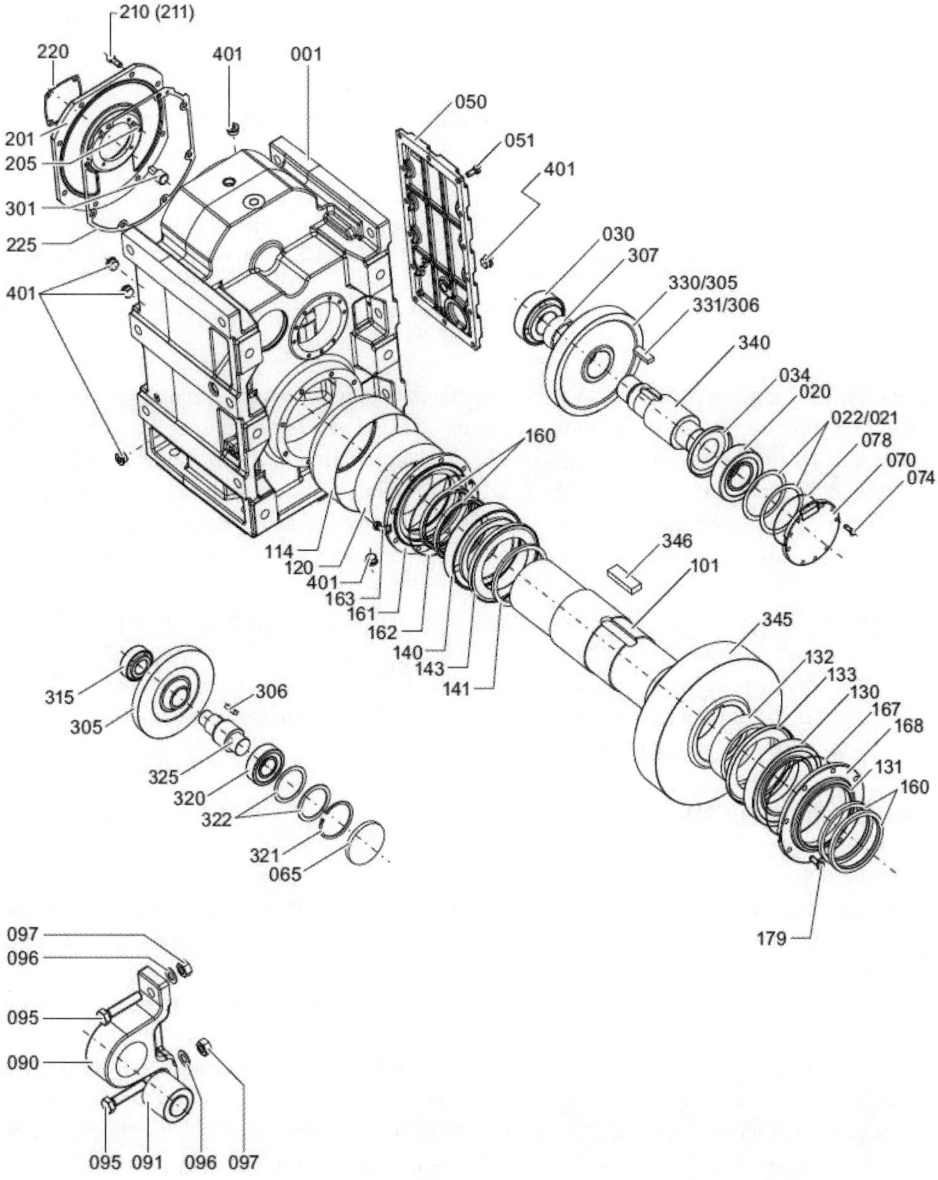

Figure 12.5—Ball Bearing and Shaft Assembly

ture and its use, you develop a design logic process map for the part and its neighbors to clarify the situation. You would not map the whole gearbox unless it is necessary. For straightforward engineering designs, a person experienced in the engineering and construction of the equipment would know how it operates and could identify its critical parts from the general assembly drawing and the bill of materials, like the parts list in Figure 12.6 for the gearbox of Figure 12.5. Rather than leave a critical part out of the analysis, add parts that you are not sure about to the spreadsheet and let the Physics of Failure Reliability Strategy Analysis determine its criticality.

You don't need to collect failure history in order to choose the right maintenance strategy for critical components. We can understand a lot about our equipment risks from the manufacturer's general assembly or exploded parts drawings and from the bill of materials. Competent peoples' past industrial experience and good engineering design knowledge can be used to select sound maintenance and reliability strategy for critical components in the parts list.

Steps to Follow in a Physics of Failure Reliability Strategy Analysis

Using a Physics of Failure Reliability Strategy worksheet, systematically work through each requirement listed. The numbers correspond to the steps noted in the Physics of Failure Reliability Strategy Analysis and Development worksheet in Table 12.1, included in the downloadable materials onlline.

1. Select the asset to be investigated and number and name it.

2. Identify all operating processes to which the asset belongs. This information comes from plant drawings, process and instrument diagrams, electrical and control cabling drawings, and equipment asset lists. Draw process maps of the operation showing the equipment to be analyzed. Ensure that every item used to run the asset is on the map. This includes power supplies and ancillary items used only at start-up or shutdown. Specify the full range of necessary functions, their service duties, and required availability for the asset and each ancillary item.

3. Determine the prior and remaining years of service life required from the asset. Identify how long the parent equipment is required to run without unplanned downtime, safety issues, production slowdown, or product quality problems. This allows later measurement of the effectiveness of the risk control strategies and provides a means for prioritizing reliability improvements. Set the target for breakdown-free years of operation.

4. Get all of the technical details, assembly drawings, and bills of materials covering all the asset's mechanical, electrical, and electronic items.

5. Conduct an Operating Criticality Analysis to determine whether the asset has components that are critical to the operation. Estimate the biggest TDAF cost for each assembly. If an asset is part of a duty-standby configuration, indicate that in the spreadsheet. For the analysis of operating risk, presume that the standby unit is available when determining criticality; however, also identify the full TDAF cost if the duty unit fails when the standby is not available. From CMMS records and operating records, identify the failure frequency of each major assembly. You will need to use a representative time period that reflects the effects of the operation's culture and management practices; at least five years of data is ideal. If the plant was upgraded or a process was changed, look at the extent of the change and records from the date of commissioning the change to determine its impact on failure frequencies.

001	Gearbox housing
020	Bearing
021	Supporting disk / shim
022	Supporting disk / shim
025	Locking ring
030	Bearing
034	NILOS ring
040	Output flange
045	Screw
050	Housing cover
051	Screw
055	Seal
065	Sealing cap
070	Sealing cap
074	Screw
078	O-ring
090	Torque arm
091	Rubber bush
095	Screw
096	Locking ring
097	Nut
101	Input shaft
109	Locking ring
110	Locking ring
114	Protection cover (optional)
120	Shrink disk
128	Supporting disk / shim
129	Supporting disk / shim
130	Bearing
131	Supporting disk / shim
132	Spacer / bush
133	NILOS ring
135	Locking ring
140	Bearing
141	Spacer / bush
143	NILOS ring
144	Supporting disk / shim
146	Locking ring
160	Shaft sealing ring
161	Bearing cover
162	O-ring
163	Screw
167	O-ring
168	Bearing cover
179	Screw
201	Adapter plate
205	Screw
210	Screw
211	Screw lock
220	Seal
225	Seal
301	Pinion
305	Helical gear wheel
306	Parallel key
307	Spacer / bush
310	Locking ring
311	Supporting disk / shim
315	Bearing
320	Bearing
321	Locking ring
322	Supporting disk / shim
325	Pinion shaft
330	Helical gear wheel
331	Parallel key
340	Pinion shaft
345	Helical gear wheel
346	Parallel key
401	Screw plug
420	Vent filter

Figure 12.6—Bill of Materials for Reduction Gearbox

6. List all critical major assemblies that make the asset operate.

7. Identify the vital subassemblies and parts in the bill of materials for each critical assembly item that make it work. When necessary, develop a reliability block diagram to understand the logic of the design and identify the influence of parts and interacting components.

8. Determine the likelihood frequency of failure for each vital subassembly or part from history data in operation and maintenance records. The likelihood scale of the risk matrix provides suitable frequency designations. For the analysis, presume that any standby unit is available when determining operational impact. If necessary, research the typical failure events that each component can suffer. When technical and failure information is not available in house, search the Internet for similar events, contact similar operating sites, or ask the original manufacturer for estimates.

9. Transfer Operating Criticality 1 and 2 values for the item from the Operating Criticality Analysis.

10. Do a Physics of Failure Factors Analysis for each vital assembly or part with a criticality above acceptable to identify all possible human error, degradation, and deformation failure mechanisms for all of the material-of-construction factors. If a part has multiple materials of construction, do an analysis for each material (e.g., a flexible drive coupling connecting two shafts is made of steel and elastomeric parts, and each material will fail from different mechanisms). Get a competent and experienced engineer who knows the equipment and its operation very well to do the POFFA (an electrical engineer for electrical equipment, a mechanical engineer for mechanical plant, and instrumentation and control engineer for control systems), or team with competent and career-experienced operators and maintainers and do the POFFA together. Transfer the POFFA results to the Physics of Failure Reliability Strategy worksheet.

11. Identify all of the life-cycle phases when each failure cause mechanism can occur, including during the remaining service life of the asset.

12. Select error elimination, microstructure degradation and deformation prevention, opportunity elimination, material reliability improvement, or consequence reduction strategies for each mechanism to eliminate the possibility of failure or to minimize the chance of a failure starting. This will produce engineering, reliability, operational, and maintenance strategy and activities that reduce a critical component's operating risks. Choices include using life-cycle process changes, preventive maintenance, condition monitoring, renewal, or refurbishment. Set the frequency or a trigger for each activity. These plans and actions can apply to any and all issues related to the organization's design, engineering, supply chain, operational, and maintenance practices. You aim to drive a part's total operating risk so low that its contribution to the total parent asset risk provides no more than one chance of failure of the parent asset during three times the parent asset's service life. Any mix of economically viable solutions across the life cycle that delivers that operational outcome is acceptable.

    Confirm on a calibrated risk matrix or by calculation that the intended activities will significantly reduce risk. The reliability improvement strategies chosen for a component must deliver substantial risk reductions to the equipment. At a minimum, the total effect of all risk mitigations used on a component must drop risk to the acceptable level. Perform a Three-Factor Risk Analysis to identify alternative or more viable asset life-cycle risk control strategies when known risk mitigations are ineffective. If no arrangement of mitigations achieves the required service life for a part, the remaining option is to reengineer the component so the microstructure of its new design can comfortably handle all service life stresses and contact conditions.

13. Set the ACE 3T (Target, Tolerance, Test) quality limits for each Physics of Failure mechanism. These are the quality standards established throughout the asset life cycle that, when achieved, prevent each cause of failure arising. A quality standard tolerance value is the lowest value specified by either the parent asset's manufacturer or by a component's manufacturer. For example, the manufacture's specification table for a flexible drive coupling used as a component by a pump-set maker to connect a motor to the pump will show the coupling can accommodate millimeters of misalignment, but the manufacturer of the pump-set permits no shaft misalignment more than a few microns. In this case, the tolerance value to use for the shaft alignment quality standard is the maximum value for misalignment specified by the pump-set manufacturer. When setting the target value range, the quality standards to use are those that deliver world-class reliability from the part's microstructure.

14. Select on-condition inspection tasks or do operating performance monitoring to detect changes in the mechanisms affecting component health, when proactive failure cause prevention or economic preventive maintenance cannot deliver failure-free service lifetimes. Proactive cause prevention is the total removal of a failure mechanism so that there is no cause that can generate a defect. Preventive maintenance is scheduled replacement of parts before failure. For the causes remaining, condition monitoring is used to look for evidence that a part's health is within its targets, or performance monitoring is done to monitor the stability of variables that raise the risk of failure. When operating variables stray out of control, that is a trigger to rectify the problem before risks grow too great. When possible, use visual condition checks against the target value; if simple observations cannot be safely and economically done, introduce suitable technological means to check component health conditions. You are not looking for evidence of failure initiation; rather, you want proof that the part is in fit condition and will remain so. If you detect adverse conditions, look for the cause(s) and remove them.

15. Set the operating and maintenance ACE 3T quality limits for each on-condition variable being monitored. Typically, the range of variables used are those that represent the stresses and health of the item being observed (e.g., operating loads, alignment, distortion, temperature, pressure, looseness, balance, lubricant chemistry, etc.). Exceeding a target value standard is a warning level alarm, and exceeding the tolerance value is the trigger for immediate rectification to return the component to within all of its target values.

16. Determine the most economical breakdown operating and maintenance activities to be done in order to recover and return to operation if a vital part should fail during the asset's service life.

17. Assign a person to be responsible for implementing each chosen error elimination, degradation, and deformation strategy and training users. This includes developing new documents and updating existing documents with relevant tasks and ACE quality controls. Update process maps; develop error-proof procedures with relevant tasks and work assurance including the required quality standards; review and update or remove existing work orders; include new, fully detailed risk elimination work orders; train and educate all relevant personnel in the new procedures and explain why they are necessary for the operation's future success; update the CMMS planned work schedule with the latest work orders; develop the future maintenance resource demand

into an overall maintenance resource schedule to understand the manpower load and the full range of skills needed in your crew.

Catalog and cost the spares identified in the Physics of Failure Reliability Strategy Analysis and/or the Reliability Growth Cause Analysis as critical requirements for plant and equipment consequence reduction. Detail the spares required for planned maintenance activities each financial year for inclusion in the annual financial budget. Update the critical spares list and order any missing spares in a controlled and financially responsible manner.

Financially model the new strategy using Reliability Growth Cause Analysis and compare the new TDAF costs to the current TDAF costs to provide economic justification for changing strategies. (This analysis is optional if you are satisfied with the estimated risk reduction and financial gain shown in the risk matrix.)

18. Assign a person to be responsible for implementing each chosen on-condition maintenance and operating task. This includes developing new documents and updating existing documents with relevant tasks and ACE quality controls, and training users.

19. Assign a person to be responsible for implementing each vital part's chosen breakdown operating and maintenance activities (this is necessary in order to minimize the total consequential losses to the operation in case of failure).

20. Assign each Physics of Failure Reliability Strategy action to all relevant phases of the asset's life cycle and implement the actions. Using the updated maintenance schedule, estimate the maintenance budget for the next two years, factoring in the reliability improvement effects on equipment as a result of the new reliability strategies. Submit the forecast maintenance budget for review by senior managers and answer their questions. Monitor and track monthly each equipment's reliability performance measures and maintenance cost to ensure they are improving.

When there are multiple identical parts configurations for an asset—for example, each piston and cylinder of an internal combustion engine—you need only analyze one set of common components.

You will find that most equipment in industry is made of the same components using the same or similar materials of construction—about 100 types of parts make up nearly all industrial machines and equipment.[1] The results of a Physics of Failure Reliability Analysis will apply to many other items of equipment. Once you have a database of past analyses, you can mostly copy and paste an existing analysis into a new analysis. Be careful, however, to detect the exception to the rule. For example, in operations with multiples of the same bearing number in machines, you will be tempted to analyze one bearing and say the analysis represents all identical bearings across your operation. But if one bearing is inside a clean room and another is outside in full weather, the situations are not comparable, and each bearing will have different causes of failure.

Selecting and Implementing High-Reliability Strategy

Keeping components in their precision zones is the universal purpose of the reliability, maintenance, operational, engineering, project, and asset management efforts applied in the Plant Wellness Way.

With that aim in mind, you establish appropriate strategies, plans, and activities throughout the organization and across the life cycle to maximize the chance of its success.

Each of the Physics of Failure Factors listed for a component are first addressed proactively to prevent all the failure mechanisms that cause the factors. The most effective solution is to select actions that will stop a cause mechanism from being present, thereby ensuring that no defects can be created and no risks can exist. One by one, engineering, maintenance and operational decisions are made so that each mechanism is prevented with appropriate action or combination of actions to remove opportunities for the mechanism to occur during the life cycle.

Table 12.2 shows the maintenance and operating strategies used at a site that follows the Industrial and Manufacturing Wellness paradigm and practices.

Table 12.2—Plant Wellness Way System of Reliability Strategy and Practices

<table>
<tr><th colspan="5">Plant Wellness Way System of Reliability Strategy and Practices</th></tr>
<tr><th>Risk Level on Failure</th><th colspan="2">Maintenance Strategy</th><th>Lifetime Reliability Strategy</th><th>Operating Strategy</th></tr>
<tr><td>Accept</td><td colspan="2">• Do Breakdown Maintenance (BM)
• Except, do Preventive Maintenance when the life-cycle cost is less than BM TDAF Cost</td><td rowspan="5">• For each failure cause mechanism select ACE 3T precision quality values and design manufacture, assemble, warehouse, install and operate so components live within their precision zones.</td><td rowspan="5">• Operate for degradation management so those factors and process conditions that initiate component failure cause mechanisms are kept within operating precision zones that maximize the component's reliability.</td></tr>
<tr><td>Low</td><td rowspan="4">• Do Preventive Maintenance (PM) where component useful life is confidently known e.g. filter, oil, brake pad, etc., coupled with defect-free Precision Maintenance, i.e. positional accuracy, no distortion, right fit and tolerance, right tightness, etc., (see Chapter 14) to ensure components are in their precision zones.
• Do early "economic" PM when cost of replacement is minor compared to the cost of a future failure, e.g. gear box, drive belt, separation screen, etc., coupled with defect-free Precision Maintenance</td><td rowspan="4">• If a failure by a cause mechanism is possible, e.g. from changed component quality parameters, component distortion, operating variable surges, etc., but when the failure occurs is a random event, then schedule component "health monitoring" (CHM) of failure cause mechanism to confirm compliance to its ACE 3T quality values (see Chapter 13), coupled with defect-free Precision Maintenance to return components to precision zone.</td></tr>
<tr><td>Medium</td></tr>
<tr><td>High</td></tr>
<tr><td>Extreme</td></tr>
</table>

Doing a Three-Factor Risk Analysis

Like all risk analyses, Three-Factor Risk Analysis starts by identifying what situations or events could arise in your operation to cause the failure of a critical part in an asset. Once potential problems are identified, strategies to eliminate or substantially reduce the risk are selected. In a Three-Factor Risk Analysis, potential risk controls are divided into three categories: consequence reduction, opportunity reduction, and reliability improvement. In each category, there are methodologies and techniques to reduce or eliminate the risk under consideration.

To confirm the effectiveness of the selected mitigations, they are either mapped onto a risk matrix, or the risk formula is used to calculate the new risk level and checked that the mitigations will reduce risk to the acceptable level.

Three-Factor Risk Analysis can be applied to address the risks in the following:

- A complete asset
- Subassemblies
- Parts and components
- Business processes and their individual steps
- Work procedures and work instructions

Table 12.3 provides a sample Three-Factor Risk Analysis worksheet for a machine component.

Table 12.3—Sample Three-Factor Risk Analysis

Event Risk = Consequence x [Opportunity x Probability (Chance of Event Happening at This Opportunity)]

Asset	Assem-bly	Part	Failure Event	Failure Cause	Opportunity to Fail	TDAF Cost Conse-quence ($)	Frequency of Opportunity (/yr)	Existing Mitigations/ Controls	Likelihood (Chance) of Failure Cause	Existing Risk Level
Pump set	Motor	Shaft NDE bearing	Seized	Interference fit tight	Every time it is machined	$50,000	1 in 20 years	Standard machine shop work quality control	Possible	Low
				No Lube	Every time it is greased	$50,000	12 per year	Greaser selects own grease	Likely	Medium
				Wrong Lube	Every time it is greased	$50,000	12 per year	Greaser selects own grease	Likely	Medium
			Spin in housing	Outer ring interference fit loose	Every time it is machined	$10,000	1 in 20 years	Standard machine shop work quality control	Possible	Low

Table 12.3, continued

Additional Mitigations/Controls	Chance Mitigations are Done Correct	New Likelihood	Potential Future Risk Level
Proof test journal and bore are to bearing OEM tolerance and form	High	Unlikely	Acceptable
Scheduled lube route, ultrasonic headset recording	High	Unlikely	Acceptable
Separate grease guns for each lube type clearly marked and colour coded. Grease points color-coded, ultrasonic headset recording	High	Unlikely	Acceptable
Proof test journal and bore are to bearing OEM tolerance and form	High	Unlikely	Acceptable

Organizational Factors of Critical Parts Failure

It is well documented that around 80 percent of equipment failures can be traced to human error and unintended business process "traps" that cause calamity.[2] Notice how many human and business factors in the "Cause Prevention Strategy" column of Table 12.1 in the online download are related to human intervention. When you put parts through a Physics of Failure Reliability Analysis, you soon collect a list of the human and business process drivers of failure in your operation. You'll see the same issues repeating.

The analysis warns that there are systemic human-related causes that create equipment problems. You must confront the reality of the three organizational factors questions below to expose the business-caused failures that have worked their way into your company's processes.

1. What human factors make an equipment part fail?
2. What business processes make an equipment part fail?
3. What engineering design issues make an equipment part fail?

Allocating Responsibility for Strategy Requirements

You are building the business-wide processes that will change your operation to deliver world-class equipment reliability. You are designing the future business and operational system that will take your company to reliability, maintenance, and operational excellence. The necessary engineering and precision standards to be met throughout the life cycle; the supporting documents needed in each department; the technical knowledge, skills, and competence needed by your people; the training of your workforce; the recording and reporting systems that confirm that strategies are being delivered correctly in your business units;[3] and so on throughout the organization need to be installed and put into use.

The Physics of Failure Reliability Analysis makes clear what needs to be done to get high reliability from your processes. All of those tasks are assigned to the people most appropriate to do them. Those people have the competence to do the requirement correctly and do it well. Duties can be shared across groups if a requirement needs multiple disciplines and skills to complete it. A column of the spreadsheet is used to identify the people responsible for implementing a strategy. You could instead color-code a strategy to represent the people responsible for the associated work. A simple color-code arrangement clearly differentiates the tasks assigned to engineering (yellow), operations (amber), maintenance (green), and so on.

Confirming the Economic Value of New Strategy

It is vital to prove that the risk controls you apply have a great chance of delivering the risk reduction needed. Be very suspicious of any maintenance strategy selection methodology that cannot prove that doing the maintenance strategy it suggests will bring real value to the company.

As part of the Physics of Failure Reliability Analysis, you compare the current risk level in the business for the asset under review to the risk that will result when the reliability improvement changes identified become standard operating practices. The proof of risk reduction can be done using a risk formula calculation in the Operating Criticality Analysis or shown on a risk matrix. In order for mitigations to be suitable, your selection must satisfy the criteria for an acceptable risk mitigation (see Chapter 4).

References

1. A. A. Hattangadi, *Plant and Machinery Failure Prevention* (New York: McGraw Hill, 2005).

2. Federal Aviation Administration, "Aviation Maintenance Technician Handbook–General, Chapter 14 Addendum/Human Factors," accessed at http://www.faa.gov/regulations_policies/handbooks_manuals/aircraft/media/amt_handbook_addendum_human_factors.pdf, August 5, 2015.

3. Bruce McLaughlin, "Comprehensive Risk Abatement Methodology as a Lean Operations Strategy," accessed at http://www.igi-global.com/article/comprehensive-risk-abatement-methodology-as-a-lean-operations-strategy/127540, August 24, 2015.

13

The Accuracy-Controlled Enterprise

Variation and risk thrive in processes that allow imprecision. The outliers of life-cycle process variations become the defects and failures that cause plant and equipment breakdowns and necessitate repairs. To get control over variation and stop defect creation, standards of quality must be met. Quality management systems are developed to combat and control variation and risk in business processes.[1] But using a certified quality management system does not give you quality performance. It isn't an International Standards Organization quality accreditation that produces excellent results.

Look carefully at how experts—total masters of their craft—work. There is confidence and certainty in every activity. They know how to do excellent work. Each action meets specific requirements with great accuracy. They continually look for evidence that they are producing the right results. A master craftsman controls variation within a narrow span of quality outcomes. By being precise in every aspect of the job, he does wonderful work. The controlled accuracy of a master craftsman needs to pervade your business if you want world-class equipment reliability. By using the accuracy-controlled methods, skills, and beliefs of the expert in your organization, you can minimize risk, control variation, and slash enterprise-wide costs as failures plummet. The organization becomes an Accuracy-Controlled Enterprise (ACE). The focus in an ACE is not product quality. It is doing a job—every job—masterly by being accurate in every respect.

The Precision Principle

You prevent failure and error by using methods that guarantee precision and accuracy. This is the Precision Principle: set quality standards for every step of a process and measure the outcomes to prove that those standards are being accurately met. If a process step cannot reliably meet the standards, you change its design until it delivers the required result. Figure 13.1 shows what happens when the Precision Principle is applied: first quality standards are set, and then the process is improved until performance meets the standard.

The Precision Principle is a rule to help you design successful processes. Having standards is the key; process improvement needs a target to hit. Start by developing appropriate standards with specific targets, tolerances, and measures. Performance and quality will follow because the process is changed until the standards are met. Once variation naturally stays within the standard, quality is consistently achieved because the process is designed to do so. There are far fewer problems and less waste when processes are built to ensure the presence of the methods, skills, equipment, tools, and know-how to produce masterly results.

Highly reliable equipment minimizes production costs and maximizes throughput. Machines require quality manufacture and precision maintenance coupled with precision operating practices

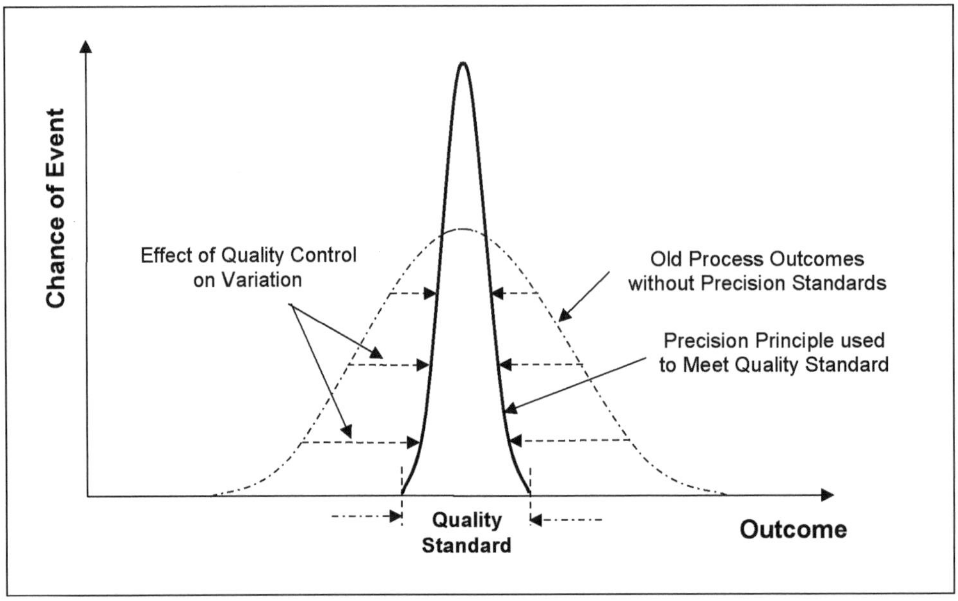

Figure 13.1—The Effect of Controlling Process Step Quality on Variation

to produce the right conditions for high reliability. You get highly reliable equipment when designers make the right choices, installers and maintainers do their work to precision quality, and operators run equipment with least operating stresses under stable production loads. There are no breakdowns when the equipment design is right for the duty, with working parts' microstructure at low stress in healthy contact environments. If an operation's equipment is not performing reliably, then something is very wrong—but not with the equipment! The problem is with the business processes. They are causing wide variation and creating risks. Your challenge is to identify the process failures that are causing defects and stopping equipment from delivering its design performance. Then you must intentionally remove them with effective risk controls and quality assurance.

Often the fault for poor reliability lies with the equipment design. Parts are not suited for the service—the materials of construction are not strong enough for the loads, the chosen manufacturing process installed weaknesses, or the material is incompatible with its contacting environment and degrades. Other reasons equipment isn't reliable are that it is wrongly installed, it is poorly assembled or rebuilt, or it is overstressed during operation. Usually these happen because the people involved in the equipment's design, installation, running and care do not know the right ways to minimize and control parts' microstructure deformation and degradation. Although designers, fabricators, operators, and maintainers are trained, they can never know enough to handle all situations competently (nor can anyone know it all). In uncertain situations, they use the knowledge they have to make a decision. If what they do fixes the problem, even if it is the wrong choice, that is how they will solve the same problem in future. Regrettably, many decisions of this type do not have an immediately negative impact. If the danger were visible, the person would instantly self-correct and get it right. But most errors of choice do not have a negative impact until long after they are made.

When a chosen action has no obvious bad consequences and things are still running fine, people start to believe they've made the right decision. A classic example from industry of falling into an ignorance trap is using the wrong valve to control product flow through a pump. Closing a pump inlet valve will reduce liquid flow, but it will also destroy the pump as a result of cavitation and vibration. When you control flow with the suction valve, you get a flow rate change, and your aim is achieved. You think what you did was right because it worked. Now a bad practice becomes set in place through ignorance and misunderstanding, and the pump will always have random failures. The only correct valve to adjust for pump flow control is the discharge valve.

There is nothing wrong with making a wrong decision if no harm is done and it is fixed immediately. Bad things happen when wrong actions progress over time to their natural and final sad conclusion. Unfortunately, there are very few decisions that have instant-replay options. If it is important in your company to have highly reliable production equipment with low maintenance costs, then the organization's work processes and business systems must support and produce that outcome. Your quality management system needs to ensure that all work is done right first time. There is great value in developing a quality system encompassing the standards and instructions that will help everyone deliver masterly results.

Creating Standard Operating Procedures to Become World Class

Human skills, talents, and abilities are typically normally distributed. If we graph the abilities of a wide cross-section of humanity to do a task, we will end up with a normal distribution bell curve. Secondary and higher learning institutions use a normal distribution curve for grading students because it reflects the spread of outcomes on exams. Figure 13.2 shows a normal distribution bell curve, or Gaussian curve, of a talent in a large human population.

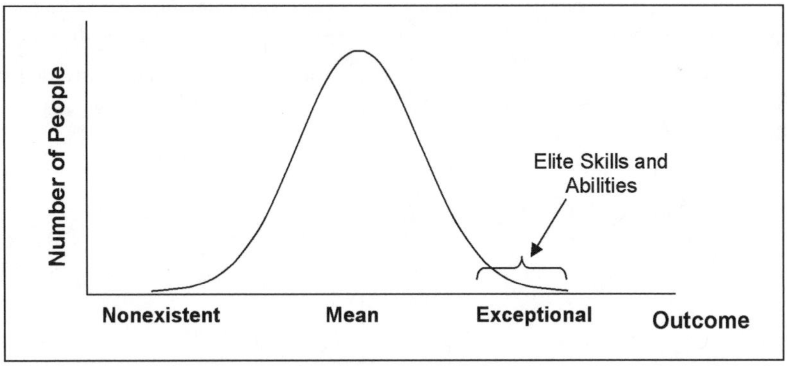

Figure 13.2—Distribution of a Talent in the Human Population

The upshot is that for human skills and talents, there are a few exceptional people, a few with astoundingly poor ability, and many in between clustered around the middle or mean. If a workplace requires highly able people, the distribution curve of human talent warns that it will be hard to get exceptional employees. The talent distribution curve also explains why continuous training of the workforce is so important to a company's long-term success. If the available skill clusters around the mean performance level, then to get better results, additional training and practice are needed to develop higher-quality skills. Training and practice have the effect of moving average performers toward the elite portion of the population,[2] as shown in Figure 13.3.

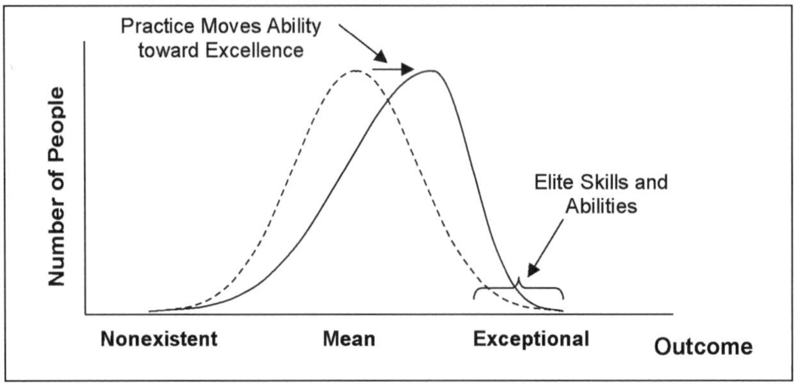

Figure 13.3—The Effect of Higher Skills Training on Developing a Talent

Adding Job Accuracy Controls

Accuracy is the degree of conformity of a measured or calculated value with its actual or specified value. Accuracy first requires setting a target value and a tolerance for outcomes that are acceptably close to the target. Figure 13.4 shows what accuracy means.

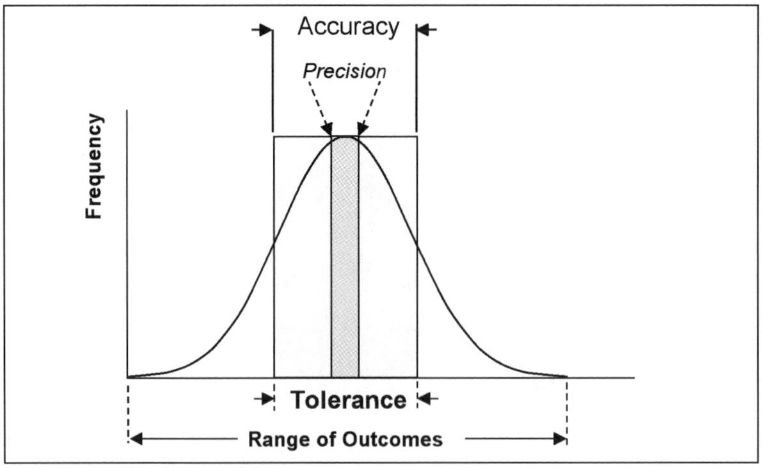

Figure 13.4—Accuracy and Precision

The problem with targets is that they are not easy to hit dead-center. It is not possible to be perfect. If a task must meet an exact value, that is an unrealistic and virtually impossible outcome. A target requires a range of outcomes that are considered acceptable. There must be upper and lower limits on allowed performance. For example, the bull's-eye on an archery target is not a dot: it is a circle of sizable diameter. You can see that the bull's-eye in Figure 13.5 is not the size of a pin head. Top marks are possible by hitting within a fair-sized bull's-eye.

Figure 13.5—
Archery Target with Bull's-Eye and Tolerance Bands

"Good, Better, Best" Quality Banding

You can drive continuous improvement in job quality by dividing your tolerance for outcomes into "good," "better," and "best" quality bands. The bands specify degrees of accuracy. Figure 13.6 shows how tolerance banding can be used to challenge people to deliver high-quality work.

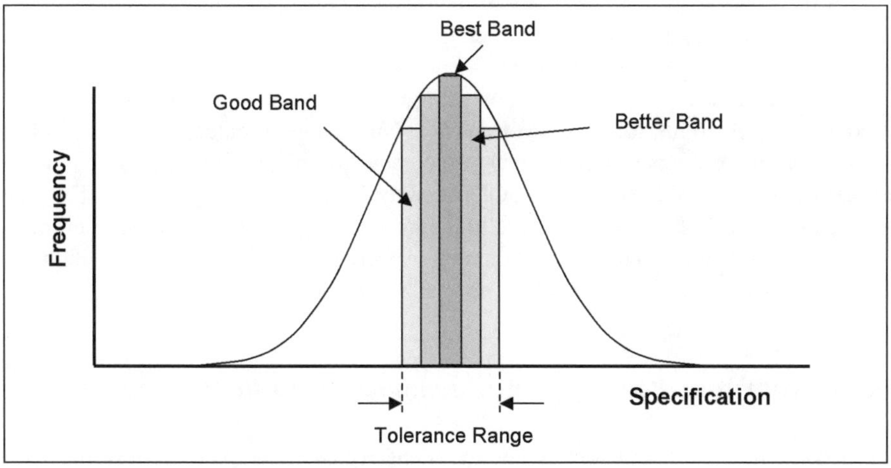

Figure 13.6—Controlling Work Quality with "Good, Better, Best" Tolerance Bands

Experts consistently achieve best-quality results. Competent people do a task to a good degree of accuracy. In between is a level of performance better than that expected of all who do the task. Using tolerance banding provides a clear indication of what is regarded as high-quality work and recognition of its achievement. Application of good, better, and best scales naturally challenges people to try to be the best. It is a simple psychological tool to improve work results.

For a standard operating procedure (SOP) to have a powerful positive effect, it needs to have clear and precise targets, tolerances, and tests—the 3Ts of masterly work—which, if faithfully met, will produce the required outcome. High equipment reliability and production performance naturally follow when work is done according to procedures written to the 3T rule. The 3Ts incorporate Series Reliability Principle 1 and the use of parallel arrangements to make work highly reliable. Tasks are made accurate through the use of masterly specifications and exact descriptions, and then a proof test is added parallel to the task to confirm that the outcome is right. Should the activity be done wrong, the proof test will identify the error. The use of the 3T rule creates statistical process control over human activity. It controls variation in work quality by focusing on task accuracy and reliability. It standardizes expert performance and ensures repeatable quality outcomes from procedures. By using the 3Ts, you limit the range of acceptable results to those you specify. Doing so gives managers and supervisors a sure way to improve workplace practices and create masterly success.

Consider, for example, this poorly defined instruction. "Inspect drive linkage for looseness." If we apply the 3T method, we might revise this instruction to read, "With a sharpened, pointed pencil, mark a straight line on the coupling and shafts of the linkage, as shown in the accompanying drawing/photo." (A sketch or photo would be provided, and if necessary you also describe how to mark a straight scribe mark.) "Grab both sides of the linkage and firmly twist in opposite directions. Observe the scribe marks as you twist. If they go out of alignment more than the thickness of the scribe mark, replace the linkage." (A sketch would be included showing when the movement is out of tolerance.) The procedure would then list and specify any other necessary proof tests and resulting repairs. With such detail provided, it is no longer necessary to employ highly qualified persons for the inspection. Anyone with mechanical aptitude can do the work reliably once they are trained. Like an automobile manual for novice mechanics, top-class procedures are written with detailed descriptions and plentiful vivid images. When do-it-yourself mechanics have such manuals in hand, they can do a lot of their own work with certainty of job quality. Once procedures contain all the information and measures necessary to rebuild an item of equipment or run a piece of plant accurately, then even people with average skills can do the job well.

The accuracy of a task can be improved by using well-formulated, clearly understood instructions that contain targets to hit, tolerances for an acceptable range of outcomes, and tests to prove that the work has been done to the required level of accuracy. When there are high-cost consequences, the first thing to do is to introduce improved SOPs to control work variability and risk. The inclusion of "target, tolerance, test"—the 3Ts of defect elimination—in all procedural tasks is the first rule of human-error failure prevention. The only solution that is better is to mistake-proof the item's design so that the design prevents mistakes from becoming inherited defects.

Train and Retrain Your People to Your Standard Operating Procedures

Having a procedure full of the best content and excellent explanations for your workforce is not, in itself, enough to guarantee accuracy. How can you be sure that users comprehend what they read? Not all people are literate, nor will everyone understand the true intent of all the descriptions and words used in a procedure. To be sure that your people know what to do and that they do it right, they need training and practice in the procedure. They need to know how to do the work correctly before they are allowed to do it unsupervised. Later, they will need regular refresher and reinforcement training. The amount and extent of training varies inversely with the frequency which with a procedure is used (tasks done often are fresh in peoples' minds and so they need less retraining), directly with its complexity (more difficult work needs more practice), inversely with the starting skill and knowledge levels of the people involved (the more you know, the less you need to be trained), and inversely with past practical experience in successfully doing the work (once you're competent, you need less retraining because you know how to do the task well).

Human memory degrades rapidly. Procedures that are correctly performed each day by the same people usually do not need retraining unless the method changes. Procedures that are done less often will need refresher training at a frequency commensurate with the risk from an error. If a procedure is dangerous and done infrequently, then retraining and practice are needed before the work is done. Training and retraining may seem like an unnecessary burden on an organization. Managers will say, "If the work is done by qualified people why do I need to train them? They have already been trained." The answer to that question is, "What risks are you willing to carry in your operation—how many defects, errors, mistakes and harm are you willing to pay for?" If an organization's risk management system uses procedures to protect the organization from trouble, then the risk protection barrier must always be working properly. It is a requirement to continuously check and prove the protection layer to ensure that the right knowledge and skills are in place and operating correctly. The training, retraining, and observation of hands-on performance helps keep that protective layer intact.

Assuming that people are "trained once, trained for life" is a serious error in judgment. For example, suppose that a connecting flange in a machine or on a pipe leaks soon after the job is completed. Flanges that are in good condition, properly rated for the service, and squarely mounted will not leak if they are bolted up right. A leak is a sign that you may need to retrain your people in the correct bolting of flanges. When a repair recurs, often on perfectly good equipment, it warns you that an SOP does not exist, or the SOP that is being used doesn't contain targets, tolerances, and proof tests, or retraining to the right procedures and practices is needed. Or maybe the procedure is sitting at the back of a shelf somewhere because no one knew what it meant.

Making Your Organization an ACE

A classic example of the value that an accuracy-controlled SOP can bring to machinery is the story of a forced draft fan bearing failure. The rear roller bearing on the fan never lasted for more than about two months after a repair. The downtime was expensive and a great inconvenience. To prevent breakdowns, the bearing was replaced every six weeks during a planned outage. It was also put on vibration analysis observation. After several replacements, enough vibration data had been collected to diagnose a pinched outer bearing race. The rear bearing housing had been machined oval when manufactured, and it squeezed the new bearing out of round every time it bolted up. You could say that the vibration analysis did wonderfully well. But the truth is, the repair procedure failed badly. If there had been a task in the procedure to measure the bearing housing roundness and compare the dimensions with the allowable target measurements, the oval-shaped hole would have been found at the very first rebuild. A badly written procedure failed the organization. An accuracy-controlled procedure with targets, tolerances, and proof tests would have found the problem on the first repair, and it would have been fixed permanently.

An accuracy-controlled procedure clearly explains each job task; there is a measurable result that is observable by the user, a range for best quality and a range for minimum quality, and a test to check the result's accuracy. With each new task allowed to start only once the previous one is within specification, you can be sure of doing a quality job. With targets and tolerances written into the procedure, the user is obliged to perform the work so that the result is within the required tolerance. Having a target and tolerance forces the user to become significantly more accurate than without them. When all of the targets have been hit, the job is done right, and the result is high quality work. If tasks are done only within tolerance, the result will be acceptable, but you will know that it is not top quality and cannot kid yourself that it is. The 3Ts automatically build defect elimination into any job. Existing procedures can be easily converted to an accuracy-controlled format, requiring only the effort and time to determine and set standards and then write clear, complete instructions for each task. The inclusion of a target with tolerances and a proof test to give feedback to confirm that each task is right before the job progresses to the next task makes top-class work quality the designed result that people deliver.

Once a procedure consistently delivers its purpose, you have developed a human-error prevention system. Job steps with target, tolerance, and tested activities are error-proofed to prevent failure and stop the introduction of defects. The work is correctly done when individual tasks are all within their tolerance limits. No longer will unexpected defects happen if work is done accurately to the requirements of the 3T procedure. Using the 3T methodology in procedures results in quality control and training documents of outstandingly high value. An organization that uses sound failure control and defect prevention systems based on proof-tested, accurate work is transformed from a quality-conscious organization to an Accuracy-Controlled Enterprise—an ACE organization. With accuracy in maintenance, operation, and engineering tasks, getting outstanding equipment reliability and consistently high production performance becomes normal.

The Value of Precision Quality

The need for precision and accuracy to control variability dominates asset-intensive industries. Precision quality is the most critical requirement for high plant and equipment reliability. Industries using machines need them to run reliably (no failures or unplanned stoppages) with high availability (ready for immediate use) and high utilization (continuously in use) for their entire working life. Outstanding reliability, availability, and utilization come from being precise and accurate in equipment assembly and operation. Precision and accuracy in equipment design, construction, use, and maintenance is a sure way to achieve a lifetime of high equipment performance and service with low operating costs. But this achievement requires the patience to develop top-class craft and operator skills and the dedication to continually apply accuracy-controlled quality assurance.

Precision is the lifeblood of equipment reliability. It results from controlling accuracy to within high quality standards. An example of assembly precision is the alignment between the two rotating shafts, shown in Figure 13.7. When two shafts are offset to each other, they run out of true and distort one another, causing massive forces to be loaded onto the bearings and drive coupling. Eventually, the bearings, coupling, shafts, and machine internals are destroyed because of the inaccuracy in the alignment. The two shafts must align with sufficient precision that they generate no destructive forces when they rotate at operating speed. When a 3T accuracy standard is set, a requirement is established that must be confirmed by measurement. An alignment standard for two shafts rotating at 1,500 rev-

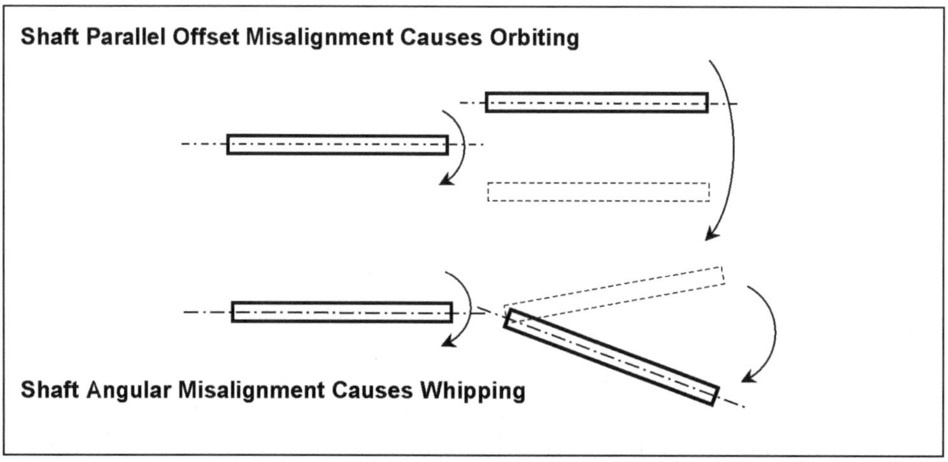

Figure 13.7—Inaccurately Aligned Shafts Destroy Machinery

olutions per minute (rpm) is to require straight shafts with the axes parallel offset aligned to better than 0.025 mm (0.001 inch) per 100 mm of coupling separation, with angular alignment of better than 0.06 degree.[3] The standard specifies how accurate the shafts need to be in line to meet high reliability requirements. The positions of the shafts can now be measured and adjusted until they are precise. Introducing 3T accuracy into workplace methods ensures everyone knows the precision that prevents defects. When achieved it translates into highly reliable equipment with outstanding availability and performance.

ACE Is a Business Culture and a Personal Philosophy

An Accuracy-Controlled Enterprise is not the same as an enterprise with a quality management system. Quality management imposes control over the processes, people, and equipment that affect the quality of a product. The ACE concept is subtly different in that it is about instilling excellence into work; it's about helping people be great at what they do. The philosophy requires people to know what an excellent outcome is for every task they do by knowing what expert results look like. From the board room to the shop floor, all roles and jobs throughout an ACE organization follow procedures with clear targets, tolerances, and tests. 3T procedures drive reliability improvement by making each person responsible for the standards of quality they deliver in their performance and guiding them to do top-quality work. An ACE workplace also permits and encourages people to improve their skills and even to change and improve the job design to make it simpler and easier for the work to be done right. Figures 13.8 and 13.9 represent the business aims of using the ACE performance quality assurance system.

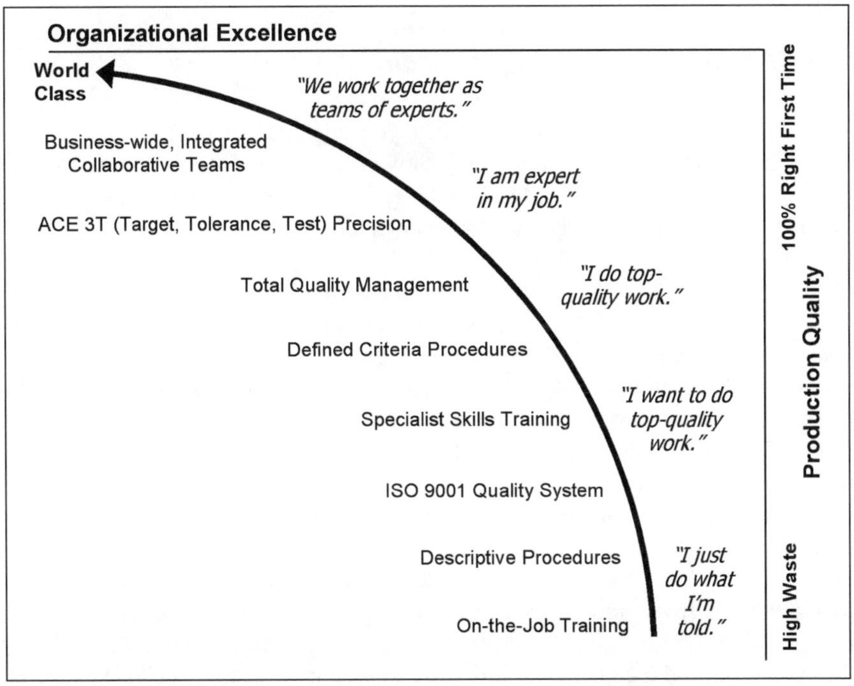

Figure 13.8—The Quality Culture of Plant Wellness Way

Figure 13.9—The People of Plant Wellness Way

Using ACE 3T Standard Operating Procedures

By eliminating human error with robust procedures, you stop equipment problems and deliver immediate positive benefits to your operation. You get lower costs, fewer repairs, fewer operating problems, fewer breakdowns, less time lost, less rework (or even no rework), no wasted materials and parts, higher availability, more throughput, less maintenance, less stock holding, and lower spares inventory levels.

The ACE 3T technique ensures that work is done correctly and accurately time after time by preventing the cause of errors. An ACE focuses on defect elimination and failure prevention. It stops operating risks dead in their tracks by making things go right so that the causes of trouble cannot exist. When you create, build, and use the right systems and techniques, you consistently get the right results by design. It is not good luck, or fate, or the "right people" that gets equipment running reliably and producing at full capacity day after day. Those results happen because the right processes and methods are in place and in use.

The statistical principle of the ACE 3T work process control method is shown in Figure 13.10. The ideal outcome of a process task is to hit its world-class target value. The poorest result allowed is the worst tolerance value. Until the measured result from the test is within tolerance, the task is repeated until it is right. In our minds, the 3Ts challenge us to become world class. It is the way of the expert.

The 3Ts go far beyond the quality control method used in inspection and test plans (ITPs). An ITP only gives you a single-sided limit—it tells you how bad you are allowed to be. With ACE 3T,

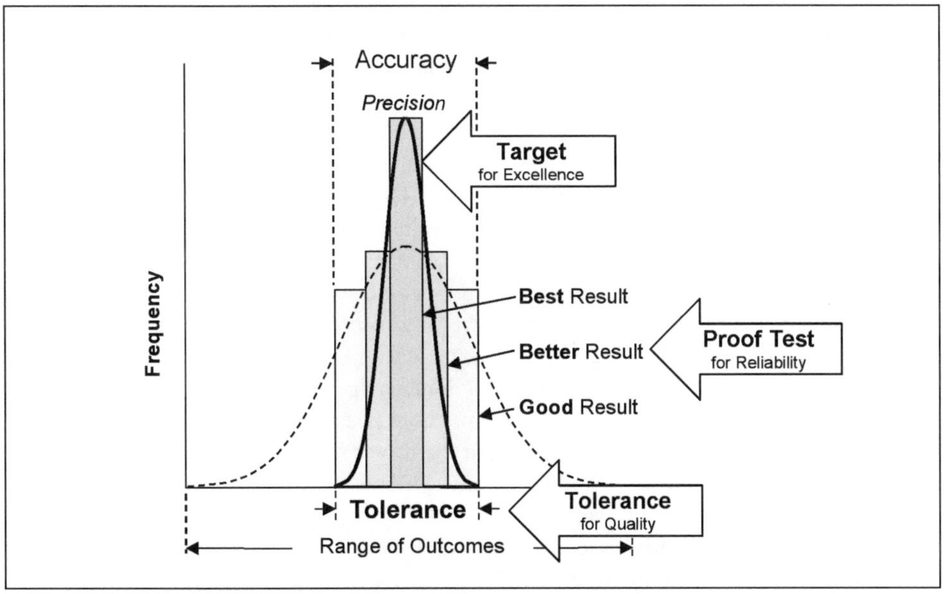

Figure 13.10—Quality Control Principles of ACE 3T Work Quality Assurance

you have two-sided limits. You have a bull's-eye that tells how good you need to be to deliver world-class results, and you have a tolerance limit on inaccuracy, after which you must correct the defect. Figure 13.11 shows the difference between using 3T and ITP to get top-quality results. Inspection and test plans never tell you what world-class performance really is! They only tell you the worst defect you can get away with in your work. Of course, you have not gotten away with it—poor quality is transferred; it just waits in your machinery and equipment for the next opportunity to fail.

The ACE approach to task process control makes work into a game. The work procedures are written to challenge people to lift their skills to the "master craftsman" level of ability. Once all work

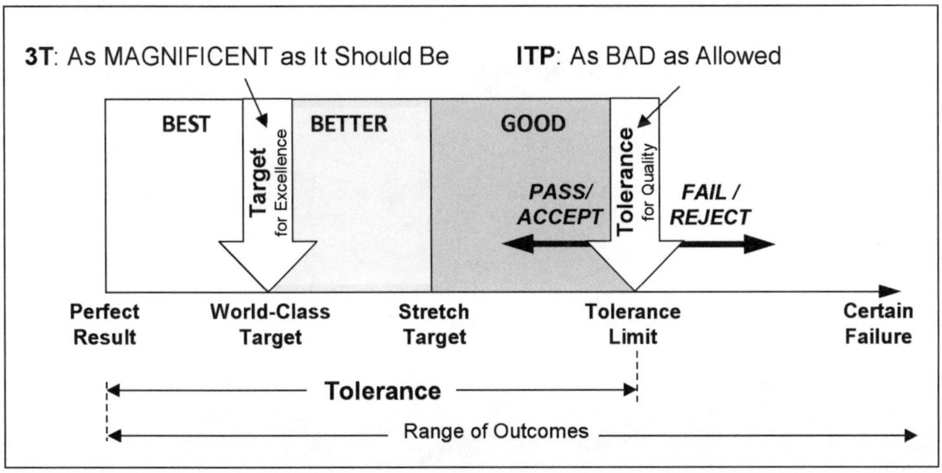

Figure 13.11—ACE 3T Procedure Encourages Masterly Work

tasks are done correctly, job success naturally follows, and your people know they have done great work. Human beings will always make mistakes. What the ACE 3Ts add to your standard operating procedures is the protection of both work quality control and quality assurance. More and more world-class outcomes are delivered as your people start hitting 3T bull's-eyes. There is a simple logic to the success of ACE procedures—be exact with your quality requirements, and your people will learn and enjoy delivering great work results.

The king of a country needed a driver for his car. The time came to interview the three short-listed candidates and make his choice. The monarch wanted to make a wise choice, so he asked the first candidate, "How close to the edge of a cliff can you drive my car at 100 kilometers per hour?" The first prospect told him, "Your Majesty, I am sure I can drive the car one meter from the edge of a cliff at 100 kilometers per hour." The king thought for a second and then told him he would not get the job and waved him away. The king posed the same question to the second prospect, "How close to the edge of a cliff can you drive my car at 100 kilometers per hour?" The second candidate said, "Your Grace, I can drive the car at 100 kilometers per hour just 150 millimeters from the cliff edge." The king looked at the candidate and told him that the driver's job would not be his and dismissed him. The last candidate met the king, and he, too, was asked, "How close to the edge of a cliff can you drive my car at 100 kilometers per hour?" The last prospect replied, "Your Royal Highness, when you are in the car, I will drive it as far from the cliff edge as is possible."[4]

The moral of the story is to not take crazy risks. Stay away from danger. If you aren't using 3T accuracy-controlled processes and procedures, you're driving your business too close to the cliff.

References

1. David Hoyle, *ISO 9000 Quality Systems Handbook,* 5th ed. (Burlington, MA: Butterworth-Heinemann, 2006).
2. Malcolm Gladwell, *Outliers: The Story of Success* (New York: Little, Brown, 2008).
3. John Piotrowski, *Shaft Alignment Handbook,* 3rd ed. (Boca Raton, FL: CRC Press, 2007).
4. Thanks to Robert Tillett of OPTIMIZE Consultants, Kampala, Uganda, for the use of his story.

14

Precision Maintenance Skills and Standards

Precision maintenance is the strict adherence to exacting machinery quality standards. It improves machines and equipment condition to a level at which no equipment-attributable quality or reliability problems occur. Precision maintenance sustains plant and equipment at the specifications that eliminate the part defects and failures that cause breakdowns and ruin production. As a consequence, it saves large amounts of money for the companies that use it.

Using highly reliable equipment with exceptional uptime that delivers unfailingly high production of top-quality product is a sure strategy for operational success. Precision maintenance brings high operating profit for the following reasons:

- Machines and equipment are built not to fail
- There is reduced need for maintenance because parts don't wear out as quickly
- It maximizes quality production and stops scrap because machines work properly
- There are far fewer stoppages and slowdowns
- Fewer spares are used because machines don't need them
- Plant availability and productivity are maximized with highly reliable machines
- Safety incidents fall dramatically because machines don't need to be touched
- Environmental incidents fall dramatically because process chemicals don't leak

Realizing remarkable machinery reliability through precision maintenance has been practiced by progressive, proactive organizations since the mid-1980s, achieving both outstanding production performance and cost reductions for all maintenance strategies.

Financial and Operating Benefits of Precision Maintenance

The two graphs in Figure 14.1 tell a remarkable story: when machine vibration levels fall, so do maintenance costs—dramatically at first, then gradually and continuously as the use of precision practices improves.[1] That means machinery does not break down. It runs for longer, and plant availability, throughput, and utilization are at their maximum. As a consequence, there is more time to make more product, at less cost, to sell for more profit using fewer people.

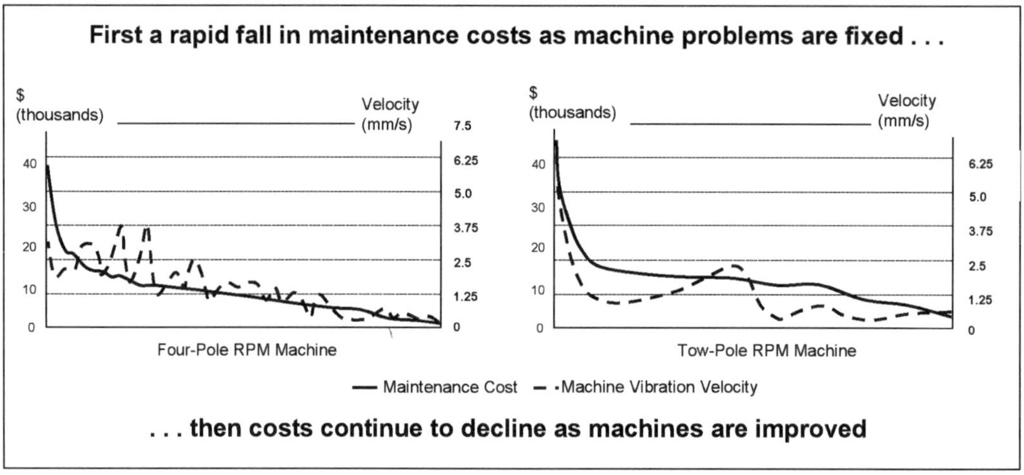

Figure 14.1—Maintenance Costs Fall When Overall Machine Vibration Levels Fall

Table 14.1 shows the results of a vibration survey in a large industrial facility for two- and four-pole electric speed machines. It records bearing vibration levels taken while the equipment is operating, along with the previous year's maintenance costs for the machines. Very low vibration for such equipment is 1 millimeter per second (mm/s) to 1.5 mm/s. If you touched the machine with your fingertips, its motion would feel like a tingling sensation. At 8 mm/s, a machine is running very rough. If you touched it, you would feel the machine shaking. At higher vibration levels, the machine would shake itself apart. The costs for equipment with low vibration are 70% to 80% less than for machines running rough. Compared with any other maintenance strategy, you will always reap the largest financial benefits when you apply precision maintenance on your machines.

Table 14.1—Maintenance Costs versus Machine Vibration

Machine Type	Highest Velocity (mm/s)	Dollars Spent Last Year	Lowest Velocity (mm/s)	Dollars Spent Last Year	Savings with Precision
Single-Stage Pumps	5.6	$3,250	2.0	$650	80%
Multistage Pumps	4.8	$6,100	1.5	$1,100	82%
Major Fans and Blowers	9.0	$900	2.8	$0	100%
Single-Stage Turbines	3.8	$8,250	1.0	$2,000	76%
Other Machines	7.8	$11,850	3.0	$3,700	69%

It is no mystery why precision maintenance lets you make more, ship more, sell more, and profit more while doing it all at less cost: it improves the operating conditions of parts within machinery and reduces their microstructure distress levels. When equipment is built to fine standards that prevent distortion and provide healthy internal conditions, it runs smoother and its parts suffer substantially less stress, fatigue, and degradation. Your maintenance people make your machines run better. This is how maintenance contributes to operating profit—by making machines run precisely so that failures don't happen and repairs aren't needed.

The Importance of Precision Standards for Machine Reliability

Exceptional equipment reliability requires a very narrow zone of high-precision quality. This level of quality puts machine parts into the "precision quality zone," where you get a massive improvement in the machine's reliability. The difference can be 5 to 10 times fewer failures over the same period, with commensurate savings in operating and maintenance costs.

Example 14.1: No One Knew How Badly the Lubricant Was Contaminated

A case study from a mineral processing refinery illustrates why quality standards are vital for machinery reliability. The refinery's worst "bad actor" machine had been a production headache for 40 years. It was critical piece of hydraulically driven equipment that stopped production every time it failed. It was so important to the business that the company paid a specialist subcontractor to be on call to maintain the asset. Regular hydraulic oil samples were sent to the laboratory for chemical analysis. Eventually, the poor reliability of the equipment could no longer be tolerated, and the company began an investigation to improve its uptime.

The latest laboratory oil condition report showed an ISO 4406 wear particle count of 23/22/19 (see Chapter 3, Table 3.4 for the range of solid particles in each value). This amount of solids contamination in hydraulic oil is so bad that it is unbelievable it could get to that point. The oil was as black as pitch. If you rubbed a drop of it between your fingers, you would feel the particles grinding across your skin. It is no surprise that the asset had been unreliable for 40 years with oil so badly polluted. One has to wonder how such a disastrous situation was tolerated for so long.

The investigation turned to the engineering procedures, which showed that no quality standard had ever been set for oil contamination. No one in the operation, nor the specialist contractor, knew that a 23/22/19 wear particle count was such a high contamination of solids that it ensured breakdowns on a regular basis. Next, the manufacturer's maintenance manual was consulted, and it, too, specified no upper limit for solids contamination in the hydraulic oil. There was one way to find out where the manufacturer might have specified an allowable ISO 4406 count. The make and model of the hydraulic pump subassembly was tracked down, and the pump manufacturer's maintenance manual was located on the Internet. The manual specified a maximum solids contamination of 18/16/13. The hydraulic oil in the circuit was contaminated more than 30 times the amount allowed by the pump maker. Forty years of equipment failures had occurred because there was no engineering standard for maximum solids contamination. Without a standard to test against, the laboratory analysis meant nothing to the people who were reading the report. As soon as the manufacturer's oil contamination limits were known, the cause of those never-ending breakdowns was obvious.

Example 14.2: Shaft Quality Control Gets Rolling Bearing Reliability

After years of suffering monthly spherical roller bearing failures on a head pulley at a steel mill, the maintenance manager brought in the bearing manufacturer to investigate why the bearing on a 150 mm diameter shaft never lasted for more than four weeks in service. The production downtime had become unacceptable, and an investigation was started to address the problem. Wisely, the bearing manufacturer was asked to be involved with the Root Cause Failure Analysis.

Spherical roller bearings of the design used come with a tapered support sleeve that sits on the shaft journal and carries the inner ring of the bearing. The pulley drove a critical mill conveyor that stopped the operation when it failed. To prevent failure during production, the conveyor was handed over to the maintenance crew every three weeks to replace the bearing. This went on for years. Roller bearing service life is measured in thousands of hours of failure-free operation, so when a roller bearing gets only weeks of service life, it is clear that something is happening to the bearing to make it fail. First, the manufacturer asked the maintenance group for the tolerance and form values of the shaft under the bearing. The measurements were not known, and it was agreed they would be recorded at the next bearing replacement.

Tolerance affects the tightness of fit between a bearing and its shaft and a bearing and its housing. A loose fit permits oscillation and vibration, while an overly tight fit squeezes the bearing's parts together and stresses the contacting surfaces. To ensure that the right fit is made for roller bearings, manufacturers publish the allowed fit and tolerance for the housing and shaft that each bearing model requires to get a full service life.

The form indicates the shape of the shaft in the region under the bearing. The shaft must be sufficiently cylindrical, sufficiently round, and sufficiently straight so that the bearing inner ring is supported at enough contact points with the shaft to keep its structural integrity and shape during rotation under maximum load-carrying service.

Both the correct tolerance and form must be present for a bearing to reach its designed operating lifetime. When the bearing was next changed, the journal dimensions were taken with a micrometer and checked against the manufacturer's requirements. Table 14.2 lists the shaft tolerances and form shape that must be met for full bearing service life. The 150 mm shaft diameter needs to be h9 tolerance—a permitted maximum size of exactly 150.0 mm and a minimum size of 149.9 mm—and also IT5 form—from Plane A to Plane C, the straightness, roundness, and cylindricity are all within an 18 μm annulus—under the bearing. The shaft diameter values measured at Planes A, B, and C are recorded in Table 14.3.

Table 14.2—Spherical Bearing Manufacturer Shaft Dimensions

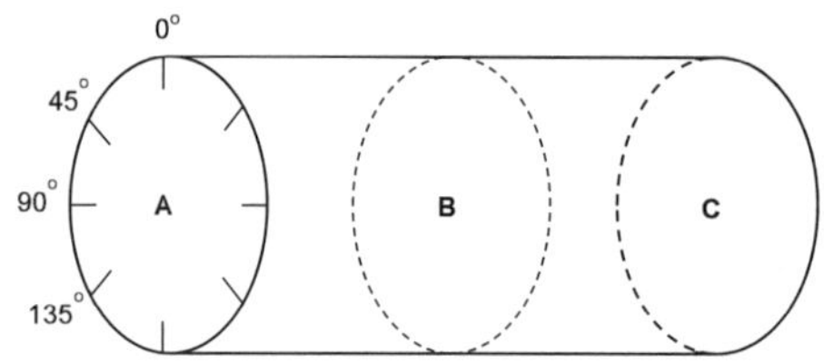

Shaft Diameter (mm)		Tolerance h9 (μm)		Form IT5 (μm)	Form IT7 (μm)
Over	Incl	High	Low	Max	Max
22	30	0	-52	9	21
30	50	0	-62	11	25
50	80	0	-74	13	30
80	120	0	-87	15	35
120	220	0	-100	18	40
220	250	0	-115	20	46
250	315	0	-130	23	62

Tolerance Evaluation

All journal dimensions were inside the required h9 tolerance. But there was still the mandatory IT5 form shape to be proven. In Tables 14.4 and 14.5, the problem with the shaft's shape is identified.

Cylindricity Evaluation

The maximum amount that the shaft is allowed to deviate from a perfect cylinder is no more than 18 μm on the diameter at any point in the region under the bearing. Table 14.4 shows that the difference between the maximum and minimum measurements along the 0 degree line on top of the shaft at Planes A, B, and C were within 10 μm—well inside the maximum 18 μm tolerance. But that was not the case for the corresponding three points along the 45-, 90-, and 135-degree lines. Along those

Table 14.3—Measured Shaft Dimensions

	0°	45°	90°	135°	Required Tolerance h9
Plane A	149.98	149.99	149.98	149.99	> 149.900 < 150.000
Plane B	149.97	149.94	149.98	149.95	
Plane C	149.98	149.98	149.95	149.99	

positions, the form is well outside of tolerance, and the shaft is not cylindrical. The measurements prove that the shaft beneath the bearing was undulating in many places and could not fully support the inner ring. There were gaps between the shaft surface and the sleeve supporting the inner ring that allow the ring to move up and down as load-carrying rolling elements move over the cavities. The flexing of the inner ring rapidly fatigues the metal, and the bearing failed as a result.

Table 14.4—Shaft Cylindricity

	0°	45°	90°	135°	Plane Average	Required IT Grade 5	IT Grade 7
Plane A	149.98	149.99	149.98	149.99		< 0.018	0.040
Plane B	149.97	149.94	149.98	149.95			
Plane C	149.98	149.98	149.95	149.99			
Max-Min	**0.01**	**0.05**	**0.03**	**0.04**			

Roundness Evaluation

The form requirement for shaft roundness can also be identified from the measured values. Table 14.5 indicates that the shaft at Plane A was out-of-round by 10 μm, at Plane B by 40 μm, and at Plane C also by 40 μm. Planes B and C were out-of-round by more than twice the manufacturer's worst allowance.

Table 14.5—Shaft Roundness

	0°	45°	90°	135°	Plane Max-Min	Required IT Grade 5	IT Grade 7
Plane A	149.98	149.99	149.98	149.99	**0.01**	< 0.018	0.040
Plane B	149.97	149.94	149.98	149.95	**0.04**		
Plane C	149.98	149.98	149.95	149.99	**0.04**		

The shaft passed on tolerance but failed on form shape—it was not cylindrical enough, it was not round enough, and it tapered toward the center. The roller bearing could not reach full service life because the journal was badly misshapen. The shaft under the bearing mounting sleeve did not support the bearing sufficiently to prevent flexing from the load of the rollers. Figure 14.2 exaggerates the problem to help explain what was wrong with the shaft.

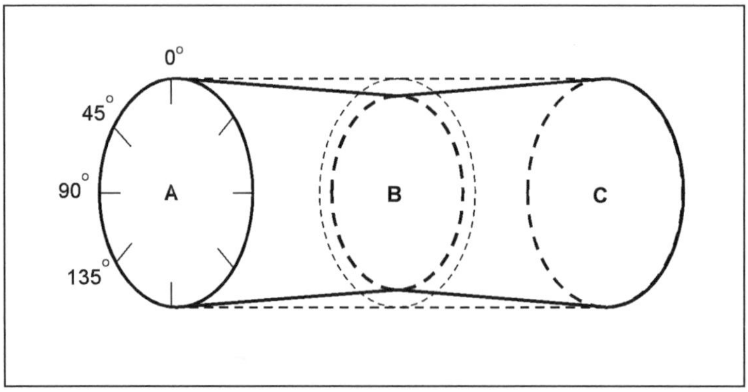

Figure 14.2—Roller Bearing Reliability Compromised by the Shaft Condition

The shaft condition became evident once the right quality assurance checks were done. The bearing manufacturer's standards for shaft form were not being met, but no one in the steel mill knew that because it was not a required maintenance inspection. The form shape checks immediately found the problem, and its solution was clear: replace the defective shaft with a new one machined to within the required tolerance and form.

The tolerance and form values for roller bearings have been published for generations. The information that would have exposed the defective shaft has always been available, but it was not a part of the repair procedure, and the shaft was never checked for its actual shape. Had the shaft tolerance and form check been included as a quality characteristic inspection every time the bearing was replaced, the maintainers would have spotted the problem and replaced the shaft many years earlier and each time the journal wear became excessive.

Your machinery parts must always be in a least stress, contaminant-free environment, or else high reliability is impossible to achieve. It is necessary to know all the vital factors that affect a machinery part's reliability and the exact values for the quality standards needed to put a part into its precision zone. Setting those values is the foundation of enterprise asset management success because with those values, you have a target at which to aim your asset management system and its life-cycle processes. They are published in your ACE 3T engineering, operating, and maintenance procedures, and your people use them to prove that equipment health and operating conditions meet the level of quality that produces the intended reliability. If a required standard is not met, you and your people immediately know there is a defect and can act to solve its causes.

The ISO vibration limits for machinery are given in ISO 10816 and its subparts. Each subpart of the standard is specific to a particular type of machine and a range of service duties. ISO 10816-3 relates to general industrial machinery, such as gearboxes, electric motors, and bearing housing. It provides warning and alarm vibration velocity values for working equipment. Table 14.6 is compiled from ISO 10816-3.

Most companies take the vibration level values literally and base maintenance interventions on the vibration range that damages equipment. They set warning limits at the restricted operation value and action limits at the prefailure level. As a result, they guarantee that they will have occasional breakdowns and that their mechanical plant and equipment will regularly undergo randomly sched-

Table 14.6—ISO 10816-3 General Machinery Vibration Levels

ISO 10816-3	Medium-Sized Machines		Large Machines	
Machine Type	Group 2		Group 1	
	15 kW–300 kW		300 kW–50 MW	
	Motor Shaft Center 160 mm–315 mm		Motor Shaft Center Above 315 mm	
Foundation	Rigid	Flexible	Rigid	Flexible
Velocity mm/s rms	Machine Damaged			
11.0				
7.1				
4.5	Restricted Operation			
3.5				
2.8	Unrestricted Operation			
2.3				
1.4				
0.7	New Machine			
0.0				

uled maintenance and suffer higher maintenance costs. Because the maintenance philosophy is to wait until the machine condition is poor and needs to be rectified, these companies incur greater equipment damage, with shorter time between failures and higher production costs. In an Industrial and Manufacturing Wellness site, you adopt the wise adage "a stitch in time saves nine" and use a proactive approach so that vibration levels never get beyond new machine condition values.

In order to get machine vibration values below 1.5 mm/s, it is necessary to guarantee that you have only miniscule misalignment between equipment components and assemblies; that there is barely detectable imbalance; that there is no body or frame distortion when mounting the equipment; that frame and body expansion or contraction do not distort internal components; that all parts have precision flatness, precision straightness, precision roundness, precision squareness, precision form, and are precisely located; that the equipment is fixed on substantial, strong mountings to which it is correctly and accurately fastened; that shaft revolutions per minute are not at resonating speeds; that bearings are selected to carry the correct load so that rolling elements roll properly and do not skate over the lubricant; that journals are sized and in tolerance to get the correct bearing clearance at operating duty without looseness or causing oil whirl; that housings are rightly sized and in tolerance; and the lubricant is impeccably clean and its chemical composition remains at specification. Once you have all that right, there are no vibration problems. It's a daunting list of precision requirements that must be achieved. Starting from the design phase of the life cycle and extending into the operating phase, it is necessary to control dozens of failure mechanisms. But until you control the outcomes of all the issues listed above, there will regularly be machinery vibration problems and no certainty of reliability.

In the Plant Wellness Way, because the causes of machine vibration are removed or prevented, you do not need to monitor vibration levels, and the predictive maintenance program only consists of lubricant analysis when it is financially justifiable. Instead, you monitor the variables that cause the established quality conditions to change. You observe the operation of the machine and process conditions for circumstances that will produce vibration if not corrected—such as the causes of material buildup on rotating equipment that unbalance the rotor, process conditions that attack or corrode microstructures, severe speed and momentum changes, abnormal temperatures causing expansion and contraction of components and the subsequent misalignment of shafts and other internal parts, pressure "hammer" in pipework, along with monitoring all other applicable Physics of Failure cause mechanisms—and then immediately correct the situation back to the stable design envelope quality range. The Plant Wellness Way gives you incredibly low maintenance costs and high uptime because your machinery is always kept in precise, good-as-new condition. Your maintenance and operating strategy is to create and sustain the precision quality needed to produce plant and equipment parts health and wellness for the whole service life of the machine.

In 2002, I visited a Sumitomo Chemical process plant in Japan on behalf of my employer. The host took me on a tour of the operation, and as we passed a pumping station, I could hear the rumbling sound of centrifugal pump cavitation. The process used hot caustic soda solution at elevated temperatures near the chemical's boiling point. The chance of pump cavitation is high at such temperatures, and it takes only a small blockage in the suction pipe to lower the pressure enough to make the pump cavitate.

At the conclusion of the hour-long tour, we went past the pumping station again. A maintenance technician was packing up his tools after swapping out the cavitating pump with a replacement. Curious to learn about the predictive maintenance employed on site, I asked my host, "What type of condition monitoring technology is used on the pump?" His answer surprised me.

Apart from the operator using his physical senses to observe pump performance, the pump was never condition monitored. The plant employed precision maintenance for all rebuilds, and its machines were installed in perfect condition. When the pump cavitated, it was caused by the process, not by poor maintenance work quality or component degradation. Putting the pump into a condition monitoring program was pointless because it would never prevent the cavitation. There was nothing more to be gained by instigating a condition monitoring program for the pump. His explanation was technically sound and financially sensible. It also made clear to me how precision maintenance brings worth and profit to an industrial operation that uses it.

Precision Maintenance Program

The following list includes the important requirements for a successful precision maintenance program:

1. Accurate fit and tolerance at operating temperature
2. Impeccably clean, contaminant-free, chemically correct lubricant lifelong
3. Distortion-free equipment for the entire lifetime
4. Shafts, couplings, and bearings running true to center
5. Forces and loads into rigid mounts and supports
6. Exact alignment of shafts at operating temperature
7. High-quality balancing of rotating parts
8. Low total machine vibration
9. Correct tension in all fasteners
10. Correct tools and test equipment in the condition to do the task precisely
11. Only in-specification parts installed
12. Application of precision skills and techniques

13. Creative disassembly to find and remove defects and failure causes
14. Proof tests that precision is being achieved
15. Business processes to consistently apply these requirements successfully

There is nothing in this list that should not already be standard practice in every industrial operation. But that hardly ever happens. The exacting standards required to deliver excellent equipment health that produces failure-free operation are not specified in many company quality systems. They are not taught to managers and engineers or to operators and maintainers. Worst of all, they are not seen as important to those supervising work quality, and they are of no interest to the people running industrial companies, despite the fortunes in operating profits that are made through the achievement of precision quality targets. Without precise standards everyone works to their own degree of accuracy, which, as a result of misunderstanding and confusion, introduces wide variation in work performance that causes inaccuracy and creates errors. Thus, defects are generated all the time, eventually triggering operating problems and equipment failures. The cause-and-effect sequence is as predictable as night following day. But it does not need to be that way.

Creative Disassembly and Defect Removal

For the thousands of defects existing within plant and equipment waiting to become failures, your maintenance and operations people use creative disassembly to fix them.[2]

Creative disassembly requires the identification of flaws in machinery and their immediate correction. It operates at the equipment part level and strikes at the heart of the thousands of defects making up the base of the equipment failure pyramid (see Chapter 10). By reducing the number of defects in machinery, fewer opportunities present themselves for catastrophic failure. The plant operator and maintenance technician become the root cause analysts for their operating plant and equipment. Instead of operators only running the plant and maintainers only replacing parts and doing maintenance, they team up and become responsible for finding the cause of a failure and correcting its root cause. They are given the authority to follow through and do all of the necessary work, including scheduling production outages to do the precision maintenance needed to prevent repetition of a problem.

There are three phases to creative disassembly analysis: before shutdown, before stripdown, and during stripdown. As the work is done, comparisons are made against specified ACE 3T precision standards. When defects are detected, they are removed or corrected during rebuild, and the equipment is returned to the right precision standards for high reliability.

1. **Pre-shutdown data collection.** Before an item of equipment is worked on, its history, information, and service condition are checked for telltale signs of potential trouble.

    - Collect records from the CMMS, parts usage, repetitive maintenance, and operating problems
    - Collect condition monitoring data such as vibration and bearing characteristics, thermography, oil analysis, etc.
    - Check for running soft foot and machine distortion while operating; identify resonance problems and poor supporting and hold-down structures

2. **Measurements and detailed observations taken at shutdown but before stripdown.** Once you've got safe assess, look over the equipment in detail, noting what is there, what is not there, and what should be present. This is when engineering, technical, and operating quality control standards are valuable for comparing real conditions against design requirements.

    - Where thermal growth occurs, collect hot growth and alignment readings
    - Identify witness marks showing relative movement between parts
    - Notice the presence of unusual deposits from wearing parts such as drive belts and couplings
    - Take lubricant samples for analysis and patch testing of wear particle count while still hot
    - Check for static soft foot distortion problems

3. **Stripdown measurements and investigative observations.** As the equipment comes apart, inspect for evidence of both poor and good operation. Take important measurements of components and compare them with the manufacture's limits to identify problems.

    - Look for witness marks and telltale evidence of incorrect operation and behavior
    - Mark relative positions of bearings to confirm correct location
    - Inspect bearing wear patterns for evidence of spalling and other failure modes
    - Note incorrect roller or race motion, cage damage, fretting corrosion, out-of-roundness, shaft straightness, etc.
    - Inspect for damage and wear patterns on moving parts such as gear teeth, pulleys, belts, etc.

Take time to do the job of creative disassembly and precision rebuilding well. It will lead to world-class equipment performance as more and more defects are removed from your plant and machinery.

Introducing a Precision Maintenance Program

Once maintenance, operations, and production managers learn about precision maintenance, they will acknowledge that it is a great concept and totally valid—although few implement it. Yet there is no other way to get lasting world-class industrial plant and equipment reliability.

Becoming a precision-focused organization starts by determining and setting 3T quality standards for every aspect of plant and equipment operation and care. Each standard is set to prevent or control the Physics of Failure mechanisms that affect the equipment. These standards should push your people to become better and your business to become a world-class performer. Every aspect of operating machine life gets measurable values to define its precision zones. Equipment is then run within the specified precision zones. Maintenance is done so that parts function within their precision zones. Rebuilding machines is done to precision specifications. You make sure your business processes deliver equipment that meets the precise standards that will always produce outstanding production success.

One way to start a precision maintenance program is to introduce precision requirements into everyday workplace practices. Everything that relates to plant and equipment health will need to meet those standards. For the organization, the symbiotic reality is that eventually its influence needs to extend to controlling the quality from the original equipment manufacturers, setting the project engineering and design specifications, ensuring parts reliability in procurement and storage, and quality-assuring plant and equipment installation, operation, maintenance, and all outsourced subcontract work. It requires confirming that the accuracy of the work performed is to precision standards. You will start keeping maintenance records, measurements, and photographs to prove how well equipment was built and rebuilt, what was used to build it, the exact conditions in which it was built, and how it was operated and maintained during its service life. You keep and use health condition records to prove that components are well and fine and need no further work to be done on them. You monitor operating variables for trends and changes as evidence that your production processes are in control or that they are not. Nothing during the life cycle affecting the health and wellness of your equipment is left to chance. Every critical part on every machine and every piece of equipment needs to meet engineering, maintenance, and operating standards that guarantee they are in great condition. When those standards are not being met, you want to know why and what will be done to return the parts back to health. The standards are measurable: they define the "engineering numbers" that are proof of compliance with requirements. With measurements to prove that the ideal conditions are present, you have the best chance of getting top-quality machinery. You can be confident that the work done on your plant is right and that your equipment will run precisely because the parts are operating in their precision zones.

All critical parts in your machines, whether they be mechanical, electrical, control system, or structural, need quality standards addressing all vital health requirements:

- Distortion
- Looseness
- Lubrication
- Cleanliness
- Shaft alignment
- Balancing
- Temperature range
- Vibration
- Manufacturing accuracy
- Surface finish
- Installation accuracy
- Tools and their condition for use
- Skills and the necessary competency
- Job history records to keep
- Calibration accuracy of test equipment
- And everything else your equipment parts need for a lifetime of low stress, health, and wellness

The original equipment manufacturer's standards spell out the minimum quality you must achieve. But they are not world-class standards. Precision maintenance standards are typically a magnitude better than what the manufacturer gives you. You need to research and agree on the precision zone quality values for your equipment because those values will be exacting of your people and service providers. Their continuous achievement requires a different approach to doing work than what was done in the past. The following is a short list of some of the necessary mechanical quality controls that should be set in a precision maintenance program. Each one must have a value that can be tested so that workplace results can be proven to either meet, exceed, or fail the standard.

- Correct tension for every fastener
- Most and least number of threads protruding from a tightened nut
- Maximum size and amounts of contamination that are acceptable in lubricant
- Position along a shaft from a datum to place a component
- Size, dimensional tolerance, and shaft form at a bearing location
- Amount of damage to a part that is acceptable before it is replaced
- Distortion in a gearbox case before misaligned holes are re-bored
- Flatness of a baseplate to ensure there is no soft foot
- Exact alignment accuracy between shafts in a drive train
- Cleanliness of workshop benchtops and atmosphere so that parts are not contaminated
- And everything else your equipment parts need for a lifetime of low stress, health, and wellness

To those who are uninitiated in precision maintenance, the task of setting up a program seems enormous. It is not so. Many standards are universal, and the necessary standards that produce outstanding reliability are already known. What needs to be done is straightforward and methodical. Training and practice in precision techniques is required, but even that is readily available.

Successfully applying precision maintenance requires competence in best-practice precision skills, supported in the workplace with the correct tools and measuring devices, along with a top-class engineering body of knowledge about the necessary machinery and maintenance standards. If you want equipment in your operation at consistently high reliability, your maintenance and operations people need to develop the higher work skills and quality practices that produce that reliability. They may not yet have such expertise and proficiency. Getting those skills requires setting high levels of excellence and then training people to meet them. Many managers, operators, and craftspeople will not believe they need such high skills in their operation. This, of course, is a fundamental error in their thinking, and it explains why many businesses undertaking reliability improvement efforts still suffer poor availability and breakdowns. They do not realize that reliability results from meeting precision quality standards. If the work processes in use are incapable of delivering that quality, they must be changed to those processes that naturally produce the required precision. Every practice and technique you use can only deliver the accuracy it is capable of producing. You cannot use imprecise practices and expect precise outcomes every time. That is impossible. Companies will always fail in their improvement efforts until they change to practices that naturally guarantee success every time they are used.

Setting Precision Quality Standards for Your Equipment

The solution to equipment reliability problems starts with setting reliability standards that bring sure reliability. The standards you need already exist and have existed for decades. Your challenge is to make them standard practices achieved all the time in your operation. The following list is an example of some of the books and international publications where you can find the necessary information and guidance.

- Accurate fits and tolerance—*ISO/ANSI Shaft/Hole Tolerance Tables*
- Clean, contaminant-free lubricant—*ISO 4406*
- Distortion-free equipment—*Shaft Alignment Handbook by John Piotrowski*
- Shafts, couplings, and bearings running true to center—*Shaft Alignment Handbook*
- Forces and loads into supports—*Shaft Alignment Handbook*
- Accurate alignment of shafts—*Shaft Alignment Handbook*
- High-quality balancing of rotating equipment—*ISO 1940 and Parts*
- Machine vibration—*ISO 10816 and Parts*
- Correct torques and tensions—*ISO/ASME Bolt, Stud, and Nut Standards*
- Correct tools in correct condition—*"As-New Specification"*
- Only in-specification parts—*OEM specifications, Machinery Handbook*
- Application of precision skills and techniques—*Precision Skills Training and Practice*
- Failure cause removal—*POF based Creative Disassembly*
- Proof test—*Precision Measuring Tools; Condition Monitoring Technologies*
- A system to use the standards successfully—*ACE 3T*

This list of publications and techniques may be incomplete for your operation's needs, and you may have to look for additional standards. Note that the benchmarks used in international standards are usually not of sufficient quality for a precision maintenance program. The quality parameters in international standards are minimum requirements that can be used to set the "good" boundary on quality performance in your ACE 3T procedures. But they are not world-class standards. When you set the "best" 3T standard, use the most demanding recommendations of subject matter experts in their field. When a documented world-class best standard cannot be found, then set your best quality at a value a factor of 10 (i.e., one magnitude) finer than the 3T good tolerance.

Accuracy-Controlled Maintenance Quality System

Item 15 in the list of key precision maintenance requirements is the glue that keeps the rest together. It requires installing a work quality management process to ensure that the other requirements are delivered to every machine and equipment in your operation. In Industrial and Manufacturing Wellness, you use Accuracy-Controlled Enterprise procedures to drive equipment reliability and production results by making precision quality your company's standard practice. You solve equipment performance problems forever because ACE procedures contain exactly the information you need to produce outstanding reliability. More importantly, ACE quality standards let you make precision maintenance a habit throughout your operation.

The beauty of the ACE methodology is you set the "good" band value at what your people are capable of doing today. It lets everyone start from where they are right now and move upward toward the excellent performance you set in the "best" quality target. Within your ACE 3T procedures is a realistic step-by-step plan for achieving the world-class quality that brings outstanding reliability and Operational Excellence.

In a nutshell, introducing a precision maintenance program consists of the following:

1. Corporate approval to implement precision maintenance and precision practices
2. Agreement across the operation on the plant and equipment to be precision maintained

3. Agreement across the operation on the precision standards to meet
4. Agreement across the operation on the best practices to be applied to meet the standards
5. Agreement across the operation on the measurement methods to prove compliance with standards
6. Written ACE procedures for all operating, maintenance, and inspection activities on the selected plant and equipment
7. Necessary test equipment, specialist tools, and facilities
8. New skills learned through on-the-job training and expert support
9. ACE procedures applied and refined
10. Monitoring of the effect of the program on plant performance and operating profits
11. Continuous improvement in the use of precision skills and practices
12. Extension of the program to other plant, equipment, and operating sites

You will only have done the job of introducing precision maintenance well when you've done the following:

- Published precision standards for equipment operation and maintenance company-wide
- Held seminars to explain and discuss the standards with all the people who need to know and use them
- Purchased the measuring and testing equipment needed to prove compliance
- Written ACE procedures for all activities related to achieving needed precision quality
- Trained people to the standards, and ensured that they can achieve them competently
- Deployed a document and record management system to collect all important information about the equipment that allows everyone fast access to the knowledge they need to make right decisions

Too few companies are that good, but this is what the Plant Wellness Way is designed to do for organizations that adopt it as their reliability creation methodology.

Engaging the Workforce

What matters most in achieving reliability success is the skills and knowledge of the shop floor people working on your plant and equipment. If you want reliability, you need to bring the engineering knowledge and maintenance work practices of the crew and subcontractors maintaining the equipment up to a level at which they can deliver world-class machinery performance. For precision maintenance to work, it needs maintenance crews and their supervision to want it and to master the necessary new skills. It requires the right engineering know-how and hand-skill competence to be used by the maintenance technician. Precision maintenance knowledge is needed, but precision maintenance skills are tactile and people must work on machines in the right ways to get highly reliable operation from them. It requires procedures to be used in a very specific way to get statistical quality control of maintenance work. When done properly, your precision maintenance program will maximize production from the efforts of highly skilled craftspeople for the least maintenance cost.

Although your shop floor people deliver precision maintenance, it is maintenance and operations managers who start the change, sustain it, and keep improving it. The great problem for industry is to find a reliable way to introduce the necessary changes in working practices so that precision thinking becomes the natural way to work. Climbing to the heights of precision maintenance success needs a safe, sound, and encouraging method to change the way people work. There needs to be a safe approach for equipment maintainers to gain understanding of precision maintenance—its work quality requirements, the skills needed, and the procedural methods to put and keep parts in their precision zones of operation.

Starting a precision maintenance program requires a well-thought-out and structured change management process that will help your people to learn and work to new, higher-skilled practices.

You do this by using the "Change to Win" change management program, which is explained in an accompanying workbook (available online). The Change to Win program gets engineers, the maintenance crew, and maintenance supervision together in setting higher quality standards and helps the organization to recognize the need to upskill to meet those standards.

Precision Maintenance Results the Plant Wellness Way

For rapid success with precision maintenance, the approach recommended in IMW is to develop a core team of specialists who undertake the development and delivery of a precision maintenance program. These people do the precision maintenance tasks and create highly reliable equipment. The rest of the maintenance crew and the equipment operators sustain the new level of performance. This is the simplest and quickest way to deliver the most operating profit from a precision maintenance program. Few people have the capability to be naturally great precision maintainers. It takes a great deal of training to get the needed skills and continuous practice to keep them top class. It is fastest to concentrate training efforts on the few people who will become specialists rather than trying to train the whole crew and getting mediocre results. As well as doing the precision maintenance work, the specialist team introduces precision methods to the rest of the maintenance crew, the operating crews, the engineering group, and the supply chain vendors that service or support your plant and equipment. These precision maintainers write procedures, deliver on-the-job training to people on how to do precision work, ensure work quality assurance, and work with the operations teams to monitor the stability of equipment and process conditions to ensure that component materials of construction have long, trouble-free lives.

References

1. Ralph T. Buscarello, "Vibration and Its Impact on Reliability and Costs" (presentation, Update International, Lakewood, CO, 1990), accessed at http://www.tappi.org/content/events/10pirm/paper/buscarello.pdf, August 22, 2015.

2. Peter Brown and Max Wishaw, "Precision Maintenance for Engineers" (course book, Industrial Training Associates, Perth, Australia, 2000).

15

Process 4– Introduce Risk Control Solutions

Summary Description of Process 4: Introduce Risk Controls

Once a mix of operating risk controls is chosen and proven on a risk matrix to be useful and effective, those controls need to be introduced into your business. Because the current business processes are producing your current business outcomes, the new risk management controls must be put into the old processes to improve work performance and company results.

Infuse Your Processes with Defect and Failure Elimination Actions

Identify all of the business processes to be enhanced with the new defect elimination and reliability creation standards and practices. Typically, these will be all of the processes across the organization's lifetime that affect operational success and equipment reliability.

Set Operating, Maintenance, and Engineering ACE Quality Standards

Set quality standards for the reliability improvement activities identified in the Physics of Failure Reliability Strategy Analysis. Meeting those standards will produce the operating equipment performance and high reliability you want from your plant. This will require you to set "good," "better," and "best" quality standards in order to produce healthy, long-lived, trouble-free equipment. Some research will be required to determine the appropriate value ranges for job tasks and to decide which tests to use to confirm compliance with the quality standards.

Write ACE 3T Procedures

Every job requires procedures that will deliver highly reliable results. The person doing the work needs to understand the excellence and accuracy required of the job. To ensure the adoption of the correct operating risk controls and to transfer them to the company's management, staff, and employees, good, better, and best quality standards are made a part of your business processes by writing

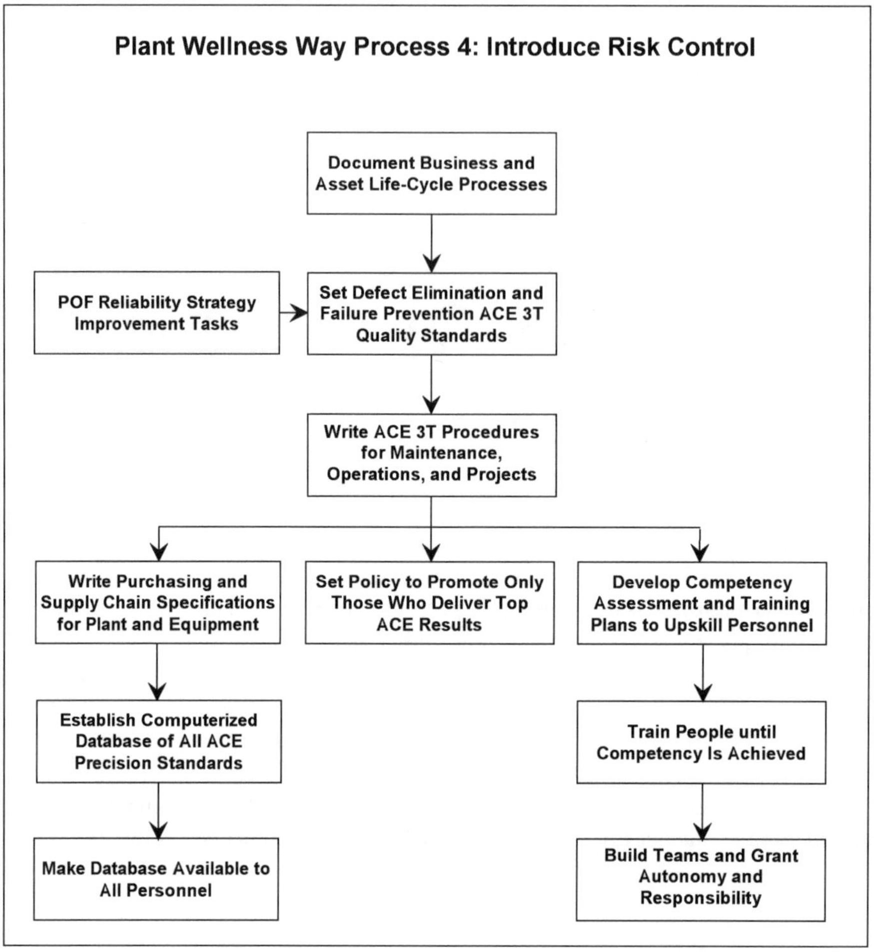

Figure 15.1—Process 4 Steps

them into all relevant documents. Doing so will make them standard operating procedures that will be taught to and applied by all users. The day will come when world-class performance is the only way the people in your company know how to work.

In an Accuracy-Controlled Enterprise, the quality standards that will take your operation to world-class performance are embedded in work tasks as target, tolerance, and test requirements. Because we seek to prevent process variation and control outcomes, a world-class target range is set for all outcomes of a process. Around that target, we identify the worst acceptable range of quality as the tolerance. Finally, you define a test to prove what quality is actually being achieved.

Write Plant and Equipment Procurement and Service Provider Specifications

The performance standards that you set for your company are the minimum equipment and service specifications that every vendor in your supply chain must meet. They define the performance that your equipment is expected to deliver and the quality standards that will deliver those results. Specific targets must be set with measures to prove that a vendor's performance meets the standard.

Develop a Computerized Knowledge Database for All to Use

The ACE standards, specifications, and procedures should be put into a database that everyone in the company can access. They are valuable best practices and important corporate knowledge to be used all the time. Your people will get the right information to run the operation in the best way to ensure least operating risk.

Make Delivery of Defect Elimination and Failure Prevention Everyone's Duty

Include in management, supervisory, staff, employee, and contractor duty statements and position descriptions the requirement to deliver ACE work quality results. Make it clear that recruits for future vacancies will be drawn from those who deliver high-quality performance. You will become a world-class operation faster and for the least cost when your staff and employees know that the best way to advance in their career is to deliver defect-free work of high reliability that eliminates the possibility of future problems.

Develop Training and Competency Improvement Plans

With performance standards and 3T procedures set, develop training plans to lift the competency of managers, engineers, supervisors, staff, and workers to meet the required ACE performance.

Build Autonomous Cross-Functional Teams

Establish cross-functional teams of people who are responsible for running a process. Subdivide large processes into smaller ones if necessary. Whether you are making a product or providing a service, use series and parallel reliability principles to build teams that have the skills and knowledge to do the required work competently. Remove all direct management supervision of the team and instead provide training to team members to develop the knowledge and skills they need to work as a team. You want to create teams with positive spirit. Let the team members share in the additional operating profits they generate above the best historical average. This will challenge them to seek better solutions while being recognized for the value their efforts bring the organization.

For more information on this topic, see the document *Organizational Structure and Teams,* which is included in the supplementary materials available online.

Setting ACE 3T Requirements

The Accuracy-Controlled Enterprise ensures defect and failure prevention by introducing quality controls for Physics of Failure cause mechanisms so that work is done correctly to eliminate each mechanism or minimize the chance that any will occur. It incorporates the "Carpenter's Creed"—measure twice, cut once—into job tasks. It uses parallel arrangements to error-proof work and lift the certainty of right performance to incredibly high levels. Process steps in an ACE have double-checks in every task. No task step is complete until it is proven to be right.

There are simple rules to use when setting the quality values in a 3T procedure. The "best" target is always the current best practice for the task. The "good" tolerance is always the most demanding original equipment manufacturer (OEM) specification. The "better" value is a stretch goal that is one-third to halfway between good and best. A test is the simplest way to check that the outcome is within tolerance. With a proof test paralleled to a work task, the two activities combined greatly lift the chance of task success. Even if a task is done wrong the first time, it is almost certain that the error will be found during the test and corrected.

The best value is typically found in the books and articles written by subject matter experts. Sometimes recognized international standards identify best values. When using published standards, pick the most demanding performance as the best outcome to achieve. When it is not possible to find the best practice for the target value, make it a magnitude better than the tolerance value.

If there is no OEM value for a task, use your industry's standard as the good value; otherwise, a suitable international standard or code of practice can be applied. You can also choose to use your organization's current standard as the good value. The "good enough" value that you use must be a fair representation of the currently accepted performance in your operation for the activity. It is the baseline from which your people will grow to become the best in your industry.

Don't set the bar low for best quality—set demanding world-class standards. The ACE 3T procedure-writing method is a psychological device to help people become great at their work. As your people learn to deliver better and then best results, you will progress faster and faster toward world-class reliability. The psychology of ACE 3T is to give your people a challenge to rise up to. They are challenged by the better and best values to outperform their past results. They will become proud of their work and of themselves. They will earn the respect of their peers, supervisors, and managers. They will be valuable, happy, highly productive people.

The practical, simple ACE method for performing masterly work with accurate results means that wasted cost and effort will become new operating profits. The ACE method of failure prevention and defect elimination permanently improves productivity and delivers new production and profits because losses disappear as job after job is done 100% correctly the first time. You can turn a poorly performing operation into a star performer simply by changing the processes and practices used by its people to ACE 3T methods. Everyone can do great work once they know the right things to do and have a sound and reliable method to help them do their job right.

Examples of Accuracy-Controlled Procedures

Accuracy-controlled procedures are simple for users but impose demanding requirements on those writing the procedures. ACE procedure writing starts by drawing a process map of the procedural steps. Each process step is given a three- to four-word description that explains the purpose of the step. For each process step, summarize what happens in that step. Boxes below each process step bullet point the step procedure and/or provide information and explanation. Above the step description, list all of the step performance indicators so that everyone knows what each step is meant to do. Using a flowchart layout of process steps with their procedures and performance indicators on a single page helps people quickly understand the process design, know what is important to monitor and measure, and find new process improvements.

Table 15.1 can be used as a template for writing an ACE 3T procedure. Tolerance is subdivided into good, better, and best ranges, with the target listed in the "best" column, and a test is specified for each task. You also advise the user what to do if the tolerance is exceeded. Tell people what they must do if they can't meet the quality standard, as they will get distressed when they find a problem they cannot fix. It's human nature to get upset when things go wrong. Simply tell them what they need to do—the "emergency procedure" they will follow when they face trouble. Procedure layouts are flexible and formats can be altered to suit an organization's existing templates. It is the inclusion of the 3Ts in each task that is the vital ingredient that must always be present in a procedure or detailed work instruction.

Table 15.1—Sample ACE 3T Procedure Layout

Task Step No.	**Task Step Owner**	**Task Step Name**	**Materials, Tools, and Their Condition**	**Full Description of Task**	**Test for Correctness**	**Tolerance Range**			**Actual Result**	**Action if Out of Tolerance**	**Sign-Off**
		(3–4 words)		(include all tables, diagrams, and pictures)	(include diagrams and pictures)	Good	Better	Best			
1	2	3	4	5	6	7	8	9	10	11	12

When writing the procedure for a job, be clear why the job is in the business. When the procedure is read by users, you want them to establish the right mind-set of wanting to do excellent work in a timely manner. Identify the process in which the job is used and its importance. Explain the purpose of the job. Indicate all of the people who are affected by the work so that the consequences of poor quality and the necessity of doing each task thoroughly and correctly are clear.

In the procedure or work instruction, incorporate the 3Ts of defect elimination into each work task. This provides accuracy control and allows users to clearly identify the quality requirements they need to meet. They must prove that they have met those quality standards before going on to the next task. Explain every step in a task in detail, using words and including images wherever possible, and even access to short videos showing correct practice. Visual explanations improve understanding, and they should be used liberally in your procedures. Define and explain the information flows, the data to be collected, and the records needed. The procedure can be used as a quality control form to record results and collect evidence of quality assurance.

Two examples of an ACE procedure follow. The first, for a clerical task, sets pass/fail criteria for each activity. In situations in which the output is either right or wrong and no tolerance range exists, you set "accept/reject" standards. A task is not complete until it meets the "accept" criterion. An accept/reject procedure can be turned into an ACE 3T procedure by introducing fair and reasonable target, tolerance, and test quality standards. You intentionally install performance requirements into a job, task, or activity by specifying factors such as timeliness, workplace cleanliness, percentage compliance achieved, condition of the equipment used, calibration of tools, or other factors that influence the quality of work results and productivity.

The second procedure, for bolting up an 80 mm pressure pipe flange, is written in the full ACE 3T format. This procedure sets the standard and quality to be achieved for each task in the job. The workmanship quality and standard of work is not left to the discretion of the person doing the work. An ACE 3T procedure clearly states the minimum acceptable outcome in the "good" column and defines top-class performance in the "best" column. This approach to error-proof work provides a practical and sure way to control task quality regardless of who does the job. Everyone knows what "good enough" is—anything less is unacceptable. Everyone also knows what top-class work is, and they are encouraged to strive for it.

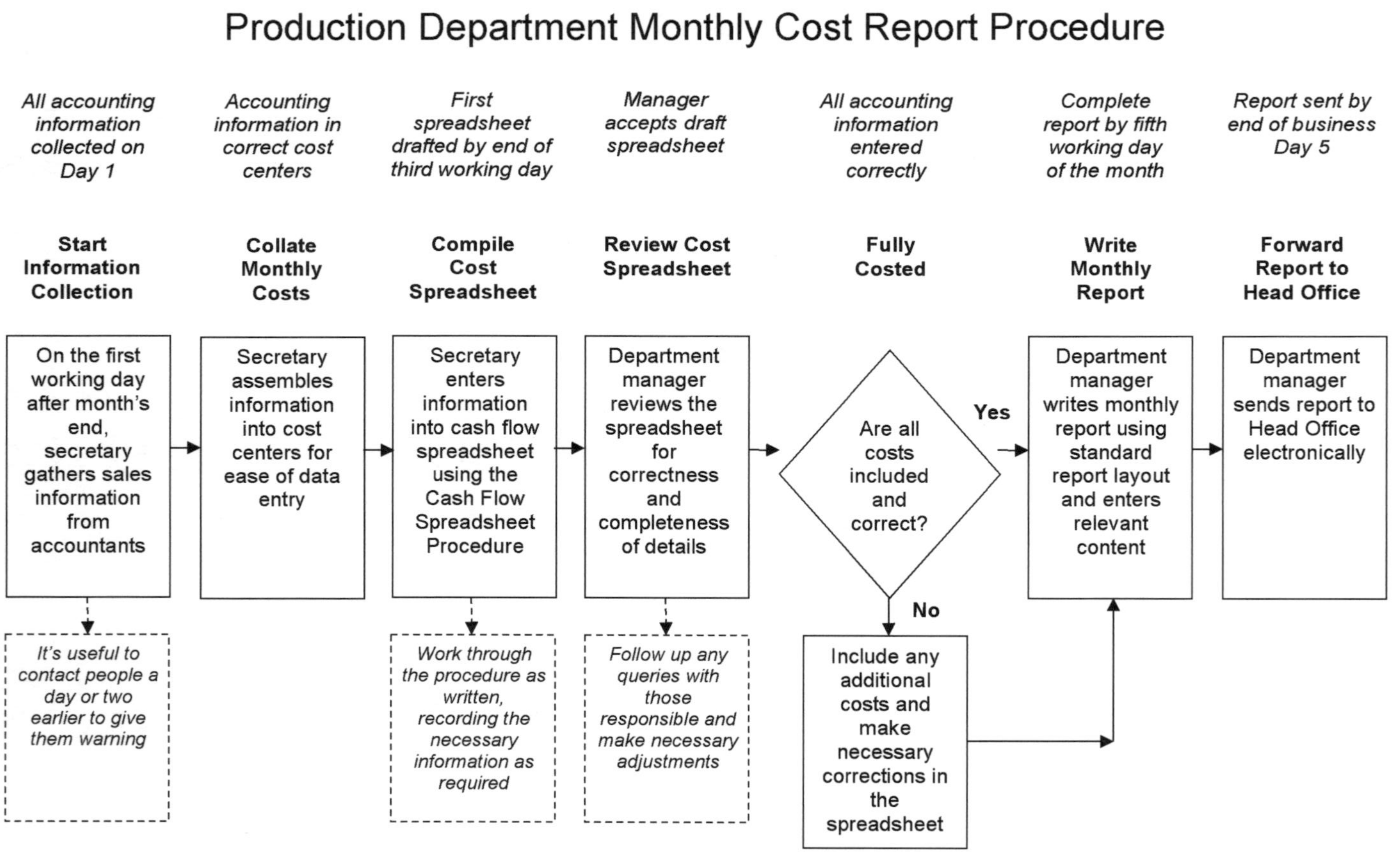

Figure 15.2—Process Map of Cost Report Procedure

Clerical Example: Cost Report Spreadsheet Accept/Reject Procedure

Figure 15.2 is a flowchart of the process steps for assembling the information required for a management report. Table 15.2 contains the procedural steps and quality requirements. The procedure opens with a statement to explain the importance of the work. The intention is to get the person compiling the report to commit to doing a good job because so much else depends on the accuracy and the quality of what he or she does.

Compile and Develop the Monthly Cost Report

Importance of developing an accurate and useful report: This procedure explains in detail how to create the department's monthly production costs summary spreadsheet. Your department manager and the cost accountants use it to make their monthly business performance reports. Any errors will flow through to the monthly report presented to the head office. The procedure is our current best practice, and you should follow it exactly. It is the result of many people's efforts over many years. It is the quickest, best way to do the job. You are encouraged to learn the job exactly as it is documented. If, after you master this procedure, you see a way to improve it, please bring your idea forward for discussion. You can test your idea and compare it with the procedure. If your suggestion proves to be better, it will become the new way of doing this job.

The people affected by the job: Department, senior, and executive management and, ultimately, everyone in the company.

Necessary equipment and tools: Computer, national monthly production computer file, national monthly production hard copy file.

Task summary: Below is a summary of the process for completing the spreadsheet. A fully detailed procedure is beneath the list. If you have a problem that you cannot solve, please see your supervisor.

1. Find spreadsheet
2. Bring up spreadsheet
3. Select worksheet
4. Get hard copy folder
5. Return with hard copy
6. Record monthly total
7. Cross-check totals
8. Totals don't agree
9. No spreadsheet error
10. Hard copy checked
11. Update spreadsheet
12. Totals agree

Table 15.2—Accept/Reject Cost Report Procedure

Task Step No.	Task Step Owner	Task Step Name	Full Description of Task	Test for Correctness	Actual Result	Initial after Complete
1.	Office clerk	Find spreadsheet	Find the shortcut on the screen called "National Monthly Production." *[You would also insert a screen image showing the item]*	See the icon called "National Monthly Production"		
2.	Office clerk	Bring up spreadsheet	Make spreadsheet ABC active on computer by double-clicking the icon. *[You would also insert a screen image showing the item]*	Note that the name on the spreadsheet is "National Monthly Production"	*(Place spreadsheet name here.)*	
3.	Office clerk	Select worksheet	Bring up the worksheet called ABC-1. *[You would also insert a screen image showing the item]*	See that the name on the SOP and actual worksheet is ABC-1.	*(Place worksheet name here.)*	
4.	Office clerk	Get hard copy folder	Get the "National Monthly Production" folder in the top drawer of the National Sales filing cabinet in the Sales Office.	Read the file name and see that it is called "National Monthly Production."		
5.	Office clerk	Return with hard copy	Return to your desk and open the folder to the Total National Production Report.	See that the page has the title "Total National Production Report"		
6.	Office clerk	Record monthly total	Total the "Purchase Price" column for the month and put into cell D8. *[You would also insert a screen image showing the item]*	Check that cell D8 has the monthly total.	*(Could also record monthly total here.)*	
7.	Office clerk	Cross-check totals	Check that the total for "Purchase Price" in the hard copy folder and the spreadsheet are the same.	Both totals are the same.	*(Record the total.)*	
8.	Office clerk	Totals don't agree	If the two numbers are not the same, check that the formula in the spreadsheet matches the correct cells. *[You would also insert a screen image showing the item]*	Check that all individual cells are picked up by the formula in the total cell.		
9.	Office clerk	No spreadsheet error	If the spreadsheet is correct, the error lies in the hard copy file. Report the error by telephone to the national production manager.	Call the national production manager.		
10.	National production manager	Hard copy checked	Confirm that the totals of individual sales are recorded correctly and ring back the correct individual production figures.	National Managers advises each figure.		
11.	Office clerk	Update spreadsheet	Correct the figures in the spreadsheet with the correct values and confirm that the totals are now correct.	Double-check the new total against hard copy file total.	*(Record the correct total.)*	
12.	Office clerk	Totals agree	If the totals in both documents agree, the job is complete. Save the spreadsheet, print a copy for the manager to review, close the electronic file, and return the hard copy file to the office filing cabinet.	See that spreadsheet is saved and the file is returned.		

80NB Flange Gasket Replacement and Fastener Tightening Procedure

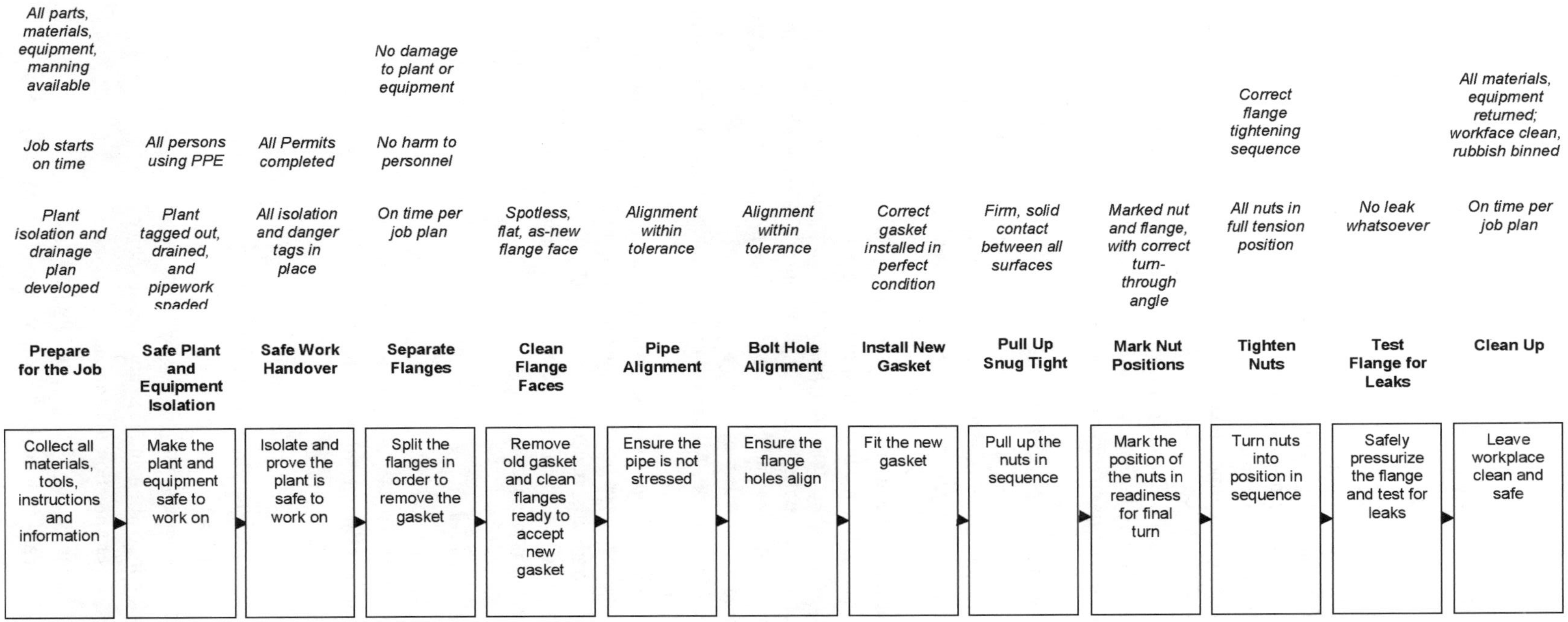

Figure 15.3—Process Map of Flange Connection Procedure

Industrial Example: Flange Connection Procedure with ACE 3T Tolerance Banding

This is an example of an ACE 3T procedure with tolerance bands for bolting together 80 NB, ANSI B36.5, forged steel, Class 150 flanges. Each task has quality ranges divided into good, better, and best performance. You must also provide instruction on what to do if the tolerance is not achieved. Figure 15.3 is a flowchart of the process steps. Table 15.3 reports the technical details applying to the work. Table 15.4 contains the procedural steps and quality requirements. The procedure opens with a statement explaining the importance of the job.

Disclaimer: The example covers the method for creating a 3T procedure and is not the actual procedure for bolting up flanges. Each organization must research, develop, and approve safe practices and procedures for bolting flanges. The use of turn-of-nut on pressure flanges may not comply with the applicable pressure piping design codes.

Pressure Pipe Flange Connection Procedure

Importance of correctly mating flanges: This procedure explains how to correctly bolt up pipe flanges on 80 mm (3 inch) diameter pipe. Pipe flanges must be bolted up so that they do not leak. Dangerous chemicals leaking from pipe flanges are a safety and environmental hazard that can lead to the death of workers and the destruction of production plant and equipment. Even a water leak from a flange can cause a slip hazard and make an unsightly mess. The procedure is our current best practice, and you should follow it exactly. It is the result of many people's efforts over many years. It is the quickest, best way to do the job. You are encouraged to learn the job exactly as it is documented. If, after you master this procedure, you believe that you can improve it, please bring your idea forward for discussion. You can test your idea and compare it with the procedure. If your suggestion proves to be better, it will become the new way of doing this job.

The people affected by the job: All operators, maintainers, and workplace staff and employees passing by the flange in future.

Necessary equipment and tools: Gasket, ring spanners (do not use adjustable shifters and pipe wrenches as they damage corners of bolt heads and nuts making their removal dangerous and unsafe), suitably load-rated studs and nuts, pencil.

Task summary: Below is a summary of the process for installing gaskets and making flanges. A fully detailed procedure is beneath the list. If you have a problem that you cannot solve, please see your supervisor.

1. Get work pack, tools, new fasteners, and new gasket
2. Get safe handover isolated and pipe drained
3. Place personal danger tags and test if drained
4. Break and spread flange safely
5. Clean up flange faces
6. Check and correct unrestrained pipe alignment
7. Check and correct bolt hole alignment
8. Mount gasket and insert fasteners
9. Pull up fasteners snug tight in sequence
10. Mark nut position and turn angle past snug
11. Turn nuts to position in sequence
12. Test flange for leakage at operating pressure
13. Safely clean up, hand back, complete job record and sign off on work order

Table 15.3—ACE 3T Industrial Procedure Engineering Information

Bolt Size	Bolt Grade	Bolt Torque	Tolerance on Torque		Engineering Standards
5/8"	A193 B7 stud and nut	201 Nm (60% Yield)	±15% with torque wrench	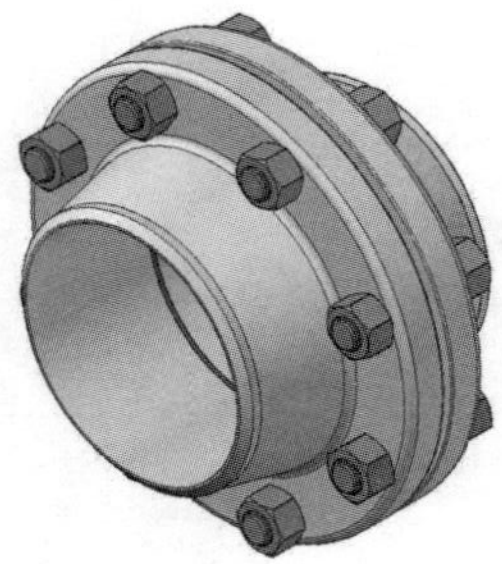	Flange Squareness: Good: Within 1 mm for every 200 mm diameter Better: Within 0.75 mm for every 200 mm diameter Best: Within 0.5 mm for every 200 mm diameter
		½ turn from snug tight	½ to 5/8 turn		Stress-Free Flange Bolt Hole Alignment: Good: Centers within 2 mm Better: Centers Within 1.5 mm Best: Centers within 1 mm
Gasket: Non-asbestos fiber, 1.5 mm thick, ring, grade as noted on work order					Bolt Lubricant: Molybdenum disulphide

Table 15.4—An ACE 3T Industrial Procedure

Task Step No.	Task Step Owner	Task Step Name (3–4 words)	Task Description	Materials, Tools, and Their Condition	Test for Correctness (include diagrams and pictures)	Tolerance Range Good	Better	Best	Reading/ Result	Action if Out of Tolerance	Sign-Off
1	Technician	Prepare for the job	Gather new studs and nuts, washers, gasket, thread paste, tools, job work order, danger tags, handover permit, special instructions, PPE	5/8" ring spanner or socket, podgy spike bar, screw driver, scraper	All materials and tools are on the job before starting the job	Request and collect issued items from store	Planner arranged all items ready for issue from Store	Planner has all items at job and job is ready to do		Only start work once all requirements are gathered together	
2	Technician	Inform operator	Contact operations personnel responsible for plant isolations and handover		Handover preparation and documents correctly done	Contact operator when ready to start job	Operator has plant off-line awaiting work	Operator has plant isolated, tagged and drained		Job can only start when Operations safely handover plant and piping	
3	Technician and plant operator	Make workplace safe	Place personal danger tags at isolation points and accept plant handover after proving isolations and drainage	Danger tags	Isolation procedure is correctly done and proven safe	Operator and repair man walk circuit and identify and tag isolations and open drains	Operator has isolated plant & tagged isolations out-of-service & drained piping	Operator provides isolation point drawing and walks circuit to show previous tagged isolations and open drains		Only start work when piping is fully drained and proven to be empty and possible gas build-up vented	
4	Technician	Separate flanges	Release tension on exiting fasteners gradually in tightening sequence. Remove one fastener at a time but leave the last fastener loosely in place if pipe springs unexpectedly, spring flanges with podgy bar	5/8" ring spanner or socket, anti-seize liquid	All fasteners removed without damage to flanges or harm to personnel or other property	Back-off all nuts half a turn in sequence and then a full turn, removing all fasteners but last one. Spring flanges with podgy	Back-off all nuts half a turn in sequence and then a full turn, catch any drops of product from flange in suitable container, remove all fasteners but last one. Spring flanges with podgy	Cover fasteners with anti-seize, back-off nuts half a turn in sequence and then a full turn, catch any drops of product from flange in suitable container, remove every second fastener and finally all fasteners but last one. Spring flanges with podgy		If flange does not spread easily review the situation and consider use of hydraulic spreader or wedges without damaging flange faces	
5		Clean flange faces	Remove old gasket and clean flange faces, remove any burrs, check face is flat with straight metal ruler and 0.05 mm shim in gaps, no draw marks, pits, or scratches allowed across flange face	25 mm wide metal scraper, 80 grit emery cloth	Flange face are totally clean and safely usable	Loose material removed, burr-free, flat face, no draw marks or pits deeper than 0.25mm	Grooves clean, face sanded, flat face, no draw marks or pits	Bright, smooth, flat face, no groove damage or pitting, as good as new		Replace or machine flange with identical rating and grade if pits are deep, or are in close clusters, or not flat *(pictures would be necessary)*	
6		Pipe alignment	Check unrestrained pipe alignment	5/8" ring spanner x 2, or socket and ring spanner	Measure misalignment with vernier callipers on flanges with studs removed	Flanges are unbolted and are in-line to within 2 mm	Flanges unbolted and are in-line to within 1.5 mm	Flanges unbolted and are in-line to within 1 mm		Cut pipe and remount flange to bring unrestrained flanges to within 1 mm alignment and 0.5 mm squareness to applicable procedure for the pipe material and grade	

Table 15.4—An ACE 3T Industrial Procedure continued

Task Step No.	Task Step Owner	Task Step Name (3–4 words)	Task Description	Materials, Tools, and Their Condition	Test for Correctness (include diagrams and pictures)	Tolerance Range Good	Tolerance Range Better	Tolerance Range Best	Reading/ Result	Action if Out of Tolerance	Sign-Off
7	Tradesman	Bolt hole alignment	5/8" ring spanner x 2	Check bolt hole alignment	Measure with vernier callipers on flanges with studs removed	Flanges unbolted and holes in line to within 2 mm	Flanges unbolted and holes in line to within 1 mm	Flanges unbolted and holes in line to within 0.5 mm		Cut pipe and realign flange to bring hole alignment of unrestrained flanges to within 0.5 mm	
8		Install new gasket and fasteners	Mount gasket and insert fasteners. Precut studs to length and de-burr so that two full threads protrude out of each nut when fully tightened. Lightly lubricate the studs and the face of the nuts in contact with the flange.	Approved new gasket, new studs and nuts, bolt lubricant, podgy bar	Only new gasket and new fastener components used	Gasket slid between flanges and centred without damage and new studs/nuts fitted by hand	Gasket slid between flanges without and centred damage and studs/nuts lightly, prelubricated and fitted by hand within 2 minutes	Gasket slid between flanges and centred without damage and studs/nuts lightly, prelubricated and fitted by hand within 1 minute			
9		Bring flanges together	Pull up fasteners snug tight in cross tightening sequence. Sung means flanges are in firm contact under about 20% of final bolt torque. It is obtained by the full effort of a well-built man pulling on a ring spanner until it can no longer be moved by hand. It can also be achieved by use of an impact wrench. When the spinning nut turns to blows, count three blows, and the bolt will be snug tight.[a]	5/8" ring spanner or socket, feeler gauges	Flanges come together square with stress-free alignment	Wind nuts onto studs by hand so studs extend equal distance either side of flange. Tighten nuts finger tight and check that flanges are parallel to an accuracy of 0.4 mm with the feeler gauges. Pull all nuts on both flanges up sung tight in correct sequence.	Wind nuts onto studs by hand so studs extend equal distance either side of flange. Tighten nuts finger tight and check that flanges are parallel to an accuracy of 0.2 mm with the feeler gauges. Pull all nuts on both flanges up sung tight in correct sequence within 5 minutes	Wind nuts onto studs by hand so studs extend equal distance either side of flange. Tighten nuts finger tight and check that flanges are parallel to an accuracy of 0.1 mm with the feeler gauges. Number the studs in the sequence of tightening. Pull all nuts on both flanges up sung tight in correct sequence within 4 minute		If flanges are not parallel, directly 180° degrees opposite widest part of indicated gap, loosen nuts one or more turns. Return to segment with gap and tighten until both flanges are in contact with gasket. This is necessary to prevent flange levering over the fulcrum formed by the outer edge of the two raised faces at points in contact with gasket. The restriction will cause exceptionally high flange to gasket clamp loading at this point, with possible damage to gasket, plus diverting necessary clamp loading bolt torque energy to correcting alignment on the opposite segment.	
10		Match mark fasteners	Match-mark nut position **on one flange only** with a pencil when all nuts on both flanges are snug.	Pencil	Scribed marks in correct position and easily observable	Match-mark the nut and flange	Clearly match mark the nut and flange within 1 minute	Clearly match-mark the nut and flange within 45 seconds			

Table 15.4—An ACE 3T Industrial Procedure, continued

Task Step No.	Task Step Owner	Task Step Name (3–4 words)	Task Description	Materials, Tools, and Their Condition	Test for Correctness (include diagrams and pictures)	Tolerance Range Good	Better	Best	Reading/ Result	Action if Out of Tolerance	Sign-Off
11		Tighten fasteners	Turn each nut **on one flange only** an extra 1/3 turn to final position in cross tightening sequence. Re-tension continuously until all nuts are equally tight. No rotation of stud is permitted while tightening the nut.	5/8" ring spanner or socket, impact wrench	Fasteners correctly tensioned to required nut position in right tightening sequence	Tighten nuts 1/4 turn in cross-sequence and finally tighten nuts to 1/3 turn in cross-sequence.	Tighten nuts 1/4 turn in cross-sequence and finally tighten nuts to 1/3 turn in cross-sequence in 5 minutes.	Tighten nuts 1/4 turn in cross-sequence and finally tighten nuts to 1/3 turn in cross-sequence in 4 minutes.		If a stud starts to rotate as the nut is tightened it indicates that the nuts were not snug to start with. Immediately stop and undo all studs and repeat nut snug tensioning procedure.	
12		Test for leaks	Test flange for leakage at operating pressure, release pressure and retighten nuts on same flange as originally tightened		No leak whatsoever at full operating pressure	No leak whatsoever at full operating pressure	No leak whatsoever at full operating pressure	No leak whatsoever at full operating pressure			
13		Clean and hand back	Safely clean up, hand back, complete job record, and sign off and record work order history		All equipment, tools and waste removed, area left spotlessly clean	Recommission procedure is written, reviewed, and correctly done and plant proven safe for continued use	Recommission procedure is written, reviewed, and correctly done and plant proven safe for continued use	Recommission procedure is written, reviewed, and correctly done and plant proven safe for continued use			

DISCLAIMER: *The example covers the method to use to create a 3T procedure and is not the actual procedure to use when bolting-up flanges. Each organisation must research, develop and approve their safe practices and procedures for bolting flanges. The use of turn-of-nut on pressure flanges may not comply with the applicable pressure piping design codes.*

a. Alan T. Sheppard, High Strength Bolting, The DuRoss Group, Inc.

The process map in Figure 15.3 showing the flange bolting procedure is not normally included in the workplace document. It is included here to show that the work is a series process and that bad quality outcomes at any stage will produce defects that cause leaks in the future.

Competency Assessment and Training Plan

Once standards of quality are set in a job, you can measure compliance of the delivered outcome. The purpose of having quality standards is to guide an organization and its people to achieve them deliberately and consistently. That includes helping those doing the work learn how to produce more successful results. If your people are not consistently within tolerance of a quality standard, they are putting defects into what they do. It's an unacceptable situation that must be addressed whenever it arises.

A fair approach when people are not meeting their standards is to ask them what can be done to get them to reliably meet all task quality requirements. Let them propose a successful solution. If they can implement the solution themselves, help them do so quickly. If they require further training to produce higher-quality results, then add courses to their training plan. If capital expenditure is needed to guarantee the specified quality then add the costs into the capital budget and get the equipment into use as fast as possible. If, after all that can be done for them has been, they still cannot consistently deliver the required quality, it is necessary to put them into a different job that they can do competently and appoint a proficient replacement.

Cross-Functional Knowledge Teams

An organization brings people together to produce outputs that are desired by its customers and stakeholders. The organizational structure connects people together in their efforts. For an industrial operation, the quality of the output is dependent on peoples' knowledge and skills, the effectiveness of business processes, and the reliability of operational assets. There is a lot to get right when world-class reliability is your aim. It's easier and faster to share the load among the people who are best able to do the work right. That's what teams are for.

The ultimate aim is to have teams of subject matter experts working seamlessly together who proactively seek to optimize their productivity. They work in partnership with all of the other teams in the business so that the organization and all of its people are continually becoming better and better at achieving its purpose.

16

Reliability Growth

These days, we know what needs to be done to make all of our machines highly reliable and reduce their rates of failure. Research conducted across the world during the late twentieth century explains why machines fail—their parts are overstressed. Furthermore, we know that equipment and machine failures are not unlucky accidents; mostly, they are failed by our "gremlins." We also know that it is possible to choose the failure rate you want for your plant and equipment and then put into place the processes and methods that will naturally deliver it.

Creating high equipment reliability means removing the risk of errors and defects in everything that impacts equipment health. Use error-proof and mistake-proof tasks and methods. Run plant and equipment so that they live well within their engineered design envelope. Provide correct education and skills covering the proper operation of the plant and machinery to all—from manager to greaser. Train operators and maintainers to understand their process engineering so that they know how their plant works. Develop fault-free, accuracy-controlled, proof-tested operating and maintenance procedures. Include independent double checks and even triple checks in tasks when the risk is great. Teach the procedures until your people are masters at doing them. Get people up and down the organization to help each other and learn from one another by applying the power of teaming up.

Failure Patterns and Failure Modes

Equipment failure follows one of the six probability patterns shown in Figure 16.1, made famous by the 1978 Nowlan and Heap study of aircraft maintenance.[1] Evidence from airline industry maintenance in the 1960s and 1970s indicated that, together, failure patterns D, E, and F represented 89% of failures, with pattern F alone, showing infant mortality failure, representing 68%. Other airlines and the U.S. Navy conducted similar studies and confirmed the patterns. Although the proportions varied between maritime and aerospace conditions, patterns D, E, and F dominated.[2] The curves show that for most equipment, once the equipment is through its early-life period, failure is not age related but is random. This does not mean there are no reasons for failure; rather, it means that the reasons arise by chance, so when the event will happen is uncertain. Nowlan and Heap questioned the practice of doing regular overhauls: if most equipment failures have nothing to do with the age of the equipment, why are parts replaced on a time basis? You could be throwing away a perfectly good part that is still suitable for many hours of service, and you are introducing the opportunity for error and defect creation with each intervention.

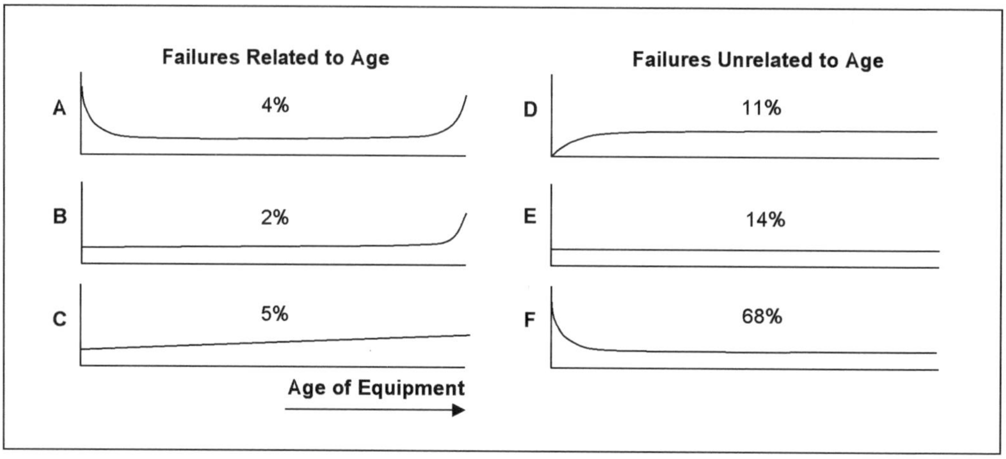

Figure 16.1—Six Failure Patterns Identified for Aircraft Equipment

The three lists in Figure 16.2 cover the vast majority of types of solutions used to prevent equipment failure. The philosophy for preventing machinery failure is explained in the figure: remove the causes of each part's failure modes. If the parts do not fail, the machine does not stop. You can create any amount of equipment reliability that you want by controlling the failure rates of your equipment parts. The failure curves are malleable by continually removing the risks that cause equipment parts to fail. They can be changed by the selection of engineering, operating, and maintenance policies and practices. The evidence of successful reliability improvement shows up as declining rates of parts failure and greater equipment uptime. Figure 16.3 shows the changed failure rate of equipment that Industrial and Manufacturing Wellness methods are designed to deliver.

Reliability Growth Cause Analysis

Reliability Growth Cause Analysis (RGCA) is a full technical and financial justification to introduce reliability growth strategies. RGCA is all about getting management support to put failure prevention and defect elimination into use quickly. Its purpose is to identify the causes of component failure and show the business value gained by preventing their initiation. You use it to fully understand how parts fail in service and to justify installing the right methods and practices in the life cycle to stop defect creation. Do that proactively, and you will get large operating profits from all the equipment failures you never have.

Improved reliability has a cause. Just like a failure has its causes, there are causes of greater reliability. You can wait for a failure to happen and then learn from the experience and change your processes to prevent it. That is root cause failure analysis. But it is not proactive behavior. The approach quickly buries you in never-ending problems, and eventually you only make time to investigate catastrophes. You fix a few causes of failures, but not the thousands of defects waiting to create the next lot of disasters. What must be done is shown in Figure 16.4. You create higher reliability in a series arrangement by raising the reliability of each step. To get high plant reliability, stop the problems throughout the life cycle that will become equipment failures. Identify all potential troubles and prevent them from starting by implementing the processes and activities that cause high reliability. A Reliability Growth Cause Analysis is used to decide how to intentionally produce and profit from world-class reliability.

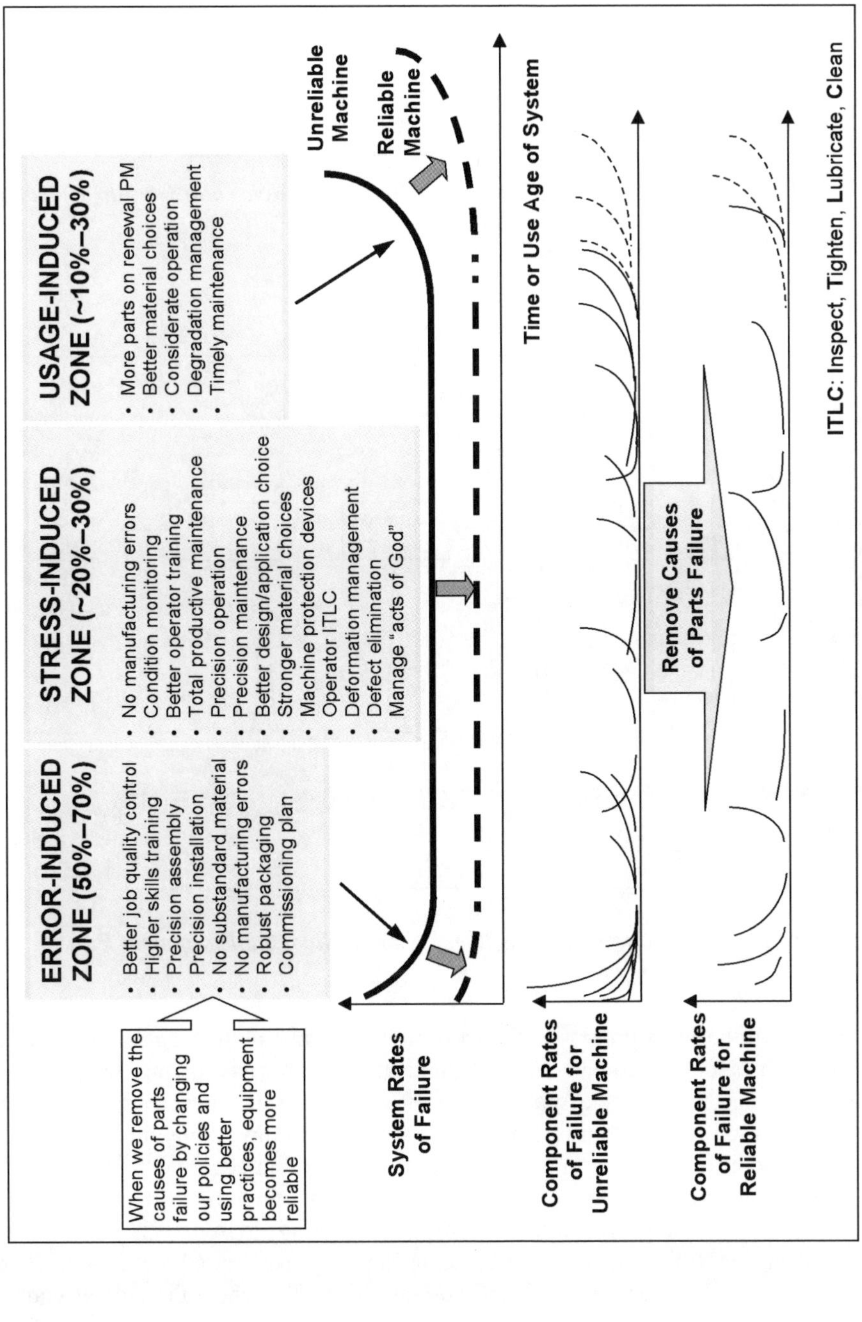

Figure 16.2—What to Do to Have Highly Reliable Machines

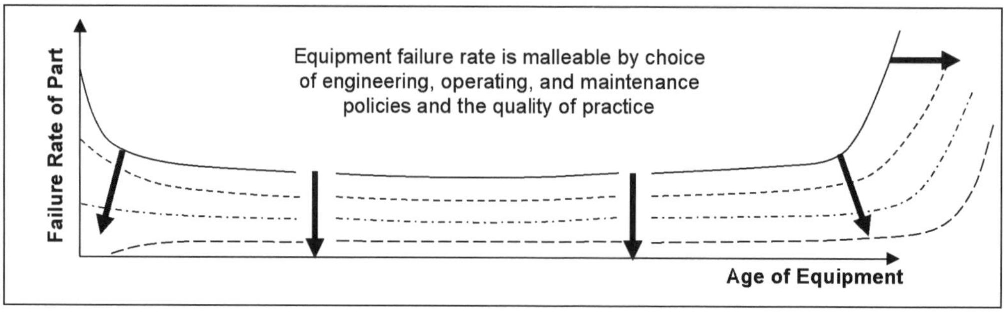

Figure 16.3—The Rate of Failure Is Malleable by Choice of Policies and Quality of Practices

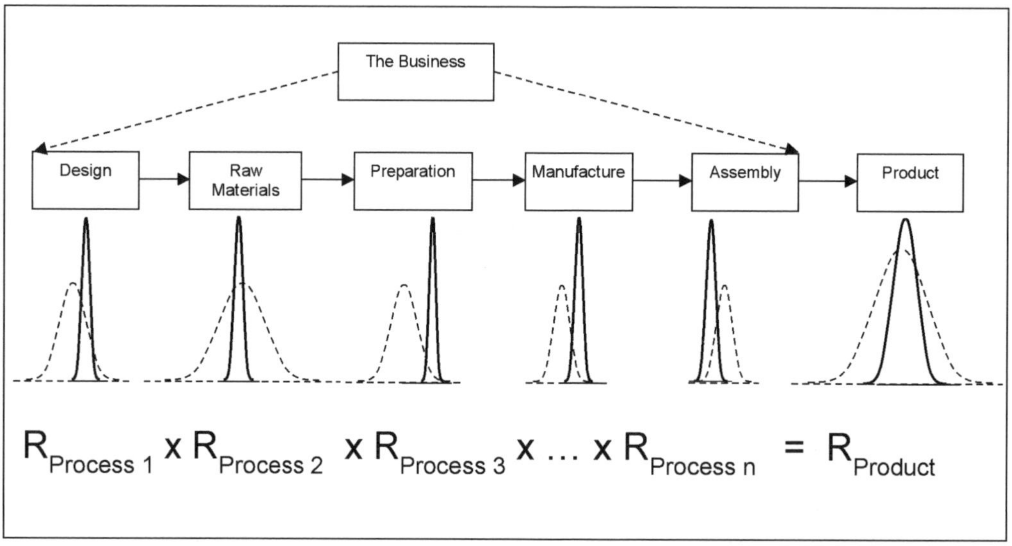

Figure 16.4—Eliminate Risk from Every Process Step and Equipment Part

The maps of your business processes, the work flow diagrams of your operating procedures, and the bills of materials for your equipment are the foundation documents for improving equipment reliability. They are respectively used to make more robust processes, to control human error, and to make a part's material-of-construction microstructure safe and healthy. Reliability Growth Cause Analysis lets you find effective ways to increase the reliability of an equipment part. It looks for what can be done to intentionally reduce stress and remove risk. Part number by part number, every identifiable way to remove and prevent stress, improve the working environment, or eliminate risk to reliability is identified. Every weakness, hazard, or danger is listed in a spreadsheet. Then the cures that bring sure reliability are selected, and you put them everywhere throughout your company.

The RGCA method adopts the same strategy for reliability growth that world-class leaders in industrial safety use for workplace safety improvement. They proactively improve safety by identifying risks and installing appropriate protection against harm before incidents happen. They don't let defects that can become accidents start in the first place. RGCA assumes that failures will happen to equipment parts as a result of defects created in life-cycle processes unless they are intentionally prevented. It requires recognizing what can cause risk during a part's lifetime and then making the necessary corrections to problematic tasks and introducing process improvements to prevent every cause from starting. Reliability increases by using the right practices and processes to prevent defects and risk alongside those that proactively promote health and wellness. RGCA requires you to identify ways that will drive reliability improvement and not simply prevent failure. Your aim is to never give a process step or part a reason to fail.

Each failure cause is analyzed in detail and its POF mechanisms addressed. Table 16.1 shows the range of operational risk, technical, and financial information complied in an RGCA for each component failure cause. The final recommendations identify the strategies, practices, and skills needed in design, manufacturing, procurement, construction, warehousing, operations, and maintenance to deliver lifetime reliability. A robust and timely plan is then developed to introduce them into the organization, including all necessary documents, training, and skills development.

Table 16.1—Reliability Growth Cause Analysis Requirements

Failure Description: ________________________________ **Cause No.:** ____ **Failure Cause:** ________________________
• Frequency of Cause:
• Time to Repair:
• TDAF Cost:
• Causes of Stress/Overload:
• Causes of Fatigue/Degradation:
• Current Risk Matrix Rating:
• Controls to Prevent Cause:
• Estimated Failures Prevented after Risk Controls in Use (/yr):
• New Risk Matrix Rating:
• TDAF Cost Savings from Higher Reliability:

RGCA requires us to ask how best to protect against failure-causing events and situations that can become future plant and machinery breakdowns. From design and capital equipment selection, through to board room decisions, and down to the operating procedures and maintenance practices, every phase of the component's life cycle is considered when looking at how to proactively prevent defects from starting. As you perform RGCA for a part, you outline its reliability creation strategy using a spreadsheet like that in Table 16.2. With full knowledge of what makes a part fail, you select the procedures, practices, and skills that prevent stress, fatigue, and degradation of the materials of construction. These are the actions that are incorporated into the organization's methodologies and practices and trained into its people. The size of the risk reduction required determines which reliability growth improvements you will use. At the completion of the RGCA, you will have a list of the necessary activities and quality standards that will deliver equipment part reliability with certainty. Those activities are put into place in your business processes, and the right skills are developed to an expert level in the people who need to do them.

Table 16.2—Reliability Growth Cause Analysis Spreadsheet

Reliability Growth Cause Analysis Spreadsheet

Business Process:
Equipment Number:

Department:
Equipment Name:

Step 1 | Step 2 | Step 3 | [ ]

Step/Item/Part Description:
Procedure Description:
Process Description:
Part Number:
Effect of Step/Item/Part Failure?
Failure of Step/Item/Part Causes System Failure (Y/N)?
Total TDAF Cost Savings Possible ($/yr):

Risks and Controls

Failure Mode 1:
Frequency of Cause 1:
Time to Repair 1:
TDAF Cost:
Causes of Stress/Overload:
Causes of Fatigue/Degradation:
Current Risk Matrix Rating:
Controls to Prevent Cause:
Estimated Failures Prevented after Risk Controls in Use (/yr):
New Risk Matrix Rating:
TDAF Cost Savings with Higher Reliability:

Failure Mode 2:
Frequency of Cause 2:
Time to Repair 2:
TDAF Cost:
Causes of Stress/Overload:
Causes of Fatigue/Degradation:
Current Risk Matrix Rating:
Controls to Prevent Cause 2:
Estimated Failures Prevented after Risk Controls in Use (/yr):
New Risk Matrix Rating:
TDAF Cost Savings with Higher Reliability:

Failure Mode 3:

TDAF Cost Savings ($/yr):

Finding the causes of reliability growth requires time, effort, and resources to do the analysis and financial justification. But without making those changes, plant and equipment reliability can never get better. If high reliability were easy, every company would already have it. But high reliability is exceptional because it is demanding and challenging to do well. Few organizations know how to achieve the exacting standards and practices of world-class performance. Without a method to find exactly what to focus on, without the financial justification that high reliability is worthwhile, and without an achievable plan to deliver it, industrial organizations waste away.

Once an RGCA is performed and the right reliability growth actions, knowledge, and practices are identified, they apply to every similar situation. Do an RGCA for one bearing, and you have done it for all similar bearings. An analysis done for a production task applies to every such task done in the operation. Take the learning from each analysis and apply it to every similar situation across your business. When you transfer knowledge, you apply Series Reliability Property 3, and rapid reliability growth can happen in only a matter of weeks as best practices are cascaded across a business. Once the reliability improvement efforts identified in a RGCA are implemented, the chance of failure-causing incidents occurring drops. You get a big reduction in the number of equipment failures because the right actions to produce reliability are done at every stage in a process, workplace task, and throughout a part's life. Your operating profits will rapidly climb, and your safety and environmental performance will become top class.

Reliability Growth Cause Analysis Example 1

The following is an example RGCA for preventing failure of the inner raceway of a single row deep groove ball bearing like the one shown in Figure 16.5. The process map for the shaft and bearing inner ring arrangement is drawn in Figure 16.6.

Figure 16.5—Roller Bearing Arrangement

In an RGCA, you are charged with finding all of the causes of reliability improvement, including what can be done earlier in the life cycle to prevent a failure. First develop a list of known and possible inner ring and raceway failures based on experience and using the Physics of Failure Factors Analysis guidewords. The causes of inner raceway failure include a cracked ring, a scoured or scratched raceway, a brinelled and indented race, a loose fitting ring, a race suffering electrically arcing, and so on. The next step is to ask what causes each mechanism. How can a particular cause mechanism arise during the life cycle? For example, how can the inner ring be cracked? A cracked ring can occur because of excessive interference fit on the shaft, because of a huge impact load, because the shaft is oval and the ring is forced out of shape, because a solid piece of material is trapped

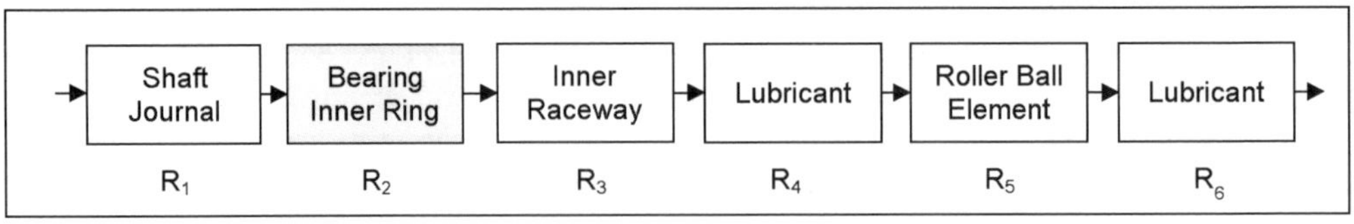

Figure 16.6—Process Map for Roller Bearing Located on a Journal

Table 16.3—Example Reliability Growth Cause Analysis on Inner Ring of a Roller Bearing

Failure Description: Cracked Inner Roller Bearing Race		
	Failure Cause 1: Excessive Interference Fit	**Failure Cause 2:** Impact to Ring
Frequency of Cause:	Early life—1 per year	Random—3 per year
Time to Repair:	5 hours	10 hours
TDAF Cost:	$20,000	$25,000
Causes of Stress/Overload:	• Large shaft • Small bearing ring bore • Tight clearance	• Hammered when fitting • Start-up with equipment fully loaded
Causes of Fatigue/Degradation:	Not applicable	• Misaligned shafts • Loose ring moving on shaft • Loose clearance
Current Risk Matrix Rating:	Medium	Medium
Controls to Prevent Cause:	• Update all bearing fitting procedures to measure shaft and bore; confirm correct interference fit at operating temperature and train people annually • Update all machine procurement contracts to include quality check of shaft diameters before acceptance of machine for delivery • Update all bearing procurement contracts to include random inspections of tolerances • Update all design and drawing standards to include proof check that shaft measurements and tolerances on drawings suit operating conditions once bearing is selected	• Update all bearing fitting procedures to include using only approved tools and equipment and train people annually; purchase necessary equipment, schedule necessary maintenance for equipment • Change operating procedures to remove load from equipment prior to restart and train people annually (Alternative: Soft start with ramp-up control if capital available) • Align shafts to procedure and train people annually • Update bearing fitting procedures to measure shaft and bore; confirm correct interference fit at operating temperature and train people annually
Estimated Failures Prevented after Risk Controls in Use (/yr):	All future failures	80% of future failures
New Risk Matrix Rating:	Low	Low
TDAF Cost Savings from Higher Reliability:	$20,000 per year	$60,000 per year

between the ring and shaft during the fitting, or because the shaft is heavily burred and when the ring is forced over the burr it is damaged in the installation process.

Once the causes are listed, they need to be prevented. For a cracked inner ring due to an excessive interference fit, you would ask, how is excessive shaft interference prevented? This problem results from incorrect tolerances between the inner ring and shaft. It is usually a manufacturing error of the shaft or the ring. Therefore, you must develop proactive measures to ensure that a ring is never fitted to an incorrectly made shaft or that an incorrectly made ring is never fitted to a good shaft. One solution is to measure the shaft and the ring with a micrometer and check that the sizes match the manufacturer's tolerance and form requirements for the model of bearing. An additional protection would be to confirm that the bearing model is correct for the service duty and operating temperatures. These checks become procedural requirements that are written into the ACE 3T procedure for the job. An example of the analysis and possible measures for preventing a cracked inner ring are listed in the "Failure Cause 1" column in Table 16.3.

Continue with the next cause of a crack in the inner ring—heavy impact—and develop preventive actions. (Impact damage can occur when a ring is fitted to a shaft with hammer blows or over-

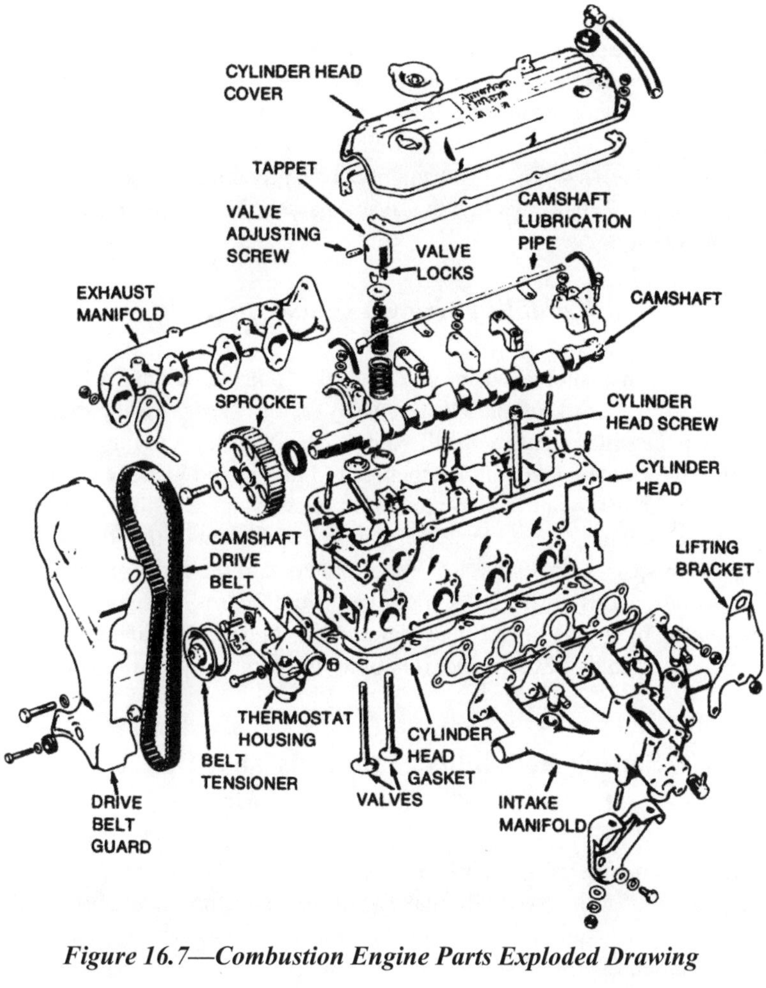

Figure 16.7—Combustion Engine Parts Exploded Drawing

loaded in a press, brinelling occurs during shipping and road transport, a badly aligned shaft cyclically vibrates the race and rolling elements together, or it suffers a huge start-up impact load.) The process continues for a shaft that is oval, for a solid piece of material trapped between the race and shaft during the fitting, for a heavily burred shaft, and so on. Each failure cause gets its own column, and the table expands until all causes are listed so that they can be addressed with effective risk elimination or prevention activities. With each preventive measure put into place and made standard practice by using ACE 3T procedures and workforce training, the reliability of your equipment parts and plant increases.

Reliability Growth Cause Analysis Example 2

This example of a Reliability Growth Cause Analysis is for an internal combustion engine driving a fire pump in a gasoline fuel storage terminal. A fire pump engine is a vital element of the firefighting service if a fuel terminal has a blaze. It acts as a backup to an installed electric motor driven pump, which runs first to supply water to the terminal tanks' spray nozzles. If the electrical power supply fails, the engine-driven pump starts up and provides the water to fight the fire.

First identify each critical part in the asset. Exploded assembly drawings like that in Figure 16.7 and bill of materials parts lists identify the components in equipment. If the critical parts are not obvious, develop a process map of the parts in use when the machine is in operation and identify each item in the chain of parts that could cause the machine to fail. Diagrams similar to Figure 16.8 show series arrangements that are high-risk locations for failure. Then put each critical part through the RGCA to identify ways to increase lifetime reliability.

Figure 16.8 and Table 16.4 provide an example of a process map for parts operation and use. Table 16.5 summarizes the Reliability Growth Cause Analysis, and the risk matrix used in this analysis is shown in Table 16.6.

Including POFFA in a Reliability Growth Cause Analysis

To improve the certainty of a comprehensive outcome when doing a Reliability Growth Cause Analysis, you can include a Physics of Failure Factors Analysis for every critical part. Once at-risk components that will stop the equipment are identified, you can use POFFA to find the cause mechanisms of stresses and the environmental conditions that lead to failure and identify where those situations arise. With the causes known, you build in protection at every stage of the life cycle to prevent the circumstances from developing. This Physics of Failure refinement of the RGCA methodology lets you create business processes across the life cycle that contain the proper actions and activities to deliver the least operational risk. An RGCA combined with POFFA helps you discover exactly what to do to produce outstanding reliability. Equally important, it gets you to identify how well each life-cycle task has to be done so that exceptional reliability is guaranteed in your equipment.

Setting Reliability Standards That Deliver Outstandingly Reliable Equipment

Because high equipment reliability and production plant availability are business and life-cycle process outputs, you need to intentionally make your processes produce those results. Reliability starts with what the original equipment manufacturer made. If the reliability of an equipment design is inadequate for your needs, you will suffer high maintenance costs and get poor production results.

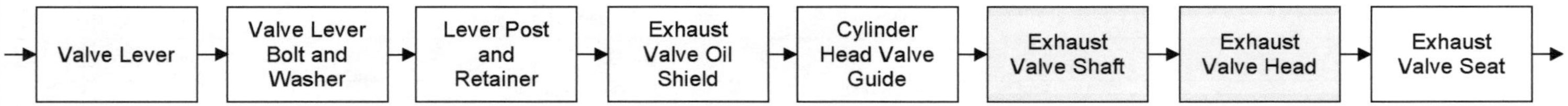

The flowchart shows the sequence of parts required to operate the exhaust valve in the internal combustion engine shown in the exploded drawing (Figure 19.7). The exhaust valve is opened by a series of mechanical parts to release the compressed exhaust gases. The hot gases (greater than 1,000° C) flow past the valve head and the valve seat embedded in the cylinder head. The timing of the opening is set by mechanical connection to the position of the crank and can be adjusted somewhat to select the optimal point to maximize the removal of burnt gases.

Figure 16.8—Exhaust Valve Operation Process Drawing

Table 16.4—Reliability Growth Cause Analysis Development

Reliability Growth Cause Analysis Development	
Step/Item/Part Description:	Item 41—Valve, Exhaust
Procedure/Drawing No. and Description:	Bill of materials and exploded parts drawing for internal combustion engine
Process Description:	Fuel terminal fire water supply pump drive motor—required operating life is 50 years (potential 100 years)
Part Number:	6505
Effect of Step/Item/Part Failure?	Engine cannot operate at full capacity because exhaust valve damage prevents compression. If valve failure occurs during firefighting duty, only the standby electric fire pump is available for backup duty.
Failure of Step/Item/Part Causes System Failure (Y/N)?	Yes—if not repaired at onset of valve damage, engine block head and cylinder block piston can be destroyed
Total TDAF Cost Savings Possible ($/yr):	A compete stripdown and rebuild of the engine costs $25,000 and 1 month downtime. Over a 50-year life, this produces an annualized cost of $500/year. Above the financial cost, the company's reputation with the regulators will suffer if the fire pump drive engine fails.
Risks and Controls	
Failure Stress Cause 1:	Exhaust valves' seat burned from normal use
Frequency of Cause 1:	The engine has a total of about 100 hours of operation per year. The motor is run for two hours each week on test and to prove that the fire water piping circuit does not leak. Valves could fail after about 4,000 hours of operation (around 40 years), although they are unlikely to fail before 3,000 hours in service (about 30 years).
Time to Repair 1:	Up to 1 month
TDAF Cost:	$25,000 once in a 50-year operating life with motor sent off site for urgent repair (annualized cost of $500)
Causes of Stress/Overload:	Not applicable
Causes of Fatigue/Degradation:	Exhausts valves are expected to degrade with use
Current Risk Matrix Rating:	This motor has a service life of 50 years as a fire pump prime mover in a tank terminal. Should it fail, the regulators will scrutinize the operation and become concerned that the company has poor maintenance practices. Likelihood 3; Consequence 3 = **M**
Controls to Prevent Cause:	Introduce planned replacement of all exhaust valves, seats, and valve guides at 3,000 hours or 25 years service, whichever comes first
Estimated Failures Prevented after Risk Controls in Use (/yr):	No failures are expected due to this mode of failure if exhaust vales are replaced every 25 years
New Risk Matrix Rating:	Likelihood 1; Consequence 3 = **L/M**
TDAF Cost Savings with Higher Reliability:	The planned refurbishment requires new exhaust valves and reseating. It is a two-day job done on site. Cost is $5,000 every 25 years.

Table 16.4—Reliability Growth Cause Analysis Development, continued

Failure Stress Cause 2:	Exhaust valve seat burned from a valve timing error
Frequency of Cause 2:	The engine is tuned once annually, or about every 100 hours of operation
Time to Repair 2:	Up to 1 month
TDAF Cost:	$25,000 once in 50 years motor sent off-site for urgent repair (annualized cost of $500)
Causes of Stress/Overload:	Localised high temperature at valve head edges and valve seat as exhaust gas flows past valve and seat due to valve closing late or opening early
Causes of Fatigue/Degradation:	Not applicable
Current Risk Matrix Rating:	The opportunity for a timing error arises annually and will not be corrected for twelve months, provided the error is then detected. In a 50 year operating life there will be 50 opportunities to mistakenly set valve timing. With 100 hours of annual service accumulated at a rate of two hours per week there is little time for the valve/seat to be burnt by one error. The valve is most likely to fail if the valve timing is not rectified for some years. Likelihood 3 ; Consequence 3 = **M**
Controls to Prevent Cause 2:	Introduce ACE 3T procedures to control tasks and to ensure a record of all valve timing adjustments is made and can be used to compare future settings.
Estimated Failures Prevented after Risk Controls in Use (/yr):	No failures are expected due to this mode of failure after ACE 3T procedures and recording is introduced
New Risk Matrix Rating:	Likelihood 1 ; Consequence 3 = **L/M**
TDAF Cost Savings with Higher Reliability:	No costs are expected in future from this failure mode when mitigation is performed
Failure Stress Cause 3:	Valve train and associated parts are wrongly installed and components come loose or break and valve falls into cylinder
Frequency of Cause 3:	This failure is only expected after a rebuild of the motor or a cylinder head is refurbished or replaced. The opportunity for this failure arises whenever other failure causes require the engine or cylinder heads to be removed. Errors in rebuilding such a complicated piece of equipment should be expected. It is estimated that once every twenty years a rebuild will arise, which is twice during the motor's operating life, and one of them will go wrong.
Time to Repair 3:	Up to 1 month
TDAF Cost:	$25,000 once in a 50-year operating life with motor sent off site for urgent repair (an annualized cost of $500)
Causes of Fatigue/Degradation:	Human error or faulty parts
Current Risk Matrix Rating:	Likelihood 3; Consequence 3 = **M**
Controls to Prevent Cause 3:	Introduce ACE 3T procedures to control engine rebuild and overhaul tasks. If work is done by a subcontractor or repair shop, ensure compliance with ACE 3T precision quality standards and implement tests and observation during rebuild to confirm compliance to quality requirements
Estimated Failures Prevented after Risk Controls in Use (/yr):	No failures are expected due to this mode of failure after ACE 3T procedures and recording are introduced
New Risk Matrix Rating:	Likelihood 1; Consequence 3 = **L/M**
TDAF Cost Savings with Higher Reliability:	No costs are expected in future from this failure mode when mitigation is performed
TDAF Cost Savings ($/yr):	Total annualized savings of $1,500 expected from the mitigations

Table 16.5—Reliability Growth Cause Analysis Strategy Summary

Reliability Growth Cause Strategy			
Failure Description: Failure of Exhaust Gas Valve in Fire Pump Drive Motor			
	Failure Stress Cause 1: Exhaust Valves' Seat Burned from Normal Use	**Failure Stress Cause 2:** Exhaust Valve Seat Burned from a Valve Timing Error	**Failure Stress Cause 3:** Valve train parts Are Wrongly Installed and Components Come Loose
Frequency of Cause:	Wear-out—1 per 50 years	Random—1 per 50 years	Early life failure—1 per 50 years
Time to Repair:	Up to 1 month	Up to 1 month	Up to 1 month
TDAF Cost:	$25,000	$25,000	$25,000
Causes of Stress/Overload:	Not applicable	Exhaust gases burn valve	Parts come loose and fail to operate properly or break
Causes of Fatigue/Degradation:	Gradual degradation from use	Not applicable	Not applicable
Current Risk Matrix Rating:	Medium	Medium	Medium
Controls to Prevent Cause:	Introduce planned replacement of all exhaust valves, seats, and valve guides at 3,000 hours or 25 years of service	Introduce ACE 3T procedures to control tasks and to ensure that a record of all valve timing adjustments is made and can be used to compare future settings	Introduce ACE 3T procedures to control engine rebuild and overhaul tasks
Estimated Failures Prevented after Risk Controls in Use (/yr):	All future failures prevented	All future failures prevented	All future failures prevented
New Risk Matrix Rating:	Low/Medium	Low/Medium	Low/Medium
TDAF Cost Savings from Higher Reliability:	$500 per year	$500 per year	$500 per year

Table 16.6—Risk Management Strategy Effectiveness

E – Extreme risk: Detailed action plan required

H – High risk: Needs senior management attention

M – Medium risk: Specify management responsibility

L – Low risk: Manage by routine procedures

Extreme or **High** risk must be reported to Senior Management and require detailed treatment plans to reduce the risk to **Low** or **Medium**

Consequence	Insignificant (1)	Minor (2)	Moderate (3)	Major (4)	Catastrophic (5)
People	Injuries or ailments not requiring medical treatment	Minor injury or first aid treatment case	Serious injury causing hospitalization or multiple medical treatment cases	Life-threatening injury or multiple serious injuries causing hospitalization	Death or multiple life-threatening injuries
Reputation	Internal review	Scrutiny required by internal committees or internal audit to prevent escalation	Scrutiny required by clients or third parties, etc.	Intense public, political, and media scrutiny (e.g., front-page headlines, TV, etc.)	Legal action or commission of inquiry or adverse national media
Business Process & Systems	Minor errors in systems or processes requiring corrective action or minor delay without impact on overall schedule	Policy procedural rule occasionally not met or services do not fully meet needs	One or more key accountability requirements not met; inconvenient but not client welfare threatening	Strategies not consistent with business objectives; trends show service is degraded	Critical system failure, bad policy advice, or ongoing noncompliance; business severely affected
Financial	<$500>	<$5,000>	<$50,000>	<$500,000>	<$5,000,000>

Likelihood – Probability:	Historical:			Insignificant 1	Minor 2	Moderate 3	Major 4	Catastrophic 5
> 1 in 10	Is expected to occur in most circumstances	5	**Almost Certain**	**M**	**H**	**H**	**E**	**E**
1 in 10–100	Will probably occur	4	**Likely**	**M**	**M**	**H**	**H**	**E**
1 in 100–1,000	Might occur at some time in the future	3	**Possible**	**L**	**M**	(1) **M**	**H**	**E**
1 in 1,000–10,000	Could occur but doubtful	2	**Unlikely**	**L**	**M**	(2) **M**	**H**	**H**
1 in 10,000–100,000	May occur but only in exceptional circumstances	1	**Rare**	**L**	**L**	**M**	**M**	**H**

The decision to buy machinery for a project is the outcome of a business process. The accuracy with which it is installed during construction is also a business process result. How well a machine is treated in operation is also a business choice. The reliability performance that you get from all your physical assets is a product of what you let your business processes do to them. To improve equipment reliability, you need to correct your business processes to make them deliver a better reliability result. You will need to work to the higher standards that produce the plant and equipment reliability you want. If you want world-class reliability, you will need to meet every quality standard that puts your machinery parts into their precision zones and then keeps them there.

Challenge Your Business to Meet High Precision Standards

Just because something is built to an internationally recognized standard or industry code, that does not make it good. Nor does using an international standard make for a risk-free design choice. For example, the specified tolerance for baseplate flatness designated in the American National Standards Institute (ANSI) pump standard is 0.375 mm/m (0.005 in/ft). The same requirement in the American Petroleum Institute (API) 610 pump standard is 0.150 mm/m (0.002 in/ft). That higher precision, with API 610 being two and a half times more demanding than ANSI, produces much higher pump reliability. API 610 pumps are designed to last many years between breakdowns; for the same service, ANSI pumps will likely last for much less time.

I began my career as a professional engineer in Perth, Western Australia, in one of the world's first fully automated breweries. A brewery is a great place for a new engineer to learn the profession because the making and packaging of beer uses a wide range of plant and machinery in diverse process manufacturing operations. It's a microcosm of the engineering world—from high-speed packaging equipment to agitated lauders, to specialty alloys for hot caustic washes, and distributed process control systems to run all the computer-controlled plants—you learn a lot in quick time. But one thing was done wrong to me.

I was told when I started as a new project engineer to use the same standards the brewery was built to. The company engineering standards stipulated the use of stainless steel ANSI pumps. So I bought ANSI standard pumps for all of many capital projects I did during the nearly eight years I worked at the brewery. I did not know, and no one ever told me, that if you buy an ANSI pump, you are highly likely to bring your company a lot of breakdowns, problems, and high maintenance costs.

The allowed ANSI baseplate un-flatness of up to 0.375 mm/m before the base must be rectified is a massive soft foot distortion problem. A new ANSI pump can be distorted so severely that you will have many breakdowns from internal components twisted out of shape by the 0.375 mm/m soft foot. When you pay to get a new ANSI pump, you carry a big risk that you will also be buying many failures because the equipment standard allows huge variation to be passed off as good quality. The API 610 pump standard instead demands that flatness be no worse than 0.150 mm/m. At that level of quality, you are forced to address soft foot and thereby prevent pump distortion. As a consequence of the better precision, pump reliability naturally increases. But you can do much better if you want to get truly outstanding reliability. The API pump flatness standard is still well short of what a world-class standard would be. A flatness of 0.05 mm/m (0.0007 in/ft) is readily achievable with modern machining equipment and practices.

This story illustrates how project engineers unwittingly destroy business profitability. Be careful what standards you select for your production equipment because that choice alone can be the cause of high maintenance costs. Once a bad machine is selected, the maintenance crew and the plant operators can do nothing to address it. All that is left for them to do during the operation phase is to keep fixing the machine when it fails.

Table 16.7—Indicative Target Values for Reliable Machines (Two-, Four-, and Six-Pole Motor Speeds)

ACE Standards for Creating Plant and Equipment Wellness and Machine Reliability							
No.	**Process Failure**	**Observation**	**Effect on Machine**	**Life Precision Requirement**	**Parameters**	**Target Value**	**Tolerance**
1	**Poor lubrication condition**	Chemistry; Contamination; Water	Short Component Life	Chemically correct, contaminant-free lubricant	Viscosity, additives, dissolved water, wear particle count	Right viscosity at operating temperature; correct proportion of additives; < 100 ppm water; ISO 4406 12/_/_ cleanliness	ISO 4406 14/_/_ cleanliness
2	**Wrong fits and tolerance**	Dimensions; Form	High Stress; Looseness; Vibration	Accurate fits and tolerance at operating temperature	Interference fit, operating temperature	Form IT5, operating temperature at design conditions	IT7
3	**Running off-center**	Dimensions; Form	High Stress; Vibration	Shafts, bearings, and couplings running true to center	Center of rotation, run-out, tolerance and form accuracy	IT5	IT7
4	**Deformed, bent, buckled parts**	Dimensions; Form	High Stress; Vibration	Distortion-free equipment for its entire lifetime	Soft foot, structural distortion	IT5	IT7
5	**Excessive loads and forces**	Vibration	High Stress	Forces and loads into rigid mounts and supports	Design load, forces into solid locations, foundation rigidity	No looseness; safely absorb/dampen forces	
6	**Misaligned shafts**	Dimensions; Form	High Stress; Vibration	Accurate alignment of shafts at operating temperature	Shaft alignment, straightness, deflection	Coupling/feet offset 10 μm/20μm	20 μm/40μm
7	**Unbalanced rotors**	Dimensions; Form; Balance	High Stress; Vibration	High-quality balanced rotating parts	Rotor balance, center of mass	G1	G2.5
8	**Induced and forced vibration**	Dimensions; Form	Vibration	Total machine vibration low	Machine vibration, machine distortion, structural rigidity	1.5 mm/s rms	4 mm/s rms
9	**Incorrectly tightened fasteners**	Tension	High Stress; Looseness	Correct torques and tensions in all components	Shank tension, looseness, fastener grade	±5% of correct tension	±10%
10	**Poor condition tools and measures**	Measurement Accuracy and Repeatability	High Stress; Looseness	Correct tools in precise condition to do task to proper standards	Good-as-new condition, reliably calibrated	As new condition/correctly calibrated	
11	**Inappropriate materials of construction**	Engineering Selection	High Stress	Only in-specification parts	Material of construction, dimensional specification	OEM-approved material and design specs	
12	**Root cause not removed**	Dimensions; Form; Precision Standards	High Stress; Looseness; Vibration	Failure cause removal during maintenance	Creative disassembly, defect elimination	Use creative disassembly and precision assembly	
13	**Precision skills and techniques not applied**	Dimensions; Form; Precision Standards	High Stress; Looseness; Vibration	Highly skilled technicians competent in precision techniques	Intelligent, competent, proactive, problem solvers	Equipment is consistently set up to world-class work quality standards	
14	**Assembly quality below standard**	Dimensions; Form; Precision Standards	High Stress; Looseness; Vibration	Proof test for precision assembly quality	Inspection test accuracy, precision standard	Ensure every activity is proven correct (apply Carpenter's Creed)	Milestone tasks tested
15	**Process out-of-control and/or not capable**	Dimensions; Form; Precision Standards	High Stress; Looseness; Vibration	A quality assurance system to make all of the above happen	Quality control standards, process in statistical control	ACE 3T procedures	ITP (Inspection and Test Plan)

NOTE: *Suggested values for 2-, 4-, and 6-pole motor speeds. These parameters are indicative and may not apply to a particular machine. Confirm actual requirements with the manufacturer or an expert.*

Set Precision Targets for Accuracy-Controlled Reliability

It is useful to know what standards will deliver high machinery reliability. Table 16.7 lists suggestions for 3T standards of machinery built to precision maintenance quality. The table is for two-, four-, and six-pole speed machinery. The values are unsuitable for high rotational speed machines. Such equipment needs even more exacting standards. The target value is the ideal outcome. The tolerance is the maximum allowance before rectification action must be taken immediately. The tolerance range is an engineering choice reflecting a balance between the consequence and likelihood of failure and the need to keep the plant in operation. An IT7 dimensional tolerance and a G2.5 balance are not precision values. But they are shown in the table as worst-case values so that machines can stay in operation until a maintenance shutdown can be planned. Sometimes you will set wider allowances and accept higher operating risk for the sake of expediency. But then you must watch the equipment condition and manage the risk well. Because machines are designed for a wide variety of duties, the suggestions in the table may not suit all operational situations. Every company has to investigate and choose the quality standards it will live by in its operation. Subject matter experts are best placed to recommend the standards that bring high reliability.

References

1. Stanley F. Nowlan and Howard F. Heap, *Reliability Centred Maintenance* (San Francisco: Dolby Access Press, 1978).

2. Some dispute exists over the curves because it was not clear in the studies which items were parts and which were equipment, or which items were refurbished and reused and which were replaced with new. See David J. Sherwin, “A Critical Analysis of Reliability-Centered Maintenance as a Management Tool” (paper presented at the International Conference of Maintenance Societies, Australia, 2000).

17

Measuring to Improve Performance

It's a mockery to use business performance indicators to manage a company or department. Using the information that shows up on the management dashboards on your screen can wreck your company. Measuring a process does not put you in control of it. A process is totally self-controlled. You always get what the process gives you. Organizational processes work well or do not because of their design.[1] A business process that cannot achieve its aims is the wrong design for its purpose.

Financial performance and productivity results are interesting for a business to know, but they are point-in-time samples of what your processes were producing at that moment. The data in weekly, monthly, and annual performance charts are the results your processes let you have. When you take another sample in the following period, you will get a different result. If a performance chart contains only natural outcomes, then that is the best the process can give, and there is nothing more you can do about it. If you don't like it, the only choice is to design and build a better process.

Plotting the frequency of process results taken over a long time gives you a distribution curve of process performance. It may look like the bell curve shown in Figure 17.1, or it may take some other shape. Every time you measure a process variable, the value will be somewhere on the distribution curve. When you take a sample and show its measure in a monthly report, it is a result from some point on the curve. One month you get one value for the indicator. The next month the sample comes from elsewhere on the distribution, and you get another result. Wait a while, and the result will change again. Variation in results is how a process works. You get fluctuating values over a period of time—some good, some bad, and many in between. If you expect great performance results every month, you are expecting the impossible because a process produces outcomes anywhere along the curve. You would get far more management value from a monthly report if it also showed the frequency distribution of the performance measure alongside its trend plot and where on the distribution the point was located.

Once you have a lousy process design, you will get lousy results until you get a better design—nothing else you do will bring process management success except having the right process design in the first place. You might think that collecting management indicators will aid in correct decision making. But if the process is already stable, every change made to it can worsen performance. Once managers and people get a result that is not ideal, they will tinker and make adjustments to the process. But it is no good to make changes unless a process is out of control. First it is necessary to check whether the process is running stably, and if it is, then leave it alone. Any tweaking will only ruin it. To be of any use for process management, your performance indicators need to measure the extent of the deviation from the ideal process outcome, its target "bull's-eye," and whether the distribution is steady. Indicators are useful when they help people make the right decisions to improve process designs or help them find out-of-control process effects and remove them. If they are used for other purposes, they will cause you to ruin your business performance.

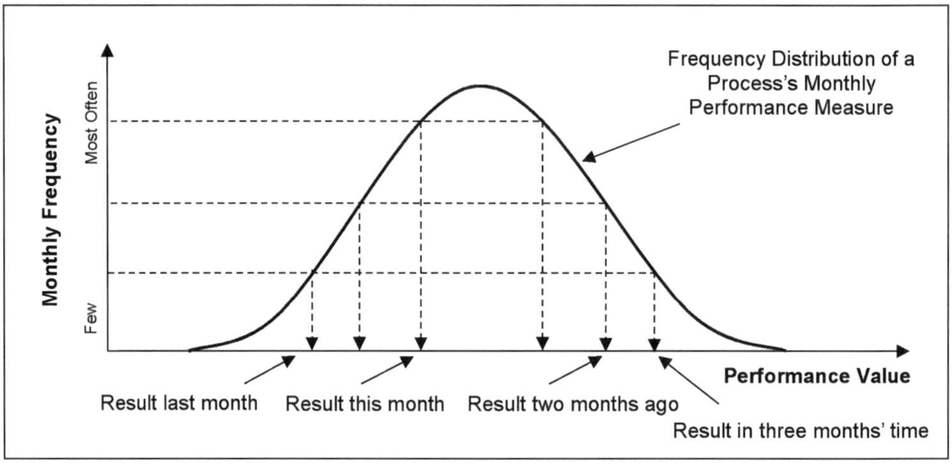

Figure 17.1—Distribution of a Process Performance Measure

Monitoring a Process and Its Process Steps

Key Performance Indicators (KPIs) are the measures used to monitor process outcomes—they show the total effect of all influences on the process. Performance Indicators (PIs), on the other hand, are used in the process steps—they tell you about a step's variability.

How do you monitor and measure a process such as production, maintenance, or a continuous improvement process so that you can make it world class? What about a job—how do you monitor and measure work and use the information to make it highly successful? What about the tasks in the job—how can you make them more efficient and effective? In Industrial and Manufacturing Wellness, every one of these questions is answered the same way: by monitoring the trend and stability of step inputs and outputs, as well as checking the trend and stability of the whole process to see cumulative effects.

Once you have the process design captured in process maps and procedures, you have everything you need to properly monitor, measure, and improve the process. On the process map, important inputs and outputs from each step are identified. In ACE 3T procedures, all variables in the process step are identified and quality standards are set for them. The ideal results and tolerances are defined using measurable performance terms with values that can be charted. The process map and supporting documents contain all the performance measures you need to monitor the whole process and the performance of each step. Useful statistical charts and graphs can be developed with the results and used to observe and compare process and step behavior.

Figure 17.2 is a process map of a purchasing process. The process is used to buy what is needed in as simple and straightforward a manner as possible. In the end, what you want from the buying process are the things you asked for. A measure of total process success is whether you received the items you wanted. A second measure of overall success is whether you got your order on time. A third is whether the order was in perfect condition. A fourth is whether the invoiced amount matched the price in the offer. These purchasing requirements are what the process should always deliver with 100% certainty. Less than 100% conformity on all measures means that the process design failed. Somewhere, one process step, or more, went wrong. If you monitor the step outputs, you will know which ones failed.

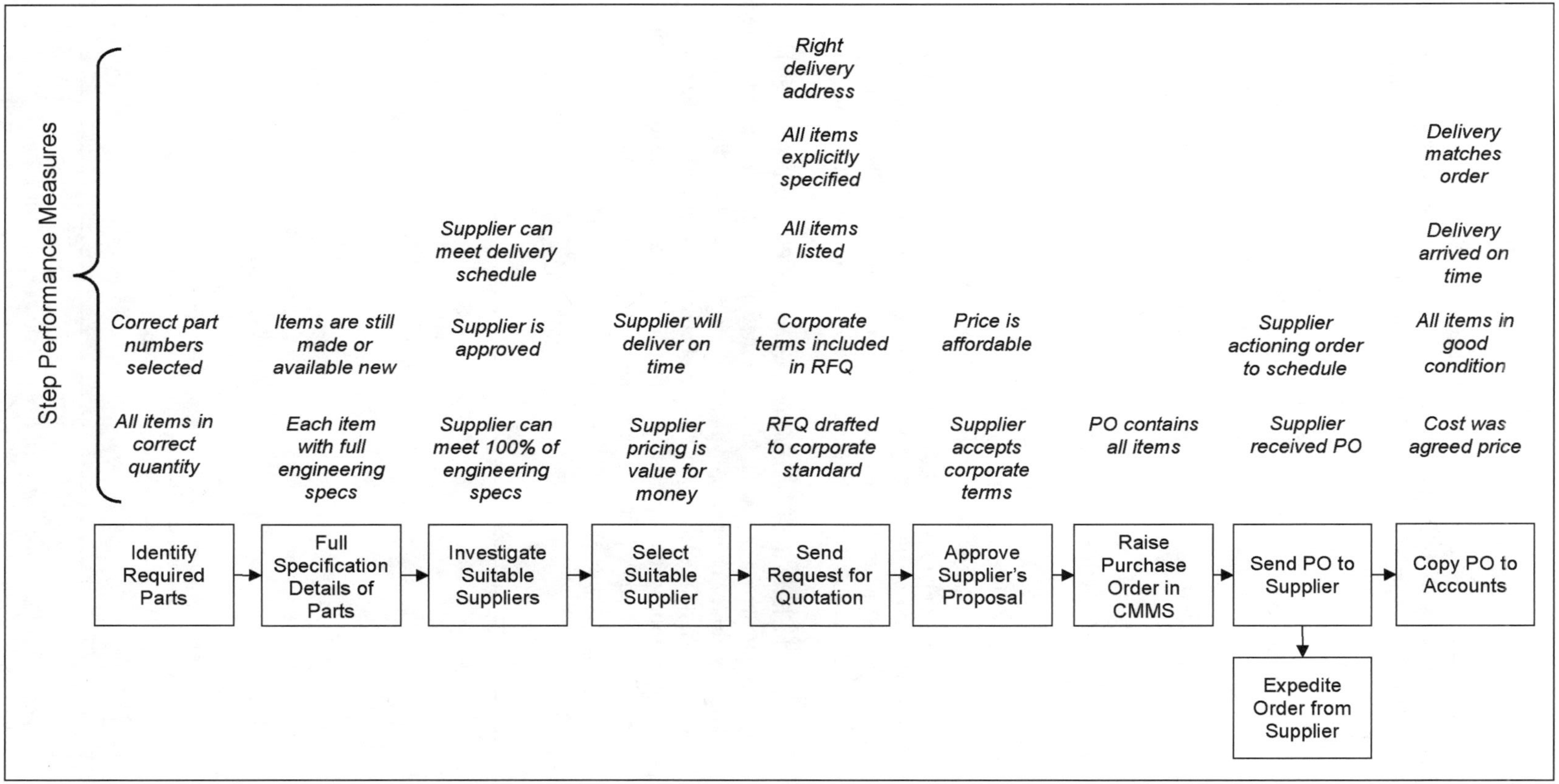

Figure 17.2—Monitoring the Performance of a Procurement Process

A process fails when a process step fails. If a result from the complete process is not acceptable, you need to find and correct the steps in the process that control the result. In order to do that, you need indicators of each step's success. The measures you use are the required step outputs noted on the process map. The beauty of using the measures written in the process map is they are what the process design is meant to deliver. They directly result from the behavior of the step and reflect the activities done in a step by those running the process. The performance values are available because the procedure asks for them to be generated as the work is done. If it is necessary to understand why and how a step failed internally, you observe and measure the variables in the step, even going as far as to statistically chart them to look for the telltale signs caused by the presence of defects. When you report on the process and talk to people about its performance, it will be with facts and answers specific to the process.

Process Performance Distribution Curves

The performance indicators you take from a process are samples of the process behavior. Each KPI and PI is a result generated as the process was being run. These indicators test the process's performance. Over time, the indicators will tell you the reliability and stability of your business processes. Look at the run charts and frequency distribution curves of your maintenance and operations processes if you want to see your company's operating future. Those graphs and plots represent future performance if all remains the same. The distribution curves of your KPIs and PIs reflect your business's outlook until you change policies, process designs, and practices to those that produce more successful outcomes.

It's interesting to watch the faces of senior managers when they first see the distribution curves for their Key Performance Indicators and you explain what they mean. For years, they have seen individual monthly KPI results plotted on graphs. Typically, the KPI graph shows the trend line across a period of time. Using a run chart alone to observe process behavior is incomplete process management. When you also show company and department managers the frequency distribution curves of the KPIs they are responsible for and explain to them that the shape indicates the process's chance of achieving the KPI, they start to understand what their business and its processes have been doing.

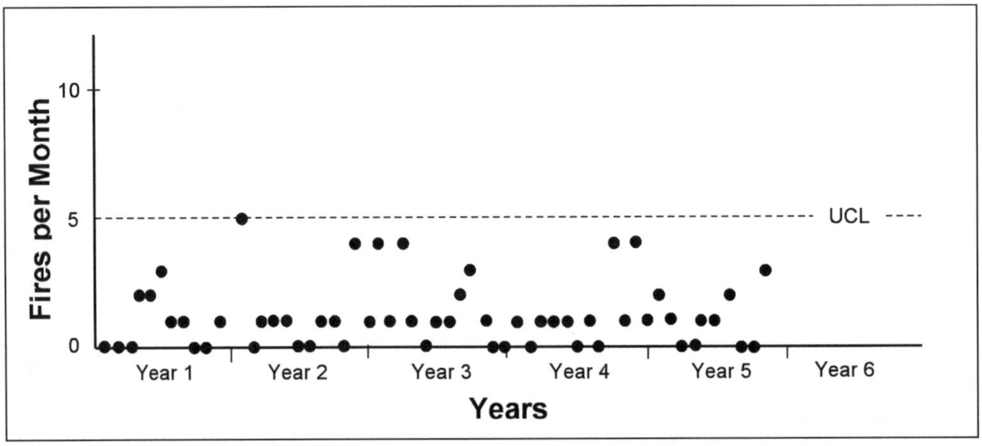

Figure 17.3—Run Chart of Fire Events in an Industrial Plant

When the curves are wide and success is uncertain, managers get a worried look. When the curve is tight but over the wrong performance value, they get a frown from wondering how they can afford to reengineer the process. When the distribution is tight and "right on the money," you get a big smile.

The run chart and the failure distribution curve developed from the chart data are two simple techniques for analyzing both system and step performance. Figure 17.3 is adapted from W. Edwards Deming's book *Out of the Crisis.*[2] It is a run chart of fire events in an industrial operation over a five-year period. Looking at the run chart, it is clear that unless something changes, Years 6, 7, and beyond will be the same as the past five years. The run chart confirms that a persistent problem exists and will continue next year, and the year after, and again the year after that unless the process is fixed. The repetitiveness of the fires indicates that a common cause problem(s) is built into the way the company operates—it gets fires by design. This place must have been endless trouble for its people and management to run.

Within the run chart is "hidden" information about the probability of the next fire event. Over 58 months, there was one or more fires in 38 months. Therefore, the odds are 38/58 (a 0.66 chance) that there will be a fire next month. Figure 17.4 is a distribution plot showing the frequency of fires per month in the months with fires. In the 38 months with fires, there were 68 fires. In 24 of the 38 months, there was one fire per month (24/38 = 0.63 chance of one fire in months with fires), and 5 months had two fires (5/38 = 0.13 chance of two fires). In fact, there is a 0.66 x 0.13 = 0.09 chance—odds of about 1 in 11—of having two fires in the next month. In 3 months, there were four fires (3/38 = 0.08 chance of 4 fires in months with fires), 5 months had four fires (5/38 = 0.13 chance of four fires), and 1 month had five fires (1/38 = 0.03 chance of five fires). No months had six fires, but from past history, the possibility existed that it would happen one day.

There is still information in the run chart. The frequency and density of months with one fire versus the density and frequency of months with three or more fires indicates that the months when one fire happened were not the same types of months as those when three or more fires happened. Something significantly different happened in the months with many fires.

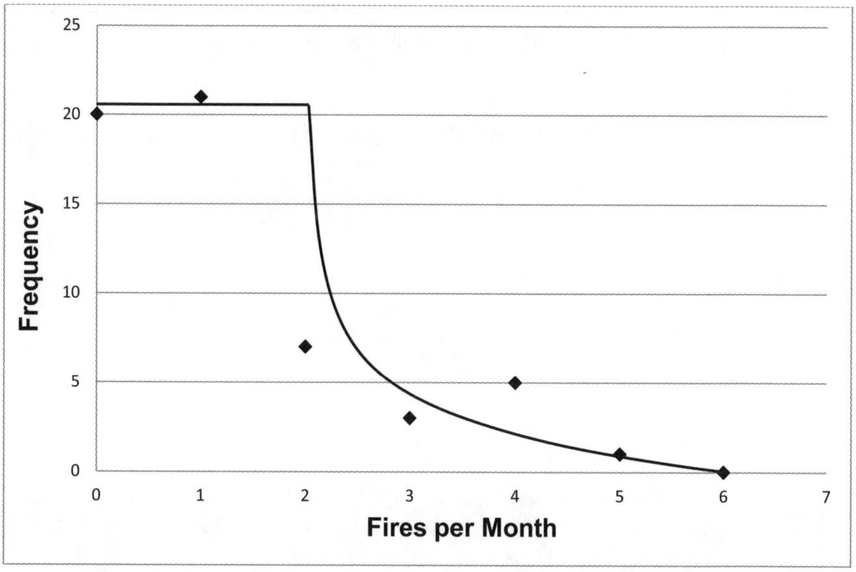

Figure 17.4—Probability Distribution Curve of Fires in Months with Fires

Another way of using run charts is shown Figure 17.5, which graphs the production downtime caused by equipment breakdowns each week in the industrial plant introduced in Chapter 3 (see Figure 3.12).

Figure 17.6 is the frequency distribution curve for the data. At first it looks like a normal bell-shaped distribution curve, but suddenly there is a discontinuity. This company has two types of breakdowns—usual and severe. The lost time from usual or standard breakdowns average 25 to 30 hours a week (consistently between 15 to 45 hours), but from time to time the company has breakdowns that are catastrophic to production uptime.

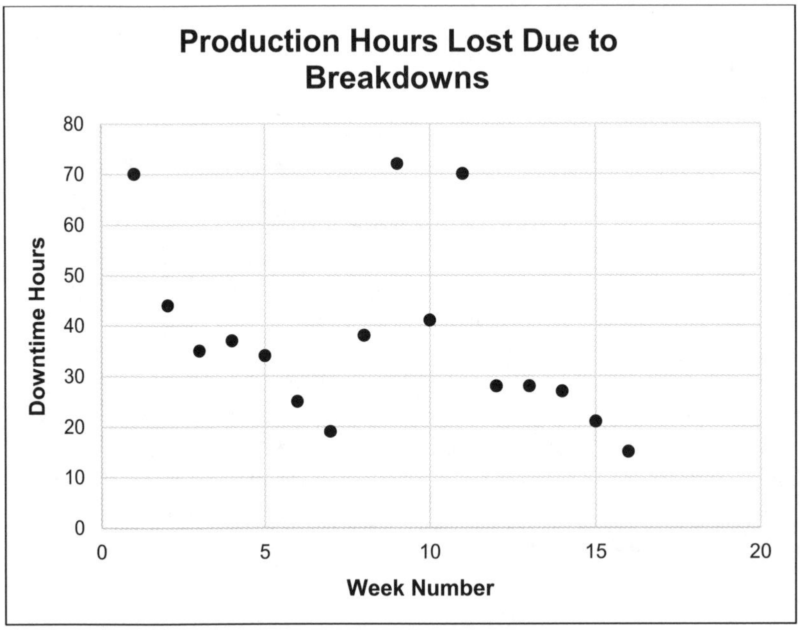

Figure 17.5—Production Hours Lost Due to Breakdowns

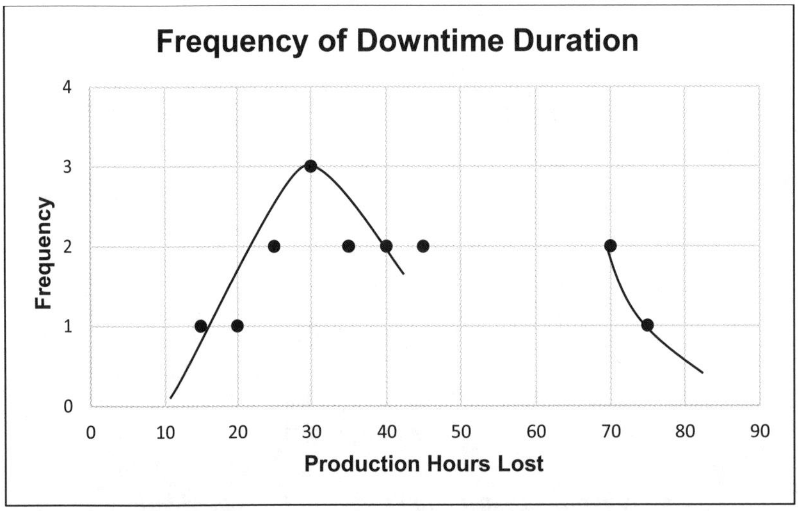

Figure 17.6—Distribution of Production Breakdown Hours

Example 17.1: Production Plant Performance

Figures 17.7 and 17.8 are examples of getting stuck trying to fix your current business when instead you ought to throw away your troubles and build a better business system. This business is a renowned company in its home country. It is well respected and profitable enough. But it could easily be much wealthier. There are vast new fortunes sitting in the business, but they will never be seen by its owners and managers. They are totally focused each day on trying to make the existing business processes and system work properly. In reality, their operational processes need to be redesigned to remove the problems stopping them from delivering the organization's purpose.

The two charts cover a period of 10 years of operation (the years and company are intentionally disguised). Figure 17.7 is a run chart showing the dates and durations of all outages. The periods between downtimes are the operation's uptime. Figure 17.8 is the uptime frequency distribution curve derived from the run chart. It is a "chance of success" chart because it shows how often each uptime period happened during the 10 years. The shape of the frequency distribution curve foretells what outcomes can be expected in the future.

Look at the density of outages in the run chart. There were years of frequent trouble and times of less. Things must have gone bad often to create the dense regions. The last two years show fewer minor trips and some of the best uptimes. Note the stratification of dots: many less than 50 hours of outage and far fewer greater than that. Having random outages of 10 to 48 hours is common in this operation.

The frequency distribution curve turns the splatter of events in the run chart into a clear message about this company's chance of uptime success. There are dozens of short-duration running periods of less than 2 days and quite a few periods from 2 to 10 days. Very few uptime periods last longer than 20 days. The story hidden in the run chart becomes obvious in the frequency distribution plot: this business is missing out on a fortune in operating profits (worth several million dollars annually) because it has so many stoppages interrupting production. Yet the same operation has several times delivered more than two months of uninterrupted production. The process design can get long production runs, but things happen in the organization to prevent success.

The whole area under the curve in the frequency distribution is the probability of plant uptime. You can estimate by eye that the area between 0 and 20 days is larger than the area from 20 to 80 days by about four times. It is four times more likely that the next plant outage will be fewer than 20 days away than it will be more than 20. It looks about three times more likely that an outage will happen in fewer than 10 days rather than longer. It is clear from the "hump" in the curve between 0 to 2 days that short-duration outages are a regular problem for the business. It would be very valuable to analyze the reasons for the outages in the hump to learn what is causing so many stoppages and eliminate them so the process can work to its best design performance.

The shape of the uptime frequency distribution curve contains information about the business process design. The clustering of results along a negative downward curve tells us there are destructive influences at work within this company. This company has an in-built outage-causing process—surely unintentionally introduced—that is producing the poor uptime results. The early failure peak is an indicator of poor business process quality control—there are a lot of defects sitting in the business waiting for the chance to go wrong. The negative slope at the back of the curve means that the company's current designed and intended processes can never get it to the production performance it wants, which is close to 60 days or more of uninterrupted production uptime between outages (the dashed curve in Figure 17.8). That's happened only five times in 10 years. Those five successes were all attributable to luck. We know it was luck because the long duration uptime points form a flat line, meaning that they were accidental results without consistency. Those 50-day or longer uptimes were chance events; they are not repeatable by intention because of the self-destroying influences within the company processes. The only sure thing in this business is that uptime has a great chance of

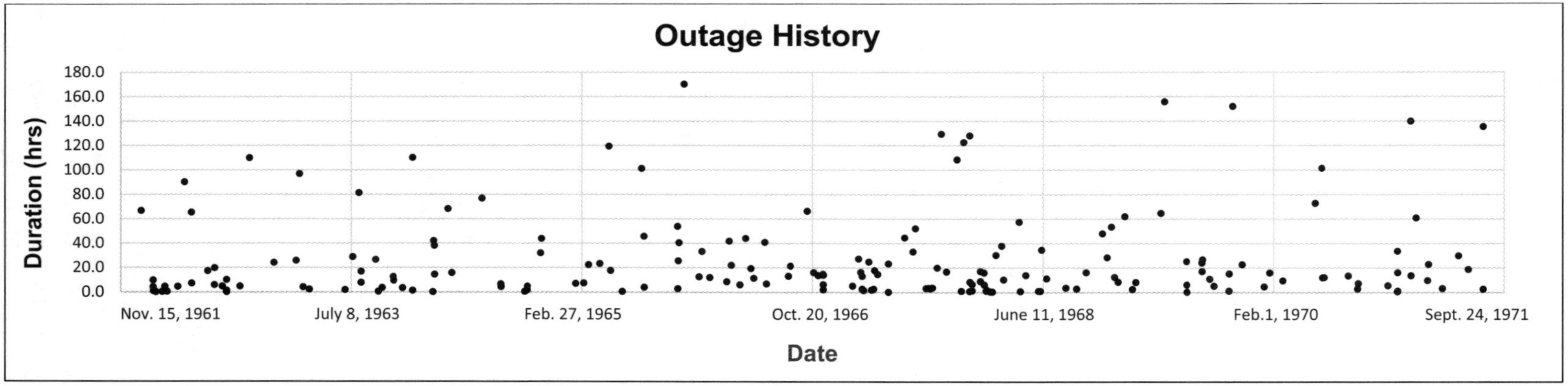

Figure 17.7—Plant Outage Duration Run Chart

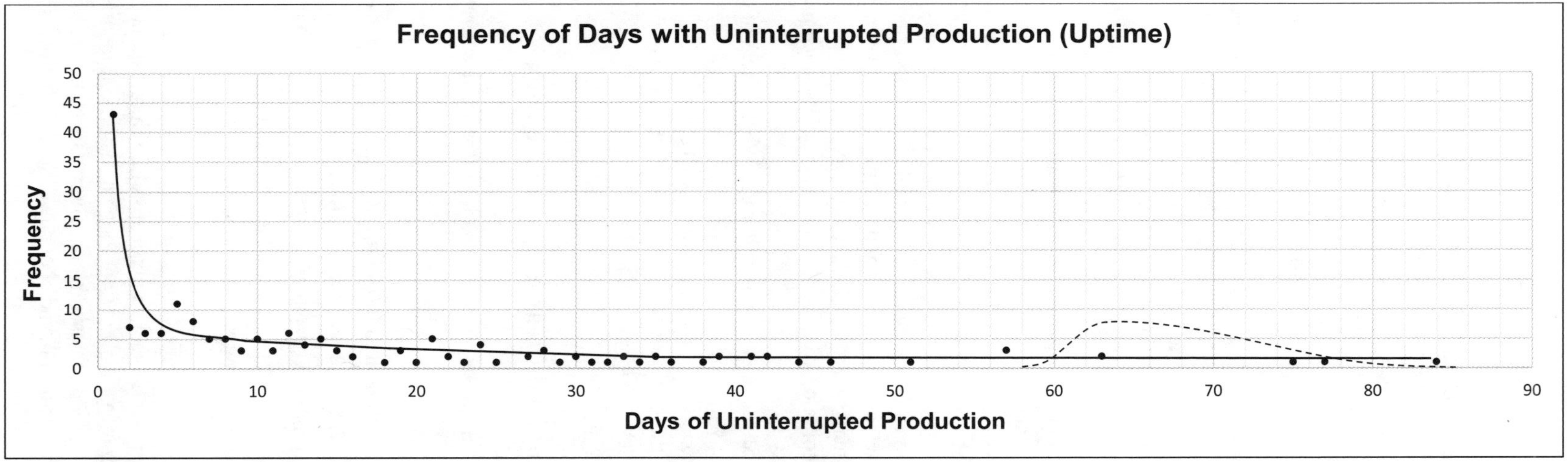

Figure 17.8—Distribution of Uninterrupted Production Days

being fewer than 20 days and some chance of being up to 30 days; the company will be very lucky if it gets the 60 days it wants (about a 10% chance). Their maintenance costs must be higher and their availability lower because of the many unwanted outages that don't need to happen.

The engineering, maintenance, and operational processes this business uses can only deliver the current uptime results seen in the plots. Remember, this is 10 years of real production data—this performance is what actually results. Until the company adopts reliability creation processes that guarantee the success it wants, future operating performance and losses will be the same as in the past.

Plant run charts are more than simply indicators of the dates when you had problems in your plant or with your equipment. They contain knowledge of the likely behavior of your operation in the future. That behavior is the result of the cumulative effects of your capital projects, asset management, operational, and maintenance management policies and practices. Once you convert run chart data into a frequency distribution, you see the natural behavior of your operation and its processes.

To do useful process improvement, it is vital to know the reason for each PI and KPI data point. A run chart of each type of event cause is then plotted, and the cause frequency distribution is graphed. Using the distribution curve of causes lets you identify the frequency of each event cause. If you then plot the cumulative TDAF costs from each event, you will understand the high cost problems you have in your company. Repeating causes are likely to be indications of common cause problems in your operation. Single events may indicate a special cause impacting the process. If the same cause reoccurs, the business-wide losses from it will justify starting an improvement project to solve the problem. Once the problem is removed, the money once lost will turn into new operating profits. Keep full and complete records of your production and maintenance problems—they are worth solid gold to you in the future.

With distribution curves, you have a means to monitor the effectiveness of changes made to a process. Each change you introduce will produce a new performance result for the steps you alter and cumulatively for the process. Once you collect about 10 to 12 results, do as shown in Figure 17.9 and superimpose the curve for the new outcomes on the past distribution curve to get feedback on the impact of the change. If the new curve does not plot within the higher performance part of the old curve, you need to keep looking for better answers.

Monitoring and Measuring Maintenance

The types of performance measures useful to the maintenance effort in an operation are those that do the following:

- Identify where your equipment failures start in your suite of business processes
- Track the influence of life-cycle factors on failure
- Direct efforts so that time and resources are used effectively and efficiently to remove the causes of failure, create higher reliability, and reduce operating risk
- Guide maintenance to deliver greater business benefits

Equipment fails because parts lose structural integrity from microstructure overstress or atomic attack. There are six major causes of mechanical equipment failure: lubricant degradation and contamination, out of balance, misalignment, working component distortion, incorrect fastening, and induced vibration. For electrical equipment, the six major failure causes are contamination, induced vibration, high temperature, moisture, distortion, and poor power supply stability. These failure mech-

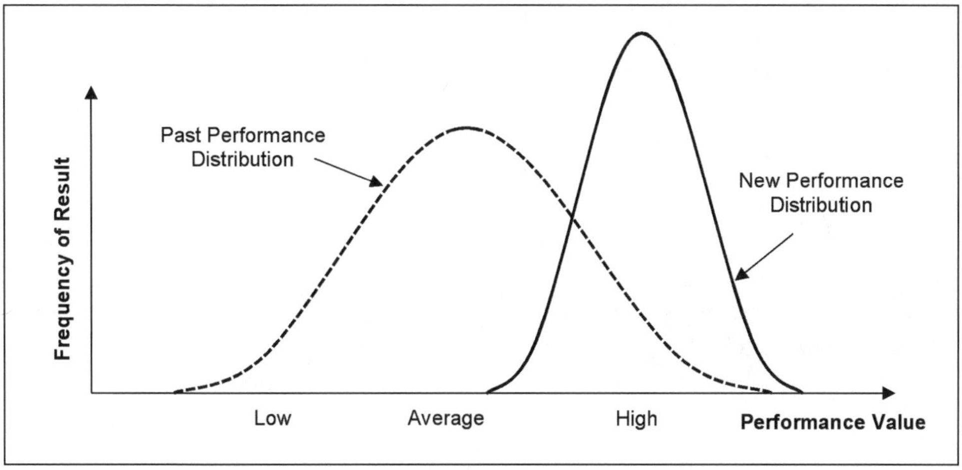

Figure 17.9—New Distribution Superimposed on the Past Distribution

anisms must be purged from your plant. This requirement goes far beyond just using condition monitoring to observe equipment health. You want maintenance to find what is creating those mechanisms in the mechanical and electrical equipment and get rid of them. You need measures to prove that maintenance is focused on successfully eliminating problems. You want the maintenance crew to be proactively creating sure reliability. To make this happen in your operation, you need to use maintenance indicators that drive reliability improvement.

It's important to know where and why maintenance allocates time, people, and effort each month. Unfortunately, maintenance resources are easily expended doing anything and everything to keep the operation running. The best maintenance strategy delivers reliability and lower risks that give back fortunes year after year. The secret is not to focus on doing maintenance but to focus on creating reliability and removing operating risk. Maintenance has the duty to stop problems from starting, and where there are problems, they permanently remove them. If you measure maintenance using the factors listed below and a significant portion of time and resources is not used on them, then the future will be the same as it is today and highly likely to get worse.

- Maintenance work orders spent on improving equipment
- Maintenance time and effort spent removing breakdown causes
- Maintenance time and effort spent improving maintenance procedures
- Maintenance time and effort spent improving maintainer skills and knowledge
- Maintenance time and effort spent reducing operating problems
- Time spent removing wasted effort and cost from maintenance processes
- Efforts spent improving stores management processes and stored parts reliability
- Maintenance work orders spent improving safety

The size and mix of your maintenance crew reflects the effectiveness of its work processes and of the equipment reliability creation processes in use. In a world-class, reliability-driven business, maintainers spend a lot of time designing and implementing productivity improvements on equipment and removing maintenance costs.[3,4] The Plant Wellness Way solution is to use a small, highly versatile maintenance crew with the skills to do high-quality, precision work that is right the first time. When maintainers have more engineering knowledge, use precision skills, create standardized work management processes, and employ work quality assurance solutions, the maintenance crew can be used in very profitable ways.

References

1. W. Edwards Deming, *Out of the Crisis* (Cambridge, MA: MIT Press, 2000).

2. Deming, *Out of the Crisis,* 324.

3. Jim Wardhaugh, extract from IQPC Reliability and Maintenance Congress presentation "Maintenance—The Best Practices," Singapore, 2005.

4. "Maximising Operational Efficiency Presentation," E. I. Du Pont de Nemours and Co., 2004.

18

The Chance of Success

What is the chance that the processes you use will successfully deliver the intended results? In a process in which one step is done after the other and the failure of one step can fail the whole process, there are many opportunities for problems. In a series arrangement, the risks add together. Your total business risk is the sum of all of the risks in all of your business processes. The total risk to a production line is the sum of all of the risks in each production line step. The total risk for an equipment item is the sum of all of the risks from each of its components. Figure 18.1 represents how the risks to the individual parts in a roller bearing add together to make the total risk for the bearing. A risk that fails any component of the bearing will fail the whole bearing. Every machine, process, and job carries the cumulative risk of failure from all of its elements.

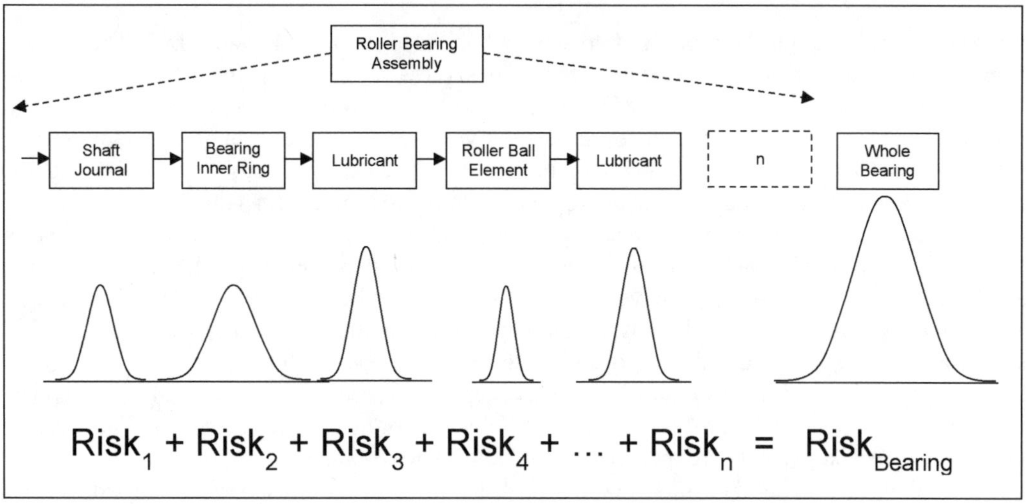

Figure 18.1—Risk from Each Equipment Part Adds Together

To improve equipment reliability, remove the risk of part failures. To improve production line performance, remove the risk of production line step failures. To improve workplace safety, remove the risks of workplace harm. You'll create a more successful business when you use processes containing fewer risks.

You can estimate the chance of successfully completing a process by gauging the chance that each step will be successfully completed. The method is called Chance of Success Mapping.

Estimating the Chance of Process Success

The four-step process for crossing a busy road shown in Figure 18.2 is a simple example of how Chance of Success Mapping can improve the process success rate. The aim is to understand how each step can be made more successful. By removing the risks of failure at each step and introducing more effective step practices, the chance of process success rises.

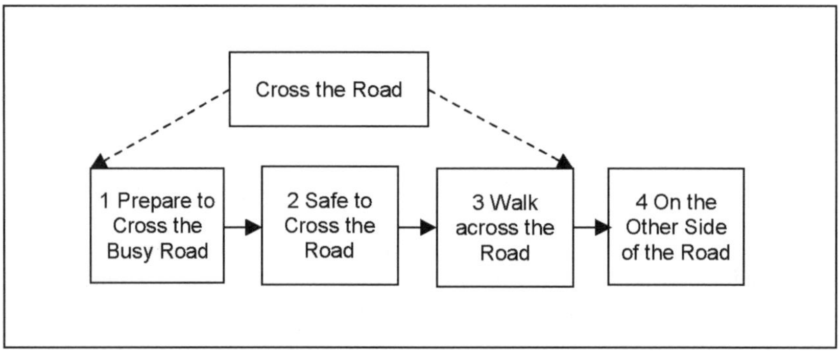

Figure 18.2—Simplified Process for Crossing a Road

The spreadsheet for the process of crossing a busy road is shown in Table 18.1. The process map goes across the top of the table, and the chance of success investigation is conducted underneath. Because the process is a series, if one step is done poorly, the whole process performs poorly—which, when crossing a dangerous road, means that you increase the chance of being run over or getting seriously injured. Success depends on preparing yourself well and ensuring that nothing bad happens as you cross the road. Your chance of successfully crossing the road diminishes with every risk you face, and it improves with every uncertainty you remove. Because the risk to a process is the sum of the risks from each step, you investigate individual steps to see how they could be failed and look at what can be done to remove the causes of failure and promote the chance of success.

A process chance of success analysis starts by replicating the process map across the top of a spreadsheet. These steps make up the process as designed, and the combination of all of the steps is your current process design. The purpose of each step in the process is explained. Every step has one or more functions in the process, and it must deliver all of its required outcomes. These are its target values. All step targets are recorded in the table. When a target is not achieved, the step has failed. Smiley faces are used to show how critical a step is to overall process success; a sad face denotes a critical step, and a normal face is an important step. A step may be unimportant, important, or critical. It is unimportant if its failure has no impact on the process outcome. A step is important if, when it goes wrong, it causes a loss of time or money but can be corrected to perform properly. A step is critical if, when it goes wrong, someone is injured, there is a disastrous product release, or a step target cannot be achieved. It is very unlikely that any step in a series process is unimportant. If that were the case, you would eliminate the step because it is a waste of time and effort.

To find the problems that could prevent process success, you identify all of the risks in each of the steps and note them in the spreadsheet. With each step's existing risks and problems known, you can estimate each step's current best chance of success and worst chance of success. This establishes

Table 18.1—Chance of Success Mapping for Crossing the Road

Chance of Success Mapping for Crossing the Road

No	*Process Steps*	1 Prepare to Cross the Busy Road	Chance of Success for this Step		2 Safe to Cross the Road	Chance of Success for this Step		3 Walk across the Road	Chance of Success for this Step		Chance of Success to this Point		4 On the Other Side of the Road
1 2 3			Low %	High %		Low %	High %		Low %	High %	Low %	High %	
4	*Step Purpose*	Check the traffic conditions			Confirm there is time to cross			Physically move the distance					Reach the other side unharmed
5	*Step Importance*	☹			☹			☹					😐
6	*Step Target Outcome*	Look both ways	99	100	No traffic too close	90	98	Traverse directly across the road	98	100	86	98	Safely across the busy road
7	*Opportunity to Do Step*	1.1 Be at roadside			2.1 When no vehicles within 200 m			3.1 When no vehicles approach during crossing					4.1 On the other side of the road
8								3.2 Surface is traversable					
9 10 11	*Step Risks*	1.1.1 Obstructions 1.1.2 On a bend			2.1.1 Speeding car 2.1.2 Reversing car			3.1.1 Speeding car 3.2.1 Uneven surface 3.2.2 Potholes					
12 13 14	*Risk Mitigations*	1.1.1.1 Find clear line of sight 1.1.2.1 Move 100 m clear of bend	100	100	2.1.1.1 Give 300 m clearance 2.1.2.1 Look both ways	98	99	3.1.1.1 Give 300 m clearance 3.2.1.1 Plan route to follow	99	100	97	99	*Alternative option:* Move to a better spot with higher chance of success

the two ends of the frequency distribution of possible outcomes for the step. The best and worst chances of success are estimated using historical records of the frequency of past bad results and the frequency of great results in doing each step.

When crossing a busy road, you will most often cross it successfully, but occasionally, you will not be able to cross because of existing dangers. Your past experience tells you that for every 100 times you arrive at a busy road, you won't be able to cross it every time. You use historical evidence accumulated over the years to gauge the lower and upper ranges for the chance of completing each step of a process. There is a degree of uncertainty in estimating the chance of success range, but you are unlikely to come up with wild, crazy values if estimates are based on factual historical information from similar past situations. The likelihood scale of a calibrated risk matrix is a useful gauge of uncertainty when there are no complete historical records available.

The individual steps multiply together to give you a low to high range for the chance of success of completing the whole process. In the table, the chance of successfully crossing the road goes from 86% to 98%. An 86% chance of success of safely crossing the road is too low. Because you know the poor risks to address, you choose mitigations to eliminate them or minimize their impact on success. The controls you will use are recorded for each step. To estimate the revised chance of success range, you presume that the new risk controls are in place and that they are being done properly. If the planned risk mitigations noted in the table have been done, the chance of safely crossing the road goes up to 97% to 99%. In order to massively reduce the chance of harm when crossing a road, you need to use traffic lights controlled for pedestrian crossing. To totally eliminate traffic risk when crossing a busy thoroughfare, take a pedestrian bridge over the road.

Estimating the Chance of Equipment Success

For your equipment, you can do risk simulations for the chance of success of its parts. In Table 18.2, a section of the chain of components that assemble together into a ball bearing is modeled using a Chance of Success Mapping spreadsheet. The spreadsheet is configured to let you record each part's risks and then add risk controls to reduce them. Following that, you gauge how much difference the new mitigations are likely to make in increasing the whole component's reliability.

The component map of the bearing design is laid out across the top of the spreadsheet. The accumulated steps make up the process design. Each part is briefly described so that you understand why it's there and what it does. Below each item is a note on the consequence or utility when it performs successfully. Next you identify the risk severity of each step with smiley faces. For each item in the component chain, clearly describe all of the required outcomes it must deliver (i.e., its targets). Under each step list the tasks to be done in order to successfully deliver each step target.

The first intent is to gauge the chance of success range for the bearing exactly as your company is running its equipment today. The estimates of low and high chance are for the operating service and care that are now being used. Beside each step target, indicate the worst odds (low) and best odds (high) that a step will deliver each of its required targets. The upper and lower chances of success allocations are based on the historical frequency of risk events over the organization's operating history. This information is typically found in the maintenance and operations management systems. If complete and accurate failure information is not available, ask for the estimates of long-serving operations and maintenance staff.

To simplify choosing the lower and upper ranges, suppose you have 100 of these parts in your operation. How many times during the service lives of the imagined 100 items could a defect arise for the issue being considered? This gives you a percentage for the low chance of success value. Then, for the same issue, ask yourself, of the 100 items, how many are likely to be perfect every time this situation arises? That is the upper value of the chance of success for that step in your process. When the chances become very small for a 100-opportunity scenario, use bigger groups of opportunities, such as the chance in 1,000 or 10,000 of the issues or items under consideration. Mass-produced parts will have a very low chance of containing defects, likely measured in a few per 10,000 items, whereas any part made one by one using job shop processes has many humans involved and therefore a greater risk of errors. The failure rates from work done in job shops would be measured in defects per 100 and even defects per dozen.

The chance of success for a step is dependent on how many risks are present in the step and each individual risk's probability of occurrence. Each risk reduces the odds of a step's successful completion. The effect of all risks in a step on the step reliability, or the step chance of success, is taken into consideration by multiplying all low chances together to determine the total worst impact on the step and then multiplying together all high chances to determine the best odds of step success.

Once low and high chance of success estimates are made for each component step, the entire assembly's chance of success range can be gauged by multiplying together the relevant values of all its steps. To get the assembly's least chance of success, all of the component's low chances are multiplied. Similarly, to get the assembly's best chance of success, all of the components' high chances are multiplied. At any place along the chain, you can gauge the success range by inserting equations to calculate the lows and highs of success to that point.

All risks to the outcome of a step need to be identified and recorded. Every situation that can fail the step needs to be assessed for its impact on reliability. For example, in the table, every risk to journal machining needs to be listed when calculating the probability of success for the journal and the subsequent impact on the life of the bearing. As the Step 1 shaft journal is made, the same machining activity produces both the journal shape and the surface finish required in Step Targets 1.1.1 and 1.1.2. Because there are two separate quality requirements for the shaft journal from one machining event, and one or the other or both can be done wrong, each machining risk from the same job activity is noted in the spreadsheet and included in the process calculation.

According to Table 18.2, the least chance of success for the assembled bearing using current practices is 69%. For every 100 ball bearings mounted on a shaft, up to 31 of them will contain defects related to the journal, the inner ring, the lubricant, or the rolling element. Doing the very best possible, the highest chance of success of all items being defect-free is 93%. With the best odds, seven ball bearings out of 100 mounted on shafts will fail before their full service life. With the worst odds, 31 out of every 100 will fail before their time. The biggest contributor to the current chance of failure is lubricant contamination.

To increase the chance of success, you do a risk analysis for each item to identify what causes the risks at each step so that those risks can be removed. In the example in Table 18.3, the causes of risks to the success of each step are listed. It is important to identify every risk that will prevent the item from working right the first time. If you do not spot a risk, you cannot protect against it. When analyzing equipment, it would pay for you to review the Physics of Failure Factors Analysis guidewords to check whether there are other risks you need to include into the spreadsheet.

After the risk identification is completed, select mitigations that remove the risks or at least control them to very low frequencies. This includes changing the way a task is done or making process design changes to increase the chance of success. In the example in Table 18.4, controls for the risks are added to the list. The low and high estimates for risk are then updated for each item on the presumption the mitigations will be correctly implemented and used. The revised chance of success from eliminating risks moves the low chance of success to 88% and the best chance of success to 95%.

Table 18.2—Chance of Success Mapping for a Roller Bearing

	1 Shaft Journal	2 Bearing Inner Ring	3 Lubricant	4 Roller Ball Element
Step Functions	1.1 Fully support bearing inner ring 1.2 Locate bearing	2.1 Allow rolling element to rotate 2.2 Take full service load	3.1 Reduce friction 3.2 Prevent surface contact	4.1 Rotate 4.2 Take full service load 4.3 Axially align shaft inside housing
Consequence of Achievement	Full bearing service life	Full bearing service life	Full bearing service life	Full bearing service life
Step Importance	☹	☹	☹	☹
Step Targets	1.1.1 Correct fit, tolerance, and form 1.1.2 Suitable surface finish 1.2.1 Suitable metallurgical properties	2.1.1 Raceway surface finish to specification 2.2.1 Ring shaped to specification 2.2.2 Take full service loads	3.1.1 Chemically correct 3.1.2 Correct viscosity 3.1.3 Contaminate-free 3.2.1 Take full service load	4.1.1 Element perfectly smooth 4.1.2 Element perfectly shaped 4.2.1 Take full service loads

		Chance of Success for This Item			Chance of Success for This Item			Chance of Success for This Item		
		Low %	High %		Low %	High %		Low %	High %	
Step Tasks for Each Target	1.1.1.1 Accurate machining	99.5	99.9	2.1.1.1 Accurate machining	99.95	99.99	3.1.1.1 Correct lube for the service	99	99.9	4.1.1.1 Accurate machining
	1.1.2.1 Accurate machining	99.5	99.9	2.2.1.1 Accurate machining	99.95	99.99	3.1.1.2 Correct formulation	99	99.9	4.1.1.2 Accurate machining
	1.2.1.1 Correct material selection	99	99.9	2.2.2.1 Fully supported on a round shaft journal	99.5	99.9	3.1.1.3 Correct additives	99	99.9	4.2.1.1 Fully supported on a round shaft journal
				2.2.2.2 Accurate machining	99.95	99.99	3.1.2.1 Correct temperature	95	99	4.3.1 Position inner ring at proper point
							3.1.3.1 Solids-free	99	99.9	
							3.1.3.2 Water-free	99	99.9	
							3.1.3.3 Product ingress free	99	99.9	
							3.2.1.1 Load within design limits	95	99	
								84.969	97.423	

Table 18.3—Process Chance of Success Mapping: Identifying Risks for Each Part

	1 Shaft Journal			2 Bearing Inner Ring			3 Lubricant			4 Roller Ball Element		
Step Tasks for Each Target	1.1.1.1 Accurate machining	99.5	99.9	2.1.1.1 Accurate machining	99.95	99.99	3.1.1.1 Correct lube for the service	99	99.9	4.1.1.1 Accurate machining	99.95	99.99
	1.1.2.1 Accurate machining	99.5	99.9	2.2.1.1 Accurate machining	99.95	99.99	3.1.1.2 Correct formulation	99	99.9	4.1.1.2 Accurate machining	99.95	99.99
	1.2.1.1 Correct material selection	99	99.9	2.2.2.1 Fully supported on a round shaft journal	99.5	99.9	3.1.1.3 Correct additives	99	99.9	4.2.1.1 Fully supported on a round shaft journal	99.50	99
				2.2.2.2 Accurate machining	99.95	99.99	3.1.2.1 Correct temperature	95	99	4.3.1.1 Position inner ring at proper point	99	99.5
							3.1.3.1 Solids-free	99	99.9			
							3.1.3.2 Water-free	99	99.9			
							3.1.3.3 Product ingress free	99	99.9			
							3.2.1.1 Load within design limits	95	99			
		98.505	99.8		99.45	99.89		86.694	97.716		98.456	98.495
Risk Causes in Step Tasks	1.1.1.1.1 Incorrect machining setup 1.1.1.1.2 Measurement error 1.1.1.1.3 Wrong drawing 1.1.1.1.4 Incorrect part size or tolerance in drawing 1.1.1.1.5 Misread a drawing measurement 1.1.1.1.6 Poor cutting tool condition 1.2.1.1.1 Wrong material chosen 1.2.1.1.2 Wrong material supplied 1.2.1.1.3 Misread catalog selection			2.1.1.1.1 Incorrect machining setup 2.1.1.1.2 Measurement error 2.1.1.1.3 Wrong drawing 2.1.1.1.4 Incorrect part size or tolerance in drawing 2.1.1.1.5 Poor cutting tool condition 2.2.2.1.1 Wrong clearnace bearing supplied			3.1.1.1.1 Service duty misunderstood 3.1.1.1.2 Misinterpreted selection table 3.1.1.2.1 Wrong lube taken from store 3.1.1.2.2 Wax, varnish, coking 3.1.1.3.1 Additives depleated during service 3.1.2.1.1 Machine overloaded 3.1.2.1.2 Machine in high temperature location 3.1.3.1.1 Wear particles in lube 3.1.3.1.2 Sand, rust, or other solids in lube 3.1.3.2.1 Water ingress 3.1.3.3.1 Product ingress 3.2.1.1.1 Overloaded during operation			4.1.1.1.1 Incorrect machining setup 4.1.1.1.2 Measurement error 4.1.1.1.3 Wrong drawing 4.1.1.1.3 Incorrect part size or tolerance in drawing 4.2.2.1.1 Poor cutting tool condition 4.3.1.1.1 Inner ring not hard against shoulder		

Table 18.2, continued (See the website for the Excel spreadsheet at full percentage)

5 Lubricant	n	Whole Bearing
5.1 Reduce friction	n.1:	
5.2 Prevent surface contact	n.2:	
Full bearing service life	Full bearing service life	Full bearing service life
☹	☹	☹
5.1.1 Chemically correct		
5.1.2 Correct viscosity		
5.1.3 Contaminate-free		
5.2.1 Take full service load		

Chance of Success for This Item			Chance of Success for This Item		**Chance of Success to This Point**		Assembly Chance of Success	
Low	High		Low	High	Least	Best	Least	Best
%	%		%	%	%	%	%	%
99.95	99.99	5.1.1.1 Correct lube for the service	99	99.9				
99.95	99.99	5.1.1.2 Correct formulation	99	99.9				
99.50	99	5.1.2.3 Correct additives	99	99.9				
99	99.5	5.1.2.1 Correct temperature	95	99				
		5.1.3.1 Solids-free	99	99.9				
		5.1.3.2 Water-free	99	99.9				
		5.1.3.3 Product ingress free	99	99.9				
		5.2.1.1 Load within design limits	95	99				
98.407	98.485		84.969	97.423	69.2	93.1		

Table 18.3, continued (See the website for the Excel spreadsheet at full percentage)

5 Lubricant					n	Whole Bearing
5.1.1.1 Correct lube for the service	99	99.9				
5.1.1.2 Correct formulation	99	99.9				
5.1.2.3 Correct additives	99	99.9				
5.1.2.1 Correct temperature	95	99				
5.1.3.1 Solids-free	99	99.9				
5.1.3.2 Water-free	99	99.9				
5.1.3.3 Product ingress free	99	99.9				
5.2.1.1 Load within design limits	95	99				
	86.694	97.716	72.5	93.8		

Same as Item 3 Lubricant

Table 18.4—Process Chance of Success Mapping: Risk Controls Added for Each Part

	1 Shaft Journal	2 Bearing Inner Ring	3 Lubricant Grease	4 Roller Ball Element
Risk Controls	1.1.1.1.1.1 Second person check machining setup prior cutting 1.1.1.1.2.1 Double-check measure with calibrated micrometer 1.1.1.1.3.1 Include drawing check in machining ITP 1.1.1.1.4.1 Second person check all critical measurements on drawing 1.1.1.1.5.1 Critical drawing measurements to be highlighted on drawing 1.1.1.1.6.1 Second person check cutting tool condition 1.2.1.1.1.1 Include material check in machining ITP 1.2.1.1.2.1 Include material check in machining ITP 1.2.1.1.3.1 Check catalogue selection against material list on drawing	2.1.1.1.1.1 Second person check machining setup prior cutting 2.1.1.1.2.1 Double-check measure with calibrated micrometer 2.1.1.1.3.1 Include drawing check in machining ITP 2.1.1.1.4.1 Second person check all critical measurements on drawing 2.1.1.1.5.1 Second person check cutting tool condition 2.2.2.1.1.1 Include bearing clearnace check in maintenance ITP	3.1.1.1.1.1 Include service duty confirmation in engineering ITP 3.1.1.1.2.1 Second person to check selection table recommendation 3.1.1.2.1.1 Include lubricate confirmation in maintenance ITP 3.1.1.2.2.1 Include chemistry condition in test laboratory report 3.1.1.2.2.2 Include test laboratory chemistry report review and action by maintenance engineer 3.1.1.3.1.1 Additives' levels in test laboratory report 3.1.1.3.1.2 Include test laboratory additives report review and action by Maintenance Engineer 3.1.2.1.1.1 Operate equipment to precision operating procedure 3.1.2.1.2.1 Include location ambient temperature range in engineering ITP	4.1.1.1.1.1 Second person check machining setup prior cutting 4.1.1.1.2.1 Double-check measure with calibrated micrometer 4.1.1.1.3.1 Include drawing check in machining ITP 4.1.1.1.4.1 Second person check all critical measurements on drawing 4.1.1.1.5.1 Second person check cutting tool condition 2.2.2.1.1.1 Include position check in maintenance ITP

Chance of Success with Mitigations		Chance of Success for This Item			Chances of Success for This Item			Chance of Success for This Item			Chance of Success for This Item	
		Low %	High %		Low %	High %		Low %	High %		Low %	High %
Step Tasks for Each Target	1.1.1.1 Accurate machining	99.9	99.95	2.1.1.1 Accurate machining	99.95	99.99	3.1.1.1 Correct lube for the service	99.9	99.95	4.1.1.1 Accurate machining	99.95	99.99
	1.1.2.1 Accurate machining	99.9	99.95	2.2.1.1 Accurate machining	99.95	99.99	3.1.1.2 Correct formulation	99.9	99.95	4.1.1.2 Accurate machining	99.95	99.99
	1.2.1.1 Correct material selection	99.9	99.95	2.2.2.1 Fully supported on a round shaft journal	99.95	99.99	3.1.1.3 Correct additives	99.9	99.95	4.2.1.1 Fully supported on a round shaft journal	99.95	99
				2.2.2.2 Accurate machining	99.95	99.99	3.1.2.1 Correct temperature	98	99.5	4.3.1 Position inner ring at proper point	99.5	99.5
							3.1.3.1 Solids free	99.5	99.9			
							3.1.3.2 Water free	99.5	99.9			
							3.1.3.3 Product ingress free	99.5	99.9			
							3.2.1.1 Load within design limits	98	99			
		99.7	99.85		99.8	99.96		94.323	98.063		99.351	98.485

Human nature is naturally positive, and so you will look at the odds of the best outcome and expect that is what you will get. The honest truth is that you should expect to get the worst outcome and not presume that anything better will result. Until your low value for the total process increases to reasonable odds, you need to keep refining your plan with better and tighter risk controls for uncertain situations.

The reduction in risk from before to after the mitigations brings financial gain to the business. The money made from the improved chance of success can be estimated using the potential TDAF costs saved. By lifting a bearing's least chance of success from 69% to 88%, potentially 19 bearing failures in every 100 bearings will be prevented. Those savings can fund the cost of the changes that need to be implemented. You have a business case to invest some of the savings to get the higher bearing success rate so that you can make new operating profits.

In the analysis table, you gauge the influence of various risks on step reliability and their consequential impact on process success. The selection of probability or chance of success values to use in the analysis depends on how much information you have about the effect of a cause on the opportunity to fail a process step. In an ideal world, the failure rate per opportunity for each process step would already be accurately collected by your enterprise resource management system. You would then feed current chance of success values into the spreadsheet. After making a change, you would quickly know whether a mitigation was successful from the new failure data collected. If you don't have accurate records to use in the analysis, it is necessary for a person who is knowledgeable about the situation being analyzed to use the organization's operating history records and their personal

Table 18.4, continued (See the website for the Excel spreadsheet at full percentage)

5 Lubricant Grease	n	Whole Bearing

5.1.1.1.1.1 Include service duty confirmation in engineering ITP
5.1.1.1.2.1 Second person to check selection table recommendation
5.1.1.2.1.1 Include lubricate confirmation in maintenance ITP
5.1.1.2.2.1 Include chemistry condition in test laboratory report
5.1.1.2.2.2 Include test laboratory chemistry report review and action by maintenance engineer
5.1.1.3.1.1 Additives' levels in test laboratory report
5.1.1.3.1.2 Include test laboratory additives report review and action by maintenance engineer
5.1.2.1.1.1 Operate equipment to precision operating procedure
5.1.2.1.2.1 Include location ambient temperature range in engineering ITP

	Chance of Success for This Item		**Chance of Success to This Point**		Assembly Chance of Success	
	Low	High	Least	Best	Least	Best
	%	%	%	%	%	%
5.1.1.1 Correct lube for the service	99.9	99.95				
5.1.1.2 Correct formulation	99.9	99.95				
5.1.2.3 Correct additives	99.9	99.95				
5.1.2.1 Correct temperature	98	99.5				
5.1.3.1 Solids free	99.5	99.9				
5.1.3.2 Water free	99.5	99.9				
5.1.3.3 Product ingress free	99.5	99.9				
5.2.1.1 Load within design limits	98	99				
	94.323	98.063	87.9	94.5		

experience to estimate the low and high values of chance. If you are not knowledgeable about the risks for a situation being analyzed, you need to get help from the people who know the process well and feed their experience into the spreadsheet.

If it is vital to be highly certain of the estimates for success, you will need to collect actual data on the steps you want to control. This requires the introduction of a data collection form or software application to gather accurate information over an unbroken period of time for each process step's failure event frequency and the causes of each failure. Once you gather enough data on a step to be confident that you have captured all of its variations, you have the historical information needed to gauge the range for its chance of success.

Risk and reliability professionals would find technical issue with the Chance of Success Mapping, or risk probability modeling, approach. It neglects mutually exclusive events, where the occurrence of one outcome precludes a future outcome from occurring. It is an imperfect methodology because risks are allowed to be informed guesswork. You can easily miss important risks, and those risks you spot require chance of success estimates based on past history that is never stable over the long term and may not be representative of the future. It does not use the correct, integrated reliability block diagram of a process.

Despite the technical impurities, Chance of Success Mapping gives you a simple, practical way to address the risks you identify and pick a better design for a more successful process. It applies fundamental risk management principles in a structured, logical way to let you improve a process by improving its individual steps' chance of success. Your model's chance of success values are unlikely to be accurate, but accuracy is not what you want from the modeling. Your aim is to find and use better solutions to reduce risk than you now have. The model gets you to look closely at what makes your processes fail and what you can do about maximizing their chance of success. In fact, it's not worth wasting a lot of time being excessively accurate with the risk probabilities, as you only seek to compare risk prevention options to pick the better choice. Even if a chance of success map has errors, provided the errors are consistently made you will still be able to identify the better choice to take. You will hardly ever be right in any single chance of success value, but you will understand what can go wrong with the process, and, most important of all, you will have a very good idea of what will or will not work to get higher odds of success. Though your risk probability model may technically be quite wrong, it will still indicate the risk mitigation that will most likely work best out of the choices you have available.

The future of your business, plant, and equipment can be as good as you want it to be if you keep spotting and removing the troubles that prevent success and build in the causes of great reliability.

19

Process 5– Control Operational Processes

Summary Description of Process 5: Risk Monitoring and Control

World-class plant reliability is a foundational requirement for world-class production performance. Least cost operation needs equipment that does not break down, is low cost to run, requires minimal maintenance, is supremely safe, maximizes throughput, and makes 100% quality product all the time. These are all life-cycle process outcomes that are achievable only from processes designed and run to produce those results. The Stress-to-Process Model of asset management, coupled with Physics of Failure Reliability Strategy Analysis, is used for life-cycle process design in the Plant Wellness Way.

Design Processes to Operate Successfully

A process should work continuously at optimal performance. The effectiveness of a process is impacted by each of its steps. The success of a process depends on its step reliabilities, and they, in turn, depend on the adverse risks in the step. How each step is configured to optimize the process is important. To maximize success, a process is simulated using Chance of Success Mapping to find effective ways to make process steps robust and reliable. After that, the improvements are written into ACE 3T standard operating procedures, and those involved with the work are trained on how to more successfully run the process.

Establish Process and Step Performance Indicators

On a process map, specify the performance indicators used to monitor the process and its steps. Key Performance Indicators are required for the whole process, and intermediate Performance Indicators are used for each process step. They let people see how the parts of a process are working and whether corrections are needed. Those measures at the process step level are used by the people doing the work to spot losses, waste, and opportunities for improvement. The measures at the process level are used by management to guide them in optimizing the process and maximizing operating profit.

Gather Evidence and Measure Performance and Chance of Success

The range of performance measures used includes those needed to confirm that process steps have been done correctly and to ensure that future work will be done correctly. This requires a mix of

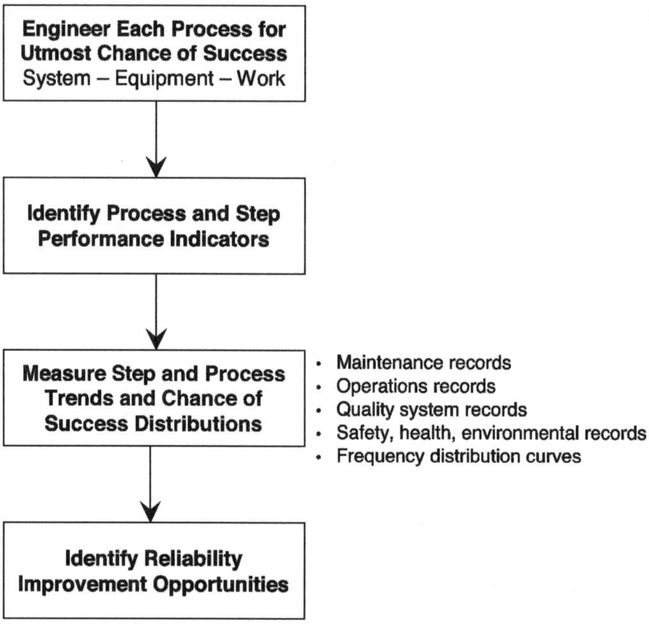

Figure 19.1—Process 5 Steps

leading and lagging indicators. All performance indicators are plotted on a run chart. From that run chart, the frequency distribution curve is developed to see how stable the values are and how successfully the process and its steps operate.

The operating procedures include the collection of data for performance indicators. Gathering, analyzing, and interpreting process data are part of what the people running and managing a process do. They use the measurements they take to check for themselves how the process is tracking. You give those with the problems the knowledge and tools they need to find the solutions for themselves.

Monitor for Reliability and Improvement Opportunities

Visual means such as diagrams, charts, and graphs are used to turn data into easily understood, useful information. When performance is not what is wanted, team up with the people in the process and plan what to do about it. Test the ideas and document, train, and then implement permanent changes to solve the problem. Use PIs to track the direction and progress caused by the improvement. Correct and improve those activities that are not yet performing well enough with the help of the people doing them. You can use the Change to Win improvement program to get ownership and involvement when solving problems.

Engineer Processes to Run Successfully

At a Plant Wellness Way site, each business process is designed and built to produce outstanding equipment reliability based on the Stress-to-Process Model of asset management. How to achieve world-class plant and machinery reliability is specified in the Physics of Failure Reliability Strategy

Analysis. It outlines what to do in your operation to prevent and control equipment failure and create lasting plant reliability. That knowledge and the required methods are put into your processes and become standard practice for everyone in your organization and its supply chains through the use of ACE 3T procedures.

You start engineering your Industrial and Manufacturing Wellness asset management system and processes by drafting process maps for all life-cycle processes across your operation. In combination, the process maps explain how the operation is to be run to achieve all its aims. A process begins as a mental construct laid out on paper. Initially, it not important for a process design to be totally right. It will be put through the six IONICS steps and come out completed in its most effective form.

A summary of the IONICS process can be found in the Appendix, which is included in the downloadable documents online.

Process Chance of Success Modeling

The future is never 100% certain, but you can influence your chance of success to a great extent. To get the operating results you want, you need to remove the effect of chance variation and adverse risk on process performance. IMW gets you to design your asset management processes for maximum success by specifying exactly how to be most successful. You can improve your chance of getting great operational performance by using Chance of Success Mapping to refine your asset management, operations, plant maintenance management, and equipment reliability creation processes.

Using a process map and ACE procedures, a review is done to build a more successful process by removing weaknesses from the original design. In a spreadsheet, you perform Chance of Success Mapping one process step at a time. When you find potential troubles, defects, and risks, you introduce more successful solutions. The effectiveness of a redesigned process step is checked by confirming that the new design delivers better outcomes than the previous version.

An example of how to use Chance of Success Mapping for work process design is shown in Tables 19.1 and 19.2. The example investigates an existing maintenance process used to select shutdown and turnaround work. A maintenance shutdown or turnaround requires a production outage, and the duration must be kept to the shortest possible time. Understanding the amount of work to be done, the type of work, and its complexity are key factors in the shutdown's success. The workload chosen will set the costs, staffing, and length of the outage. Until the workload decision is finalized, all necessary skills, resources, support services, equipment hires, and other requirements cannot be established or organized. It is an important process to get right.

All of the steps in the process map for the shutdown work selection process are transferred to the top of the Chance of Success Mapping spreadsheet. Each of the tasks in a step is considered for the ranges of success from low to high. The low estimate is based on the worst events known to happen in the company for each task, and the high value represents the best known outcomes for the task. The worst-to-best spread of odds covers the full extent of results that can occur in the task under investigation. The low probabilities of success for all steps are multiplied together, as are the high probabilities of success, and the current range of chance of success for the whole process is determined. The process performance indicated will either make you happy or get you worried. In the example, the chance of success for the shutdown work list being correct goes from an abominable 1% to as good as 90%. This process produces wide variation and has many weaknesses that need to be improved.

Table 19.1—Existing Process Chance of Success for a Maintenance Shutdown Work List—and Table 19.2—Reengineered Process Chance of Success for the Shutdown Work List—can be found in the downloadable materials online.

Creating a more successful process involves understanding where and how the current process steps fail and then designing and implementing better solutions. A risk analysis is conducted on each weak step, one task at a time, to identify the problems that can arise. Mitigations are developed for each weakness to ensure correct performance of the task. The investigation, analysis, and improvements continue step by step until the whole process is reengineered to be highly reliable and to maximize its chance of success. Table 19.2 shows the redesigned shutdown work selection process. The new range of low to high chance of success is 40% to 92%. This range is still too great, however, and too much variation remains. Further work is needed to minimize the risk of failing to properly define an accurate list of shutdown work. There is still too much opportunity to jeopardize the outage or fail to get the full benefit to future plant uptime through wrong allocation of staffing, resources, and infrastructure needed to do a successful maintenance shutdown.

Establish Process and Step Performance Indicators

Each process has a purpose, and every step in the process has a role. A process step is there to convert inputs into outputs and to move the process toward its required conclusion. How successfully a step does this can be monitored by the effectiveness of its performance and the quality of the results it feeds to the rest of the process.

In the process maps and modeling spreadsheets, you have a description of the purposes of each process step. It is necessary to set measures to check each step and indicate its success. Typically, a step's measures monitor how well its outputs meet the required quality and efficiency for the step, but monitoring can extend to checking individual task outcomes when necessary. The same applies for the complete process. KPIs monitor total process success, and PIs observe step success. Whole-of-process measures usually monitor productivity issues such as overall costs, timeliness of delivery, and quantity of saleable production.

In the Plant Wellness Way, you observe overall process and individual step performance using run charts for KPIs and PIs to track, trend, and report on a step or process. The same values are used to plot the frequency distribution curve of results. This curve shows the chance that the process or step will deliver the required performance. When it is clear that the required performance cannot be consistently achieved, the process and its steps are investigated for weaknesses and redesigned to be more successful.

Gather Evidence and Monitor Results and Chance of Success

Collect the designated process and step performance data and present it visually in run charts. For individual critical steps and the whole process performance, show each measure's trends and plot each measure's frequency distribution curve. For example, successful maintenance prevents equipment failures and minimizes production losses. It does that by keeping plant and equipment fit and in good health. Fit plant and equipment in good condition cost less to operate while making quality production to schedule. To measure the success of maintenance, you measure the share of maintenance costs in your production costs; they should be decreasing with time. The run chart should show a steady falling trend over time. When a trend flattens out, it indicates that the current process design has reached its limits of improvement and a design change is needed to get further improvement. You would also monitor individual production assets performance to ensure the maintenance efforts used on the assets reduce operating waste, costs, and losses while maximizing asset availability.

Monitor for New Reliability and Improvement Opportunities

The aim of process improvement is to enhance a process design so that its natural performance delivers results well within specification. When the performance monitoring indicates problems, you investigate the situation to determine its business impact and whether it is necessary to address the situation or leave it alone. It is not worth making changes to a process if the long-term economics of the situation does not justify the effort required and the resulting benefits. Doing a financial justification to make the business case for a change to improve asset reliability or enhance process performance is important so that you have confidence that the improvement will bring worth to the organization.

Progress and development are evolutionary processes, not revolutionary processes. Those companies that evolve fastest are more successful than those that wait for change to be imposed on them. If you want rapid evolution in your operation, help people remove defects in their processes that become problems, help them gain the skills that create success, encourage the sharing of successful knowledge across the organization, and ensure that all new successes are embedded into your procedures so that they become a habit.

20

FAILURE ROOT CAUSE PREVENTION

Highly reliable organizations proactively focus on preventing problems from entering their operation and removing those that remain. They set control mechanisms, standards, and checkpoints to spot and stop the defects that turn into future failures. They look for what can go wrong before it does and prevent its causes from happening. Instead of having problems and then investigating their causes, they imagine their problems and proactively act to eliminate their possibility. If your operation is suffering equipment and production problems, don't try to discover why they happen and figure out how to solve them. First, look at your processes. The vast majority of your production problems are caused by bad business process design. Fix your process weaknesses and do the new training,

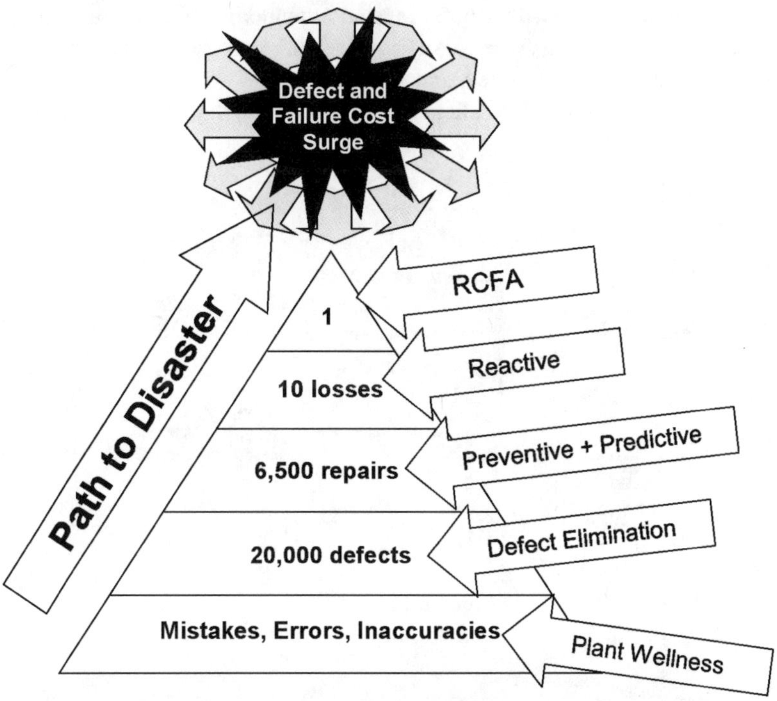

Figure 20.1—Stop the Causes of Defects at the Bottom of the Failure Pyramid

then put the answers to use. The problems disappear because they no longer can exist in your company.

Solving the real problem means finding its true causes. The technique used to investigate plant and equipment failures is known as Root Cause Failure Analysis (RCFA). RCFA is manpower and time intensive, and so it is only applied after a serious incident justifies it use.[1] Reserving RCFA for investigating major failures ensures that major failures will continue to occur. You might find and remove some contributing causes, but thousands of defects in your business will stay behind to create more future catastrophes. Failures are the result of multiple failed processes. There is never just one cause of equipment failure. There are at least four, and usually more, contributing factors to a machine failure event. Using Root Cause Failure Analysis will not discover all of the contributing factors because many of them are hidden in the distant past, while others started in other places far from your operation. In the Plant Wellness Way, the first thing you do after a failure is review the processes for the causes that contributed to the event. Figure 20.1 shows how this root cause removal strategy works to prevent the many defects at the bottom of the failure pyramid from becoming big troubles that you later must fix one at a time with RCFA.

Improve the Process Design

World-class operations recognize the interconnectivity of their processes and work hard to ensure the right results at every stage in every process. Figure 20.2 shows a failure in product assembly. The root cause traces back to its manufacture, when it leaves the process and enters another, then a second and a third. The defective item started its life elsewhere and ended up causing problems during assembly. There are innumerable opportunities for errors and defects to occur in all processes. Process after process connects with others, causing a tangled web of interaction. Errors, mistakes, and defects can come from everywhere. Any process that goes wrong has an impact on numerous others downstream. Much time, money, and resources will be wasted. If you want an operation in which good results are natural and excellence abounds, ensure that your processes permit no defects

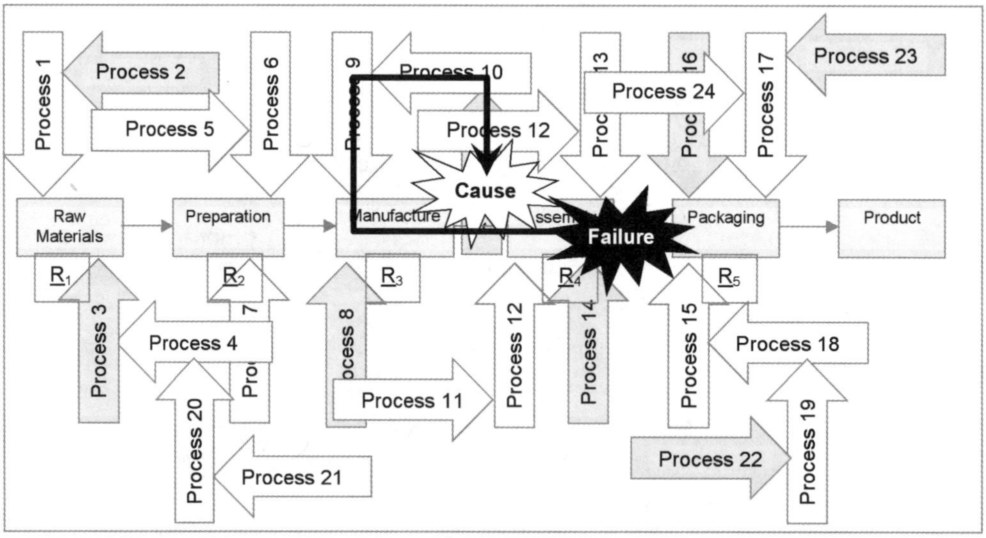

Figure 20.2—Failures Occur throughout the Process Chain

There is an insightful story told of the late Sir Ernest Shackleton, one of the great early South Pole explorers. On board his ship bound for Antarctica, he watched a man tie a knot in a rope that was holding down vital supplies. Shackleton saw that it was the wrong knot for the job. In wild seas, it would come loose and all the goods and supplies would be lost. Shackleton went to the man and asked him about his experience at sea. He learned that the man was new to seafaring. With patience and thoroughness, Shackleton taught him how to tie the correct knot, one that would be secure in all weather and sea conditions. His comment to the new seafarer is insightful for all of us who want successful outcomes: "There is always only one knot that is right for the situation."

Shackleton's method of failure prevention is the technique used in the Plant Wellness Way: do what stops the causes of failures from starting. First, put the right practices into your processes and make sure they are done right every time. In the Plant Wellness Way, when things fail, the first question you should ask is, what is wrong with the process? You can skip the RCFA, but you cannot skip finding and fixing the design faults and missing quality controls in your processes.

Prevent the Chance of Failure Starting

The necessities for high equipment reliability cannot be left to luck. If Shackleton had left it to the new seafarer to realize that he was using the wrong knot for the job, the expedition would have failed. Like Shackleton, you must find and remove the risks in your processes before they destroy your operation. Do the same for your business that Shackleton did for the Antarctic expedition: look for where troubles will start in your processes, then introduce, teach and use the right practices so that risks will never arise.

Identify Where Your Equipment Problems Begin

An important asset management indicator to collect and present is where failures arose during the equipment life cycle. Today's failures started in the past when their causes were initiated in previous processes. Tracing the parts replaced on corrective and breakdown work orders back through the processes they traveled lets you observe their life cycle. Where you find problem causes, you stop them so that they cannot arise in the future. If a part's failure was started by an error at an external repair shop, it will happen again if you don't get the shop to fix the causes. A stress-induced failure from shaft misalignment indicates that your equipment installation process is weak. You seek to understand whether your reliability troubles are in fact attributable to manufacturing defects, subcontractor mistakes, production process causes, material selection causes, equipment installation troubles, operating errors, vendor-produced causes, procurement errors, warehouse management failure, poor workmanship, and so on.

The failed item or part is used to start the review of its life. The failure is the last event in a long chain of causes and effects. The failure mode site on the part contains evidence of the causes of its failure. The causes that came together to fail the part passed through your processes undetected until they combined to initiate the failure. It is necessary find the culprit processes and fix them. This is not a root cause analysis investigation to find the actual cause of failure. It is an investigation of process design weakness to identify the presence of failure-causing steps. Typically, an experienced discipline maintenance engineer or design engineer, or a career maintenance supervisor, or maintenance planner would identify all of the processes associated with the replaced parts.

A process is weak if it does not prevent all of the Physics of Failure Factors that damage or destroy the part. Finding answers to the eight life-cycle questions (see Chapter 9) is a good place to start an investigation. Using evidence from the failure, process maps of all of the processes used during the failed part's life cycle are reviewed for risks that could have allowed defects and causes of the failure to arise and remain active. When a weakness is found, the process is reengineered to remove the opportunity for out-of-control variation. The process redesign is trialed, and the successful solution is documented and implemented. The people inside and outside the organization affected by the change are trained in the proper use of the new process.

A few examples of life-cycle process monitoring measures used to find weakness in processes by observing their effects on the operation are listed below. The indicators are simply a count of the processes used during the life of failed and replaced parts. A pie chart or bar chart of the number of maintenance corrective work orders and breakdown work orders per category for a period shows the regularity that these indicators of process design weakness arise in an operation. The measures are selected with the intention of finding the weak life-cycle processes that are making your machines fail in order to identify what more to do to make a process more robust, antifragile and successful.

- When failed equipment parts were serviced by external vendors
- Whether repaired equipment had service duty specifications
- Who did the previous repairs or replacements
- Whether repaired equipment had be run using ACE 3T operating procedures
- Count of the number of events when the equipment was run overloaded
- The equipment repairs in which parts were drawn from our store
- The equipment repairs in which parts were purchased direct
- Whether equipment repairs were done to ACE 3T procedures

These measures let you target your process redesign to build more successful maintenance and operating processes. As time goes by and data accumulate, you can develop additional subcategories within the measures to focus on finding the specific process step that starts the defects causing the repairs and breakdowns.

The lists used in the document *Plant Wellness Index* (see the supplementary material available online) contain other useful criteria that can be used as indicators of weak processes and added to the above list of process issues.

Behaviors of High-Reliability Organizations

The U.S. nuclear aircraft carrier fleet and nuclear submarine fleet are renowned worldwide as high-reliability organizations. Starting with vision and leadership, it took a lot of consistent, persistent effort, and some tragic failures to get there.

The nuclear submarine USS *Thresher* sank with 129 people on board on April 10, 1963. Although the vessel was not recovered from its resting place 2.5 kilometers deep, the naval investigation review board used photographic and retrieved evidence, along with laboratory tests, to identify failed brazed pipe joints as the most likely cause of the incident. The loss triggered a complete review of naval nuclear vessel design and operating procedures. Even though the fleet's equipment was built and maintained to high quality standards, and its personnel had specialist technical training, the quality control requirements became more demanding. Designs were simplified to remove complexity and to behave in known ways. Quality control in manufacture was improved. Operating practices became more stringent to remove the chance of variation. All crew members had to reach expert status in their discipline and equipment if they were to remain on the ship or submarine.

The organizational structure on U.S. nuclear fleet vessels is unusual. The crew members are the experts in running the ship and keeping it safe; the officers are there to support the crew in their efforts and to address issues that might reduce the crew's effectiveness.[2] That structure makes the operating crew more important to the ship's survival than the officers—a true inverted organization with managers at the bottom working for the producers at the top.

Central to the success of high-reliability organizations is the realization that everything can go wrong. The only sure protection is to know exactly what is happening with the equipment throughout the plant all the time. The equipment must be set up perfectly at the start and then monitored to ensure that it behaves exactly as it should when it is used. What you don't understand, you don't do yourself—instead, you get help from those who do know until you are trained and expert enough to do the task. Human error is acknowledged and addressed through teamwork, in which people help each other constantly and documented checks, counter checks, and double checks are a way of life.

High-reliability organizations proactively control every process and every step in those processes. Nothing is unimportant because consequential effects mean that the smallest risk can be the start of the biggest catastrophe. This requires a dedication to diligence beyond what people in commercial industry expect and are paid to do. High reliability cannot be bought with money—it lives in the hearts and minds of people who want to be the best at what they do and are respected by their peers and managers for that expertise because it is so valuable to the success of the organization.

The U.S. nuclear fleet's equipment is designed for simplicity, high reliability, and maintainability. The business systems in use demand proof of compliance with best practices. Its crews are educated to be a technical knowledge repository on their plant. Its people are trained to act skillfully in a highly reliable manner. The organization is structured to put knowledgeable experts immediately at the situation of risk and danger and bring the power of teamwork into play. Those are key reasons why the U.S. nuclear fleet is a high-reliability organization.

Limitations of Our Materials of Construction

We live in a probabilistic universe in which its physics produces divergence and sudden change in the way matter behaves when it reaches critical points.[3] Unless the physics of a situation is controlled, you can get sudden changes in the behavior of your materials of construction. The failure of equipment parts and the resulting poor reliability and safety are direct results of exceeding the physical and chemical boundaries of the materials of construction. Poor reliability and poor safety are to be expected in organizations in which people do not know the limits of their machinery and do not understand what is happening to the parts inside them.

People create high reliability when they know the engineering of their plant and process and expertly keep their equipment parts well within the capability of the materials of construction. The experience of high-reliability organizations is that equipment failure starts with poor business system process control. The necessary systems and controls that produce high reliability are not present and followed. Equipment failures then result from out-of-control variation. The organization's quality management system fails first, and then the equipment is failed by the system. To fix management system failures, it is necessary to understand how business processes can fail. By understanding how each process step can fail, you build in the correct risk controls needed to achieve high reliability at each step.

You cause your own equipment reliability through the quality management systems that you use and enforce. To get high reliability, the experience of high-reliability organizations tells us, we must put into the business the processes, the specialist technical knowledge, and the right activities done correctly that cause high reliability. Don't begrudge drawing a process flow diagram for each of your processes, and for each step in a process, to identify the hundreds of ways the processes could fail. Risks can live anywhere, and you need to see all of the places where your problems can start. Figure 20.3 traces a machine component manufacturing process step down to the fundamental tasks and actions. The work flow details expose opportunity for failure everywhere. Once you go into the details of your own processes, you'll see an enormous number of risks you were not even aware of. The presence of those risks means that things can go wrong, and they will with a frequency that is dependent on the designs of the processes used during the life cycle and whether they were constructed to stop or prevent each risk from arising.

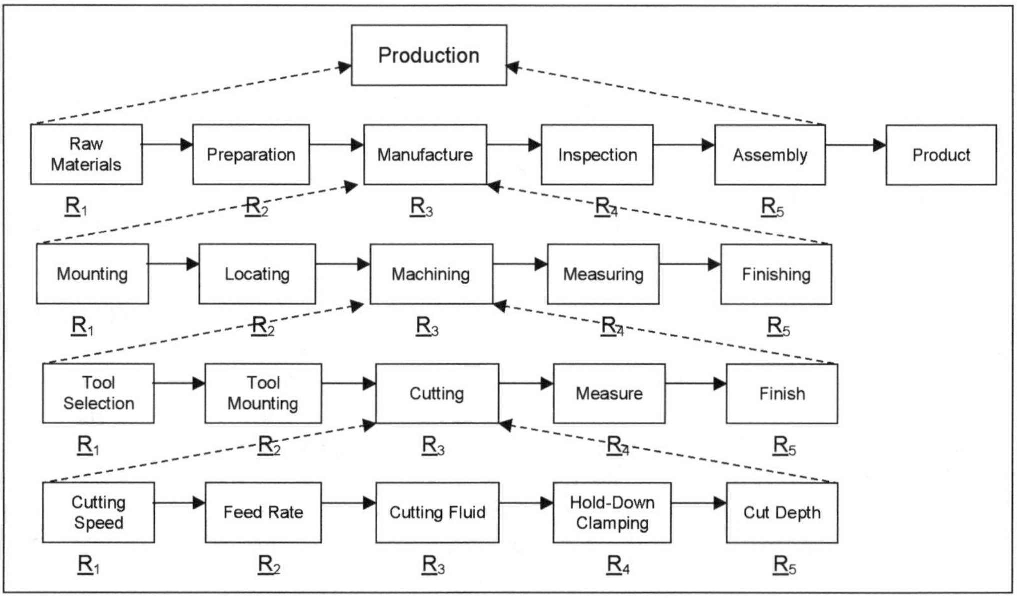

Figure 20.3—Dig into Your Processes and Their Step Activities to Spot Risks

In an organization using a Plant Wellness Way system of reliability, the problems and troubles caused by your processes are uncovered using Chance of Success Mapping to identify the risks. At each step, you list what has gone wrong in the past and what could go wrong in the future. For each risk, you develop mitigations to proactively prevent them from happening. You compile a list of the changes needed to maximize each step's chance of success, and that is your plan to create a far more successful process. In this way, you design and build highly successful operations and equipment without roots of failure inside them.

References

1. Robert C. Nelms, "The Latent Causes of Industrial Failure . . . How to Identify Them, and What to Do about Them" (Failsafe Network, Inc., Montebello, VA).

2. Charles R. Jones, nuclear safety consultant, American Nuclear Society, Nuclear Safety Culture Panel Discussion Presentation, June 2, 2003.

3. Philip Ball, *Critical Mass: How One Thing Leads to Another* (New York: Farrar, Straus and Giroux, 2005).

21

Change Management for Workplace Innovation

Change hardly ever works when it is forced on people. We need the opportunity to accept it for ourselves.[1] You have to work with human nature, not against it. People don't respond positively to being pushed around. That means letting people discover new ways for themselves. Once they find out how to do a thing better and are encouraged by their supervisors, they are highly likely to adopt the change and make it a natural part of doing their work. A workplace change management process in which people welcome innovation and positively support it is necessary.

Install Quality Management in Operations and Maintenance Processes

The late quality guru W. Edwards Deming gave us a method for addressing poor business processes. The first step is to measure the statistical performance of the current process. (He warned that appropriate statistical methods must be used if valid interpretations are to be made.) The initial analysis will show whether the current processes are statistically stable. Management can then decide to accept the natural performance of the process and stop expecting better results than the business is designed to produce. Or management can decide to reengineer the business processes so that they inherently deliver better outcomes.

Deming suggested that any new method first be tested to prove that it actually produces improvements. Using the scientific method, in which a new idea is seen as an experiment that has to prove that it is better than what came before it, the new suggestion is tracked to statistically confirm that it delivers improved results. When a technique is proven successful, it is integrated into the business processes through procedural changes and practical training until individuals are competent at the new method. If the trial is not successful, it is discarded, and the learning from the trial is used to select a new option to test. In this way, a company builds into its own design the processes that automatically deliver better performance.

"Push the Limit" Concept

Figure 21.1 shows the "push the limit" method of continuous progress and improvement. It is the remedy used by world-class companies to protect themselves from turning into average performers. They intentionally force themselves out of their comfort zone by setting higher targets and standards and then looking for ways to reach them.[2]

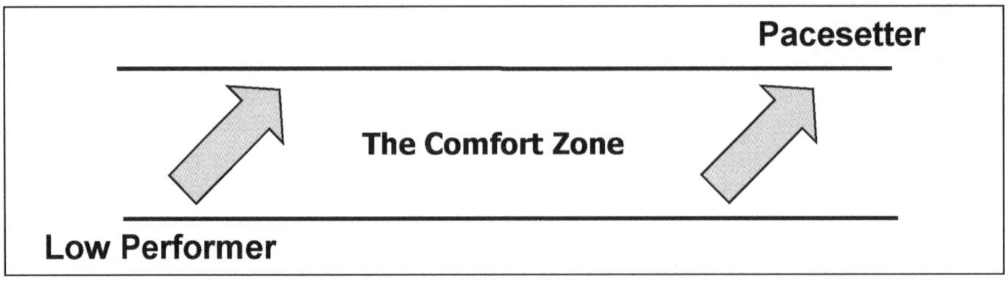

Figure 21.1—Push the Limit Improvement Model

Becoming world class means adopting the same mentality used by world-class organizations to achieve high levels of excellence. "Push the Limit" starts the planning process.

Driving Continuous Improvement with ACE 3T Procedures

Once ACE 3T procedures are developed, they become a means to push innovation and continuous improvement. Work quality and skills are advanced by resetting the tolerance range so that it is more demanding. When people consistently achieve the "good" standard, then good is no longer good enough. A new "best" practice standard is set, while the previous best standard becomes the "better" requirements, and the old better standard is reset as "good." This puts the Precision Principle into operation and harnesses peoples' desire to improve their skills and simplify their work. Figure 21.2 highlights how ACE 3T procedures can be used to drive continuous improvement.

Task Step No.	Task Step Owner	Task Step Name	Materials, Tools, and Their Condition	Full Description of Task	Test for Correctness	Tolerance Range			Actual Result	Action if Out of Tolerance	Sign-Off
		(3–4 words)		(include all tables, diagrams, and pictures)	(Include diagrams and pictures)	Good	Better	Best			
Higher Standards Drive Improvement											

Figure 21.2—Driving Quality Improvement and Innovation with Higher ACE 3T Standards

Change Your Goals to Change Your Game

When you become an IMW operation, your old capital projects, enterprise asset management, and operations and maintenance strategies and philosophies are replaced with a new business-wide system of reliability. The system of reliability is designed to help your company reach world-class performance. You do that by setting world-class standards for every step in every process. World-class performance targets require you to determine the zone of utmost success. This zone is the bull's-eye in the target. Around the bull's-eye is a region of tolerable performance. An outcome outside of tolerance is unacceptable, and it is rectified until it is correct. The bull's-eye standards become corporate, departmental, group, and personal goals to achieve. Only success is rewarded. Trying to reach the new performance levels is the behavior expected from everyone—but until that behavior delivers consistent success, the climb to the pinnacle of excellence is not yet completed, and its rewards cannot be claimed.

An example used as an analogy is shown in Figure 21.3, in which the standard soccer goal is shrunk down to one-tenth the area of the goal opening. To score within the much smaller target—the "precision zone"—requires a change in game plan. The goalie can now easily cover all of the area of the goal face. No longer can you shoot from 20 meters away and kick the ball past the goalie. The possibility of tapping a ball over the goalie's head and getting a lucky bounce into the net no longer exists. You must invent a new strategy and game plan to win the match. You need to find ways to make the goalie come out and open up the goal for a shot. The strikers will need to play as a fluid team, with one drawing the goalie out and then kicking the ball quickly to another striker who is already positioned for a one-hit flick into the net. To win the match, you will need highly skilled soccer players with outstanding ball skills who are master strategists and forward-thinking "readers of the game." You must also be totally prepared and fully practiced before a match. Your people must know what strategy to use for each scenario that may happen on the field. They must not hesitate in the few seconds they have to score a goal. Such players are among the elite in their sport.

If players who are skillful at playing the games with a goal one-tenth the size are put onto a field with a standard goal, they will score effortlessly against the "normal" players and goalies they confront. The level of excellence reached by the "precision players" will make them unbeatable in any standard game. The Plant Wellness Way of setting component health precision values and developing systems, processes, and skills to always hit the "wellness targets" is a sure way to get world-class plant and equipment reliability. Once people know what world-class performance looks like, and they are fairly rewarded for its accomplishment, it will be nearly impossible to stop them from becoming elite at doing their work.

The "Change to Win" Team-Based Business Improvement Program

"Change to Win" is a structured change management program used to bring new practices and innovative improvements to an organization. The program uses a team-based process for helping people

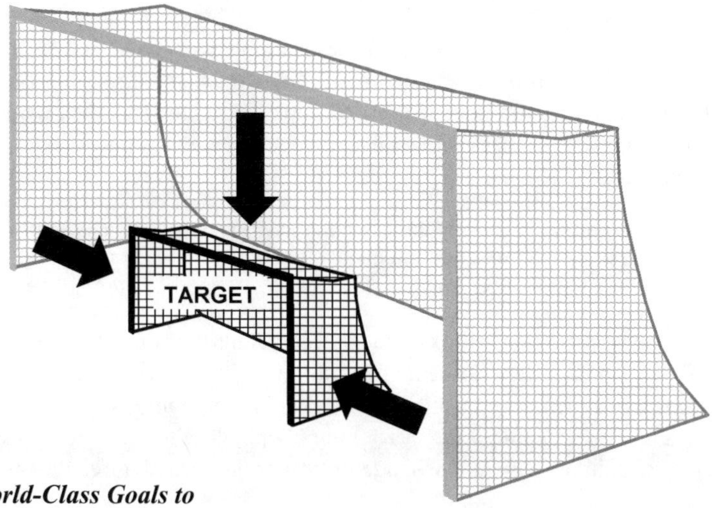

Figure 21.3—Change to World-Class Goals to Play a World-Class Game

learn better ways and better practices to include in their work. A team consisting of the manager, the supervisor, and people from the affected workplace is assembled to introduce needed changes with the help of a facilitator. They are responsible for understanding the issues, finding the solutions, and planning how the organization will adopt the changes, including testing the ideas and then, once they work successfully, making them a part of standard practice by putting them into the procedures and training the users. They bring their individual expertise together to find improvements that "push the limit" and bring better methods to the operation.

Internalized ideas and values can change when new knowledge contradicts current beliefs and causes cognitive dissidence. The Change to Win approach lets people discover new knowledge and then use it to fix their problems. To give people the chance to learn better ideas and develop ownership of them, the Change to Win program gets team members to research and investigate the range of options available to address a problem. It encourages team members to go outside their comfort zones and look for other practices and technology they don't yet know about. Driving workplace evolution is the goal of the Change to Win 100-day program and its Five Wheels of Change.

The usefulness of an innovation to a business needs to be proven. People will only change their current practices if the evidence and the support structure is in place to make the change. A non-threatening way to do that is with a pilot project to show people the worth of an innovation. The Change to Win program is a trial project that is limited to 100 days. This period is short enough for people to wait for evidence, yet long enough for the project to be done well. Once the experimental project is a success, you will have proof from within the business that the change works. With each success, more 100-day projects are started so that everyone becomes involved in making positive changes.

The Change to Win approach is not for problem solving, although it can be adapted to do so. Solving problems is done with Chance of Success Mapping to improve processes and creative disassembly to improve machines and equipment. The Change to Win method is about achieving behavioral changes in organizational performance by letting people introduce and integrate sure improvements into their workplace processes. The Change to Win program is used to prompt a company to use best practices and then adopt even better practices. Examples include introducing precision maintenance into the production workforce, introducing precision operation into an industrial site, introducing new software into a business, introducing an ACE 3T quality system into an organization, introducing safety habits into a factory, and introducing the Plant Wellness Way into a company.

The Change to Win program uses a simple workbook that each team member follows over the 100-day period. It is a friendly, low-risk, low-cost strategy to introduce changes into an operation. The teams start at the front of the workbook, and each week they work on agreed tasks until the project is complete. At weekly meetings, the team reviews progress on the action plans. When the workbook is completed, the program ends.

The current version of the Change to Win change management program workbook is included in the downloadable documents that accompany this book online. It is a part of the Plant Wellness Way methodology.

The workbook is self-explanatory. It contains the complete change management process to apply on the shop floor. It uses a team facilitator to guide the 100-day change process. The facilitator helps teams work their way through the workbook and apply the process. He or she keeps the team on track and on schedule. Like everything people do, the more often it is done, the better we become. Once a facilitator uses the Change to Win program with three or four teams, it will become second nature.

The example used in the workbook for applying the Change to Win method is the introduction of precision maintenance into an organization. Although shop floor people deliver precision maintenance, it is maintenance and operations managers who need to start the change, sustain it, and keep improving it. Starting a change initiative such as precision maintenance requires a well-thought-out and structured change management process that encourages people to want to work to new, higher-skilled precision practices. Instead of risking that your improvement project will become another failed management fad, use this practical process to help people buy in to the change—first with their heads, and then with their hearts and souls as they see the change begin to work.

References

1. Dale Carnegie, *How to Win Friends and Influence People* (New York: Simon & Schuster, 1936).
2. Jim Wardhaugh, "Maintenance—The Best Practices" (presentation, IQPC Reliability and Maintenance Congress, Singapore, 2005).

22

Plant Wellness Index

You can achieve world-class plant and machinery reliability as sure as night follows day. Getting outstanding reliability is a formulaic engineering outcome. Equipment reliability is totally dependent on component reliability (except for sabotage and acts of God), which can be fully specified by measurable engineering criteria. This makes the achievement of world-class plant reliability completely definable, as every part's necessary health conditions can be fully specified by measurable parameters. How well your business processes achieve the requirements for world-class reliability is an indicator of its effectiveness in producing highly reliable operating assets.

The results of business and workplace practice are seen in the historical evidence kept by a company. By comparing what a company does with its engineering, operations, and maintenance processes to what really creates equipment reliability, you can identify why an operation gets the plant availability it does and report with certainty on what must be changed and how to make the change to guarantee world-class production performance.

Potential for World-Class Reliability

Industrial and Manufacturing Wellness uses two indexes to focus your organization on how to achieve a world-class life-cycle asset management and create outstanding reliability.

The quickest measure for gauging an operation's potential to deliver world-class reliability is the Reliability Health Capability Index, shown in Table 22.1. Within two minutes, you'll know whether your organization has what it takes to achieve world-class reliability performance. It measures the reliability creation capability of an organization. The index comprises five factors—Processes, Quality Standards, Competence, Discipline, and Continuous Improvement—each with three measures to gauge a company's proficiency in creating and sustaining plant and equipment reliability. They are challenging and unforgiving requirements, and the insights gained during the assessment will give you correct guidance. You want to have an index score of 8 or better. A score below 5 is life threatening.

Organizational Capability to Have High Reliability

The second index used in IMW gauges the ability of your life-cycle asset management system to produce world-class plant and equipment reliability. Your operation is audited and graded based on how well its processes delivery outstanding reliability, maintenance, and operational performance. The index comprises six criteria: Least Stress in Parts, Systematic Success, On the Same Mission, Right Information, Role Competence, and Defect Elimination. The index value for each criterion is placed on the target board in Figure 22.1 to show how closely the organization's practices come to excellence. The criteria reflect their importance in the Plant Wellness Way methodology for achieving long-lived, highly reliable equipment. Assessments are made for each category of relevant skills, practices, techniques, and solutions that bring reliability success. Audit results by category are listed in worksheets that are used to assess the telltale signs exhibited by an operation.

The Plant Wellness Index (PWI) is a holistic, fact-driven enterprise asset management system assessment that assigns ratings from 1 to 10, where 1 = Excellent, 2–4 = Satisfactory, 5–7 = Poor, 8–9 = Very Poor, and 10 = Failed. The meaning of each rating is explained in Table 22.2. The higher you score, the greater the variability in your life-cycle asset management processes.[1]

The PWI measures how well a company controls the business risk hierarchy introduced in Chapter 9, extending from the Physics of Failure mechanisms impacting component microstructure health to the failure of equipment and eventual harm to the business. Once you move away from excellence, the hierarchy value representing the severity of the risks that you carry increases from 1 to 10. As you move up the hierarchy from mechanism to organization, the number of available measurements for an asset decreases and they become less refined indicators of what is driving an asset's operating condition. At any time, there can be multiple risks present, but you will see them less clearly as you go further up the hierarchy. If you use historical indicators of business performance, you will only see the effect of the final failure and never its multiple causes. To prevent the causes of asset failure, you need to measure and control how well the organization operates at the Physics of Failure mechanism level of component health.

When trying to change from traditional failure-based maintenance, the PWI naturally moves people toward a wellness strategy and the necessary parts health and work quality practices. The grading of an organization's capability to achieve world-class reliability looks into whether its processes and practices focus on doing right things and delivering the right outcomes to cause lasting component reliability.

There is a complex but predictable hierarchy of cause-and-effect failure processes traceable to the initial loss of part integrity. A misaligned shaft leads to roller bearing component overload, which leads to lubrication film breakdown, then to high metallic surface stress from metal to metal contact, then to bearing spalling and wear particles, and then to the bearing's collapse, which finally causes the equipment failure that stopped the plant.

The PWI numbers do not describe the health of equipment or even the health of a component but rather the ability of an operation to manage and control the precision of the quality results that deliver high reliability to parts, such as lubrication condition, alignment, unbalance, operational forces, distortion, and so on. It is a measure of the willingness of an organization to live with escalating risks of failure.

An assessment of "Very Poor" does not mean that equipment breakdown is imminent; rather, the extent of control over a quality parameter permits defects at the component level that will cause the parent equipment to fail in the future. When the assessment of a function is "Failed," the component is still likely operating, but a breakdown is sure to happen because the organization's processes didn't even identify a serious problem. In such cases, the timing of equipment failure depends on the severity of microstructure stresses and the rate of atomic structure degradation suffered by the component in operation.

By assessing the degree of control a company has over equipment Physics of Failure Factors and the development of life cycle failure cause mechanisms, you can gauge its ability to create outstanding plant and equipment reliability. Instead of using condition monitoring to observe failure modes from initiated failures and reporting their growing severity as the primary indicator of health, the Plant Wellness Index measures your asset management system's ability to instigate and sustain the causes of excellent microstructure strength and health for all critical components.

By making such information readily available across the organization, equipment users and maintainers know how to change their operating and maintenance practices to cause high reliability. Pre-

Table 22.1—Reliability Health Capability Index

Reliability Health Capability Index			
Criteria	**Description**	**Grade**	
Processes	Procedures and methods specifying the best way to produce all required outcomes are available	No/few documented processes, with reliance on use of human mind and memory	0
		Complete process maps with procedures for all processes, plus work instructions for critical factors	1
		Fully flowcharted, instructed in all procedural detail, and error-proofed to prevent critical step failure	2
Quality Standards	Clear criteria defining top-class outcomes, with acceptable and unacceptable results at critical points in all processes	No/few specified work quality outcomes, with reliance on human opinion for decision making	0
		Measurable quality standards set and monitored for all critical outcomes	1
		3T quality standards set, monitored, and trended for all critical process step outcomes	2
Competence	Personnel are statistically capable in all they do, with technical and emotional skills to achieve the quality standards	Qualified and trained personnel used in key positions	0
		People competent and proven capable in the role used in key positions	1
		People competent, proven capable in the role, and accomplished in the processes used in all positions	2
Discipline	Best available procedures are correctly followed by everyone throughout the organization	Most work is done as considered best by the person doing it	0
		All work is done to defined, documented instructions	1
		All work is done precisely to meet specific quality outcomes	2
Continuous Improvement	Continuously looking for, scientifically testing, and implementing better solutions	No specified or irregularly applied reliability improvement process in use	0
		Regular measuring and reporting of key process indicators and failures to identify opportunities and then doing improvements	1
		Continuous trending of process distributions and analysis of cause and effect to proactively make changes that get best results	2
		TOTAL	

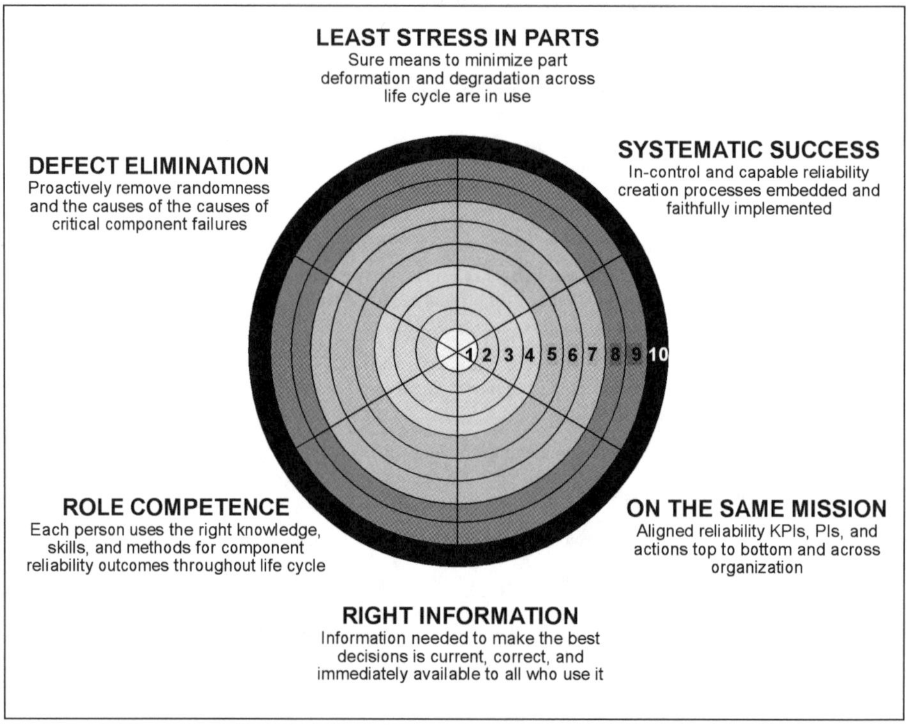

Figure 22.1—Targeted Plant Wellness Way Index

cision alignment becomes the norm because keeping shafts precisely aligned minimizes the stress suffered by the lubricant and the bearing surfaces. After a while, people start to talk about the engineering numbers that create outstanding reliability, and soon after that, everyone learns how to deliver them to their plant and equipment.

When reliability increases, the operation is seen in the new light of brightening success as the number of outages falls, the associated costs fall, safety incidents markedly decrease, and, most important of all for business success, operating profitability rapidly increases.

The use of the PWI makes it relatively easy and straightforward to assess and display the risks associated with plant ownership, which, in turn, makes the relationship between the organization and its statutory regulators more amenable because it provides the company with clear evidence that its processes monitor and control asset health in the rare event that there is an incident.

The appropriate audit values are selected following site observation and a review of historic records of the equipment, process, and system risks. For example, in an operation that has above-design pressure events in a high-pressure gas system, the stresses on all control valve shaft seals, isolation valve internals, instruments and pipe flanges increase the chance of a leak from all items put under extra pressure. The lack of operational control will immediately result in a surge in the piping system risk. The evidence in the operating records of continued lack of control over operating pressure rates as "Poor" on the PWI for both Least Stress in Parts, because stress increases with every high pressure event, and for Systematic Success, since the associated risks of loss of production,

Table 22.2—Plant Wellness Index Key

Plant Wellness Index Key			
Criteria	**Condition**	**Description**	**Activity/Action**
	Cannot be monitored	The factor cannot be monitored for assessment,	Note why in comment column. If not rectifiable, use 10.
1	Good	All required outcomes are delivered well inside specification most effectively and efficiently. Exactly what the right thing to do is clear, and there is a clear and correct explanation of the proper way it is to be done right the first time, and it is actually done that way.	Copy these best practices across the organization.
2	Satisfactory – Top	All aspects are operating well within specification, and there is a long history of controlled and capable outcomes.	Copy these best practices across the organization.
3	Satisfactory	All aspects are operating within specification with no significant problem indications.	No action
4	Satisfactory – Low	All aspects are operating within specification, but there are future problem indications.	Plan for preemptive response. Continue to monitor event initiators.
5	Poor – Top	There is a possibility of problem(s) adversely impacting the required outcome(s).	Start preemptive response. Monitor event initiators with higher frequency to observe for rapid worsening.
6	Poor	Existence of a problem is self-evident.	Rectify the problem. Department manager reviews all relevant processes for weaknesses that allowed the problem to arise and undertakes process improvement(s) until weaknesses are prevented.
7	Poor – Low	Problem is adversely affecting the outcome(s).	Too late, should never have gotten this bad. Do as per 6, but manager extends resources to use of an expert team. Undertake retraining of all involved in the correct standards and practices to be used.
8	Very Poor – Top	Problem prevented outcome(s) from meeting requirements.	Too late, should never have gotten this bad. Department manager, their senior manager and expert team redesign the process(es) to eliminate causes initiating the problem. Institute necessary capital expenditure and personnel and/or supply chain training.
9	Very Poor – Low	Problem destroyed any chance of delivering the required outcome(s).	Too late, should never have gotten this bad. Do as per 8.
10	Failed	Required outcome(s) is unachievable by this approach.	Too late, should never have gotten this bad. Do as per 8.
Notes: a. The key is used to rate the adequacy of the response to each of the requirements noted in the audit worksheets. b. Only real evidence specific to the operation being audited is acceptable proof of the existence of a requirement.			

environmental damage, and harm to personnel were allowed to repeat. The repetitive over pressure events indicate a business system failure and the inability to sustain reliable, safe production. When the problem is resolved, the PWI values revert to their normal status. The records will show the historical PWI values for reference, which will be used as evidence of lack (or not) of process control for regulatory reporting, business performance monitoring, and auditing requirements.

The Plant Wellness Index provides a means for prioritizing risk elimination and control, connecting the mechanisms that affect component microstructure to asset life-cycle success and business profitability. When an asset is perfectly healthy with its assemblies and parts at least stress right down to Physics of Failure level, it is at its lowest risk condition. In this state, asset failure becomes rare because there are few opportunities to cause degradation and deformation. At this level of performance, a company generates exceptional operating profit month after month, year after year.

Reference

1. Thank you to Paul B. Price, Reliability Consultant and Condition Monitoring Specialist, based in the United Kingdom for providing his Equipment Health Index (HI), which I have modified to create the PWI.

23

Process 6–Synthesize Ideas to Continuously Improve

Summary Description of Process 6: Continuous Risk Elimination

Bad business risks bring ruin. The causes of problems must never again enter your operation. No risk elimination secured must ever be lost. As quickly as possible, remove all remaining operating risks and drastically reduce those you cannot eliminate. Reaching the heights of reliability excellence and being a high-reliability organization is a tremendously profitable place for a company to be.

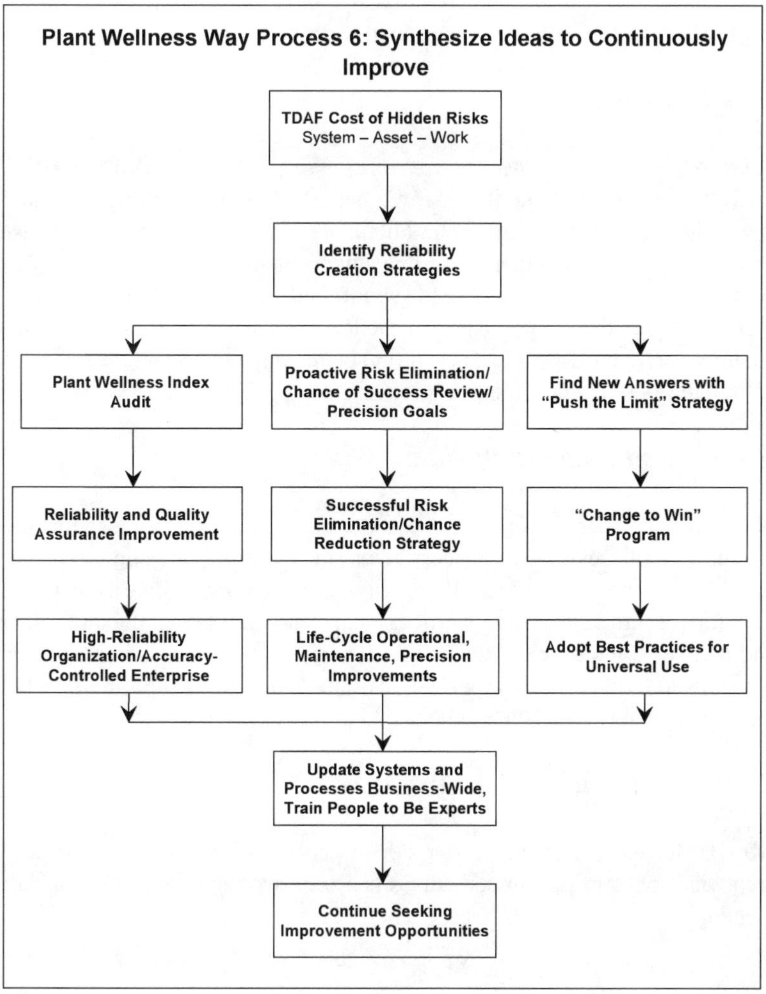

Quantify and Prioritize Remaining Risks

Find the ticking "time bombs" hiding in your business before they explode. To spot the risks you have already got in your plant and equipment, look at each equipment's critical components. Use TDAF costing of the potential business-wide losses from critical parts failure and check the odds of their failure to assess the operating risks you still carry. Use the Plant Wellness Index to clarify business system and process inadequacies. Find and fix the weaknesses in your asset management, operating, and maintenance processes using Chance of Success Mapping to discover the steps that will cause your future disasters.

Identify More Successful Risk Reduction Strategies

The best answers are those that bring sure success by design, such that the design itself protects you from having risks. For at-risk components, use Physics of Failure Factors Analysis combined with Three-Factor Risk Analysis to generate new solutions for reliability improvements. For each asset's critical parts, embed effective ways into the appropriate procedures that make them failure-free for exceptionally long times. Create lasting world-class reliability by setting "precision goals" for your equipment parts' health that your people need to achieve. Do a Reliability Growth Cause Analysis when you have to financially justify the risk reduction strategy to use.

"Push the Limit" Projects

To get world-class performance, you need to use world-class practices. Higher reliability needs higher quality standards and precision methods. Research better solutions and bring them into your operation. Make them "the way we do things around here." Don't wait for problems to justify improvements; make improvements so that there won't be any problems.

Permanent change requires the combined support and commitment of senior management, the people doing the work, and their supervisors. Introduce a never-ending Change to Win program and let your people improve their own work. They'll find better ways and simpler solutions to lift process and equipment reliability with continuous improvements.

Update Systems and Processes Business-Wide

Use your business systems to trap world-class practices in your organization so that they are always used and are never lost. Take every new success across the whole organization so that each improvement gained generates compounding benefits by using it many times. Make a change and its success permanent by putting it into your business process maps and all relevant documents. Embed the requirements and new ways into work procedures and performance measures. Make necessary information easily available to everyone so that they can make decisions right the first time. Train and retrain your people to perform the new methods and practices masterfully.

Monitor to Prove Reliability Growth

Use process Key Performance Indicators, step Performance Indicators, and frequency distributions to track the trend and speed of progress. Address and improve those activities that are not yet performing well enough.

Find the Hidden Risks and Eliminate All of Them

You know that equipment failures are caused by component failures. To climb to the heights of world-class reliability, your organization and its people need the education, skills, and processes for doing the correct quality practices and precision methods that deliver microstructure reliability.

Find All Remaining Asset Life-Cycle Risks

The causes leading to component failure can arise anywhere during the life cycle. It is necessary to remove opportunities for all defects damaging to any critical part's materials of construction throughout the asset life cycle. Use the eight life-cycle questions of the Physics of Failure Factors Analysis (Chapter 9) to regularly review your company processes and proactively eliminate the causes of the causes of failure.

Do an Operating Criticality Analysis at component level to discover the size of the risks you have still got in your operation.

Audit your organization's capability to have outstanding reliability against the Plant Wellness Index.

For an explanation, see the document *Plant Wellness Index,* which is included in the supplementary materials available online.

You will most likely miserably fail the audit and find that you can never have great asset management success using your current business processes. Be grateful that you have the truth and that you know the direction you must take to create a world-class reliable operation.

Identify More Successful Risk Reduction Strategies

For each unacceptable risk identified, seek to eliminate the risk and make reliability improvements to prevent it. Chance reduction strategies change the design of a process to make it more robust and reliable. Altering a process design to remove the possibility of defects error-proofs your business. To prove the effectiveness of chosen improvements and confirm their financial value, assess the proposed mitigations using Reliability Growth Cause Analysis so that you build a powerful business case to make the change. At the very least, plot the before and after risk on a calibrated risk matrix to show people what will be gained by the process improvement.

Keep consequence management options as a last resort because they add costs and complexity to your business. They require you to add another process that will surely fail from time to time, opening you up to the full risk that you wanted to prevent. Include consequence management methods for a complete a risk management solution, but don't use them as your only risk management strategy.

"Push the Limit" Projects

Put continuous improvement in everyone's duty statement. Every day, you could find better ways to do your job if you were asked to suggest them. This natural human inclination to want things to be simpler, easier, and faster needs to be intentionally harnessed and put into useful service.

Apply Series Reliability Property 1 and get the people who are directly working in a process to improve it. They work with the process and its problems all the time. They are best placed to understand its details and to know what changes to make to get each step to perform better and better every day in every way.

Apply Series Reliability Property 3 and get the managers of a process to find and implement system-wide improvements. System-wide changes make easy fortunes.

Where problems persist and need to be addressed in revolutionarily ways, dedicate an improvement team of subject matter experts to design and implement a solution. Where problems can be addressed in an evolutionary way, allocate them to a workplace team using Change to Win and set a deadline for their first-cut implementation.

When putting in place actions to solve problems, past experience with change management tells us that it's important to have:

- Leadership and supervisory support, as both are vital to any program or effort being continued
- Presentation of a successful business case, which is vital in an environment of scarce resources, as it justifies both the leadership commitment and resource alloca tion necessary to successfully implement the program
- Education and involvement of people, as this gives them a sense of ownership and understanding in the program, which are vital to reduce their innate resistance to change
- An effective communication plan to proactively inform people and to ensure that no one feels left out because they got no news

Make the Best Way the Only Way

Within your process documents, explain exactly how your company is run for greatest success. Keep your process documents current and accurate and train their users if they are not following the designed process. If the process is lousy, let those who use it fix it.

Unless something positive and concrete comes out of a bad event to prevent its occurrence in future, there has been no learning and the problem will reoccur. It cannot be otherwise because nothing has been corrected and nothing has been improved. Until action is taken to make necessary enhancements permanent practice, there has been no change. When you introduce improvements and changes, they must become the new way of doing the work. Only when better ways are documented and people are trained to do them right can higher performance result. If it's not written, it's not real! No new change will ever become a workplace practice until the improvement is documented in standard operating procedures, people are trained to do it right, and it is actively supported by their immediate supervisor.

Confirm Reliability Growth in Your Processes

Use trends to prove that your processes and process steps are becoming ever more successful, ever more effective, and ever more efficient. Use distribution curves to prove that they are continuously becoming optimal.

BIBLIOGRAPHY

Published Books

Arthur, Jay. *Lean Six Sigma Demystified*. New York: McGraw-Hill, 2007.

Ball, Philip. *Critical Mass: How One Thing Leads to Another*. New York: Farrar, Straus and Giroux, 2005.

Blanchard, Benjamin S. *Design and Management to Life Cycle Cost.* Forest Grove, OR: M/A Press, 1978.

Bodek, Norman. Kaikaku: *The Power and Magic of Lean—A Study of Knowledge Transfer.* Vancouver, WA: PCS Press, 2004.

Carnegie, Dale. *How to Win Friends and Influence People*. New York: Simon & Schuster, 1936.

Deming, W. Edwards. *Out of the Crisis.* Cambridge, MA: MIT Press, 2000.

Denove, Chris, and James D. Power IV. *Satisfaction*: *How Every Great Company Listens to the Voice of the Customer.* New York: Portfolio, 2006.

Fastener Handbook—Bolt Products. Victoria, Australia: Ajax Fasteners, 1999.

Feigenbaum, A. V. *Total Quality Control.* 3rd ed. New York: McGraw-Hill, 1993.

Fitchett, Don, and Mike Sondalini. *True Downtime Cost Analysis*. 2nd ed. St. Louis, MO: Feed Forward Publications, 2006.

George, Mike, Dave Rowlands, and Bill Kastle. *What Is Lean Six Sigma?* New York: McGraw-Hill, 2004.

Gladwell, Malcolm. *Blink: The Power of Thinking without Thinking.* New York: Little, Brown, 2005.

———. *Outliers: The Story of Success*. New York: Little, Brown, 2008.

Gordon, J. E. *The New Science of Strong Materials, or, Why You Don't Fall through the Floor*. 2nd ed. New York: Penguin 1976.

Hattangadi, A. A. *Plant and Machinery Failure Prevention.* New York: McGraw-Hill, 2005.

Hofstede, Geert. *Cultures and Organisations: Software of the Mind*. 2nd ed. New York: McGraw-Hill, 2005.

Hopkins, Andrew. *Safety, Culture and Risk: The Organisational Causes of Disasters.* North Ryde: CCH Australia, 2005.

Hoyle, David. *ISO 9000 Quality Systems Handbook.* 5th ed. Burlington, MA: Butterworth-Heinemann, 2006.

Juvinall, Robert C. *Engineering Considerations of Stress, Strain, and Strength.* New York: McGraw-Hill, 1967.

Liker, Jeffery K. *The Toyota Way: 14 Management Principles from the World's Greatest Manufacturer.* New York: McGraw Hill, 2004.

Kahneman, Daniel. *Thinking, Fast and Slow.* New York: Farrar, Straus and Giroux, 2011.

Mlodinow, Leonard. *The Drunkard's Walk: How Randomness Rules Our Lives.* New York: Pantheon Books, 2009.

Moore, Ron. *Making Common Sense Common Practice: Models for Manufacturing Excellence.* Revised and updated ed. Boston: Butterworth-Heinemann, 2002.

Moubray, John. *Reliability-Centered Maintenance*. Oxford, UK: Butterworth Heinemann, 1991.

Nolan, Stanley F., and Howard F. Heap. *Reliability Centered Maintenance.* San Francisco: Dolby Access Press, 1978.

Petroski, Henry. *Design Paradigms: Case Histories of Error and Judgment in Engineering.* New York: Cambridge University Press, 1994.

Piotrowski, John. *Shaft Alignment Handbook.* 3rd ed. Boca Raton, FL: CRC Press, 2007.

Robinson, Richard M., et al. *Risk and Reliability: An Introductory Text.* 7th ed. Melbourne, Australia: R2A, 2007.

Smith, David J. *Reliability, Maintainability, and Risk: Practical Methods for Engineers.* 7th ed. Boston: Elsevier, 2005.

Surowiecki, James. *The Wisdom of Crowds: Why the Many Are Smarter than the Few and How Collective Wisdom Shapes Business, Economies, Societies, and Nations.* New York: Doubleday, 2004.

Published Articles

Foord, A. G., and G. Gulland. "Can Technology Eliminate Human Error?" Trans IChemE, Part B, *Process Safety and Environmental Protection,* vol. 84, no. B3 (2006): 171–173.

Gurgenci, Hal, and Zhiqiang Guan. "Mobile Plant Maintenance and the Duty Meter Concept." *Journal of Quality in Maintenance Engineering,* vol. 7, no. 4 (2001): 275–286.

Mosaddar, Davood, and Amir Abbas Shojaie. "A Data Mining Model to Identify Inefficient Maintenance Activities." *International Journal of System Assurance Engineering and Management,* vol. 4, no. 2 (2013): 182–192.

Saldaña, Miguel Angel Mariscal, Susana García Herrero, Miguel Angel Manzanedo del Campo, and Dale O. Ritzel. "Assessing Definitions and Concepts within the Safety Profession." *International Electronic Journal of Health Education,* vol. 6 (2003): 1–9.

Wright, Nathan C. "What Is the Correlation between Transformational Leadership and the Profitability of Arizona Copper Mining Companies?" *Engineering & Mining Journal,* August 2014, 70–75.

Online Sources

Barringer, Paul H. "Life Cycle Cost & Reliability for Process Equipment." Accessed September 14, 2015. http://www.barringer1.com/pdf/lcc_rel_for_pe.pdf.

———. "Use Crow-AMSAA Reliability Growth Plots to Forecast Future System Failures." Accessed September 14, 2015. http://www.barringer1.com/pdf/Barringer-IMEC-2006-Toronto-Paper.pdf.

Block, Heinz, and Michael Hernu. "Performance Benchmarking Update: Expectations and Reality." Accessed September 14, 2015. http://webservices.solomononline.com/dbap/Data/Articles Refining/Performance_Benchmarking_Update_-_Issue_12-2007.pdf.

International Standards Organization. Accessed September 14, 2015. http://www.iso.org/iso/home/store/catalogue_ics.htm.

INDEX

Note: *f* in italic page numbers indicates a figure; *t* indicates a table

About the Author

Mike Sondalini, MBA, CP Eng., is an expert in information and knowledge management, risk management and elimination, quality management systems, asset management, and operational excellence and reliability engineering. His immersion in these areas led him to create "The Plant Wellness Way," a life cycle asset management methodology. He subsequently coined the term "Industrial and Manufacturing Wellness" to depict a holistic, business-wide system of reliability.